Textbook of Environment and Ecology

Vir Singh

Textbook of Environment and Ecology

Springer

Vir Singh
Department of Environmental Science
Govind Ballabh Pant University of Agriculture and Technology
Pantnagar, India

ISBN 978-981-99-8845-7 ISBN 978-981-99-8846-4 (eBook)
https://doi.org/10.1007/978-981-99-8846-4

This Springer imprint is published by the registered company Springer Nature Singapore Pte Ltd.
The registered company address is: 152 Beach Road, #21-01/04 Gateway East, Singapore 189721, Singapore

Preface

Welcome to the realm of *Environment and Ecology*, a captivating journey through the intricacies of our planet's delicate balance and the interplay between living organisms and their surroundings. This book endeavors to illuminate the captivating world of the environment and its integral components, providing readers with a comprehensive understanding of the dynamic relationships that govern life on Earth.

Section I: Ecosystem Analysis opens the door to the fascinating realm of ecosystems, where we explore the fundamental building blocks of life and their interactions. From the intricate trophic organization to the intricacies of energy flow and ecological pyramids, this section unravels the hidden mechanisms that sustain life and ensure the seamless functioning of ecosystems.

In Section II: Natural Resources, we delve into the invaluable treasures bestowed upon us by Mother Nature. From water to land and soil, and from forests to agriculture and minerals, each chapter highlights the significance of these resources and emphasizes the urgent need for their sustainable management to ensure a bountiful future for generations to come.

Section III: Biodiversity invites us to marvel at the awe-inspiring diversity of life forms that enrich our planet. We will journey through the various levels of biodiversity, exploring the intrinsic value of every living organism, and discussing the pressing threats that endanger this invaluable heritage. But fear not, for this section also unveils the various conservation strategies and international efforts dedicated to safeguarding our planet's biodiversity for posterity.

In Section IV: Environmental Disruptions, we confront the daunting challenges posed by human-induced disruptions to the environment. From air and water pollution to soil degradation and noise pollution, we confront the harsh realities of our actions and explore feasible solutions to mitigate these threats, striving to restore harmony between humans and the environment.

Finally, in Section V: Environmental Management, we embrace the notion of responsible stewardship as we venture into the world of solid waste management, climate change mitigation, and sustainable development. Armed with knowledge and foresight, we set out to shape a brighter, more sustainable future, where the well-being of our planet is harmoniously balanced with human progress.

Throughout this voyage, we emphasize the vital role of education, awareness, and collective action in shaping a prosperous and ecologically balanced world. Each chapter is complemented by thought-provoking exercises and suggested readings, empowering readers to delve deeper into the subject matter and cultivate a broader ecological consciousness.

It is our sincere hope that *Environment and Ecology* will not only inspire awe and admiration for the natural world but also kindle a fervent determination to protect and nurture it. As we embark on this enlightening expedition, let us remember that our actions, both great and small, can make a significant difference in preserving the harmony of our planet for generations yet unborn.

I extend my heartfelt gratitude to all the researchers, educators, and environmentalists who have dedicated their lives to unraveling the mysteries of the environment and championing the cause of conservation. Their tireless efforts continue to inspire us, as they have undoubtedly inspired you, dear reader, on this thrilling journey.

With a sense of responsibility and hope, let us venture forth into the pages that follow, and may the wisdom we gain here fuel our commitment to safeguarding the planet we call home.

Pantnagar, India Vir Singh

Acknowledgments

I would like to express my heartfelt gratitude to the remarkable individuals who played an invaluable role in shaping this textbook, *Environment and Ecology*. Throughout the writing process, their unwavering support and encouragement created a stimulating and inspiring environment that fueled my passion for the subject matter. First and foremost, I am deeply grateful to my beloved wife, Gita, whose love, patience, and understanding provided me with the space and time to dedicate myself to this endeavor. Her unwavering belief in me kept me motivated during the challenging times of research and writing.

I extend my heartfelt thanks to my wonderful daughter, Silvi, and son, Pravesh, whose love and enthusiasm for learning about the natural world always inspired me to explore new dimensions in the field of environment and ecology. To my son-in-law, Nic, I am grateful for his insightful discussions and thought-provoking ideas that added depth and richness to the content of this book. Additionally, my granddaughter, Avery, brought joy and laughter into our lives, reminding me of the importance of preserving our environment for future generations. As I reflect on this acknowledgment, I am also filled with immense joy and pride as my family welcomes another beautiful blue-eyed granddaughter, Everly, into our family. Her presence, even from the very beginning, has added a special sense of wonder and purpose to this work.

To each of them, I extend my heartfelt thanks for their unconditional love, understanding, and creativity-inducing support, without which this book would not have been possible. Their presence in my life has made this journey all the more meaningful, and I am forever grateful for their love and encouragement.

I would like to extend my gratitude to Aakanksha Tyagi, Business Manager at Springer Nature, for her invaluable support and insightful suggestions throughout the publishing process of *Environment and Ecology*. Her expertise and dedication have been instrumental in shaping the final product, and I am truly grateful for her support and guidance. Aakanksha's commitment to excellence and attention to the overall product quality have made this collaboration a rewarding and enriching experience. Thank you for going above and beyond to ensure the success of this publication.

I am immensely proud of my students, whose indescribable dedication and passion for environmental conservation have played a pivotal role in gifting the world with the book. Their relentless pursuit of knowledge, coupled with their commitment to make a positive impact on our planet, has been truly inspiring. It is through their collective efforts, insightful research, and innovative ideas that this book has come to fruition, and I am deeply grateful for their invaluable contributions. Their role in championing the cause of sustainability and spreading awareness about ecological issues will undoubtedly leave a lasting and positive impact on our global community. Together, I have embarked on a meaningful journey, and I commend each of my students for their exceptional work in shaping this significant publication.

In crafting the *Environment and Ecology*, I have had the privilege of partnering with Springer Nature, a world-renowned publishing company with a rich legacy of delivering exceptional scholarly works. I express my gratitude to Springer Nature for providing me with the platform to share my research and insights with a global audience. Their unwavering commitment to academic excellence and their dedication to advancing knowledge in the field of environment and ecology have been phenomenal in bringing this project to fruition. I am deeply honored to be associated with such a prestigious institution, and I acknowledge their significant role in making this publication a reality.

Pantnagar, India Vir Singh

Contents

About the Author

Vir Singh is a distinguished and accomplished figure in the realm of Environmental Science, with a wealth of knowledge and experience that spans over three decades. As a Professor Emeritus at GB Pant University of Agriculture and Technology, he has left an indelible mark on the field through his exceptional contributions to teaching, research, extension, project execution, research supervision, and curriculum development.

Throughout his illustrious career, Prof. Singh has taken on various significant roles within the academic and administrative domains. He has served as the Coordinator of Liberal Education, Director of Communication, and Editor-in-Chief of two esteemed monthly magazines published by the university.

Prof. Vir Singh's academic journey has taken him to numerous distinguished universities and institutes in India, where he received a comprehensive education and training in diverse fields: environmental science, animal sciences, plant sciences, and social science. Moreover, he has collaborated with prestigious organizations beyond India's borders, further enriching his vast repertoire of experiences. Notably, he has worked with the ICIMOD based in Kathmandu, Nepal, Galilee International Management Institute (GIMI) in Israel, and Friedrich-Schiller University in Germany.

As an ardent advocate of environmental science and ecology, Prof. Vir Singh is fueled by a profound passion for understanding and preserving the intricacies of the natural world. He finds himself captivated by Nature's breathtaking biodiversity and is deeply committed to fostering sustainable practices for the benefit of the planet.

Prof. Vir Singh's contributions to academia extend far beyond confines of the lecture hall. His prolific body of work comprises an impressive collection of 56 books, numerous monographs, and over 250 research papers, book chapters, and popular articles. His extensive publications have significantly enriched the field of environmental science, providing valuable insights and perspectives to scholars, researchers, and enthusiasts alike.

Abbreviations and Acronyms

3Rs	Reduce, reuse, and recycle
ADP	Adenosine diphosphate
AFOLU	Agriculture, forestry, and other land use
API	Air pollution index
ATP	Adenosine triphosphate
BC	Before Christ
BNF	Biological nitrogen fixation
BOD	Biochemical oxygen demand
CAM	Crassulacean acid metabolism
CBD	Convention on Biological Diversity
CCUS	Carbon capture, utilization, and storage
CFCs	Chlorofluorocarbons
CH_4	Methane
CNG	Compressed natural gas
CO	Carbon monoxide
CO_2	Carbon dioxide
CO_3	Carbonate
COD	Chemical oxygen demand
COP	Conference of Parties
d	Day(s)
DALYs	Disability-adjusted life years
DAP	Draught animal power
dB	Decibels
DEFC	Direct ethanol fuel cells
DNA	Deoxyribonucleic acid
DO	Dissolved oxygen
Eco-Philosophy	Ecological philosophy
EEA	European Environment Agency
EIA	Environmental Impact Assessment
EPA	Environmental Protection Agency
FAO	Food and Agriculture Organization
GDP	Gross domestic product
GI	Gastrointestinal
GMO	Genetically modified organism
GPP	Gross primary productivity
Gt	Gigatonnes
GTP	Guanosine triphosphate

GWh	Gigawatt-hour
GWP	Global warming potential
H_2	Hydrogen
H_2CO_3	Carbonic acid
H_2O	Hydrogen monoxide/water
HAPs	Hazardous air pollutants
HCFCs	Hydrochlorofluorocarbons
HFCs	Hydrofluorocarbons
HYVs	High-yielding varieties
IEA	International Energy Agency
IFAD	International Fund for Agricultural Development
IPBES	Intergovernmental Science-Policy Platform on Biodiversity and Ecosystem Services
IPCC	Intergovernmental Panel on Climate Change
IUCN	International Union for Conservation of Nature
J	Joules
kcal	Kilocalories
kg	Kilograms
kJ	Kilojoules
km	Kilometer
km^3	Cubic kilometers
LHb	Leghemoglobin
LMOs	Living modified organisms
LPG	Liberalization, privatization, and globalization
LPG	Liquefied petroleum gas
LT-LEDS	Long-term low greenhouse gas emission development strategies
m	Meter(s)
M^3	Cubic meters
MCFC	Molten carbonate fuel cells
MDGs	Millennium development goals
mg/L	Milligrams per liter
MJ/Nm^3	Megajoules per normal cubic meter
MSW	Municipal solid wastes
MW	Megawatt
MWe	Megawatt-electric
N_2	Nitrogen
N_2O	Nitrous oxide
NDCs	Nationally determined contributions
NGO	Non-governmental organization
NO	Nitric oxide
NO_x	Nitrogen oxides
NPK	Nitrogen, phosphorus, and potassium
NPP	Net primary productivity
O_2	Oxygen
O_3	Ozone
°C	Degrees Celsius
ODS	Ozone-depleting substances

OPEC	Organization of the Petroleum Exporting Countries
PAFC	Phosphoric acid fuel cells
PCBs	Polychlorinated biphenyls
PDMFC	Direct methanol fuel cells
PEMFC	Proton exchange membrane fuel cells
pH	Potential of hydrogen
PM_{10}	Particulate matter with a diameter of 10 micrometers or smaller
$PM_{2.5}$	Particulate matter with a diameter of 2.5 micrometers or smaller
POPs	Persistent organic pollutants
ppb	Parts per billion
ppm	Parts per million
ppt	Parts per trillion
PV	Photovoltaic
RNA	Ribonucleic acid
SDGs	Sustainable Development Goals
SF_6	Sulfur hexafluoride
SLM	Sound level meter
SO_2	Sulfur dioxide
SOC	Soil organic carbon
SOFC	Solid oxide fuel cells
SOM	Soil organic matter
SPM	Suspended particulate matter
Syngas	Synthesis gas
TWh	Terrawatt-hour
UK	United Kingdom
UNCCD	United Nations Convention to Combat Desertification
UNESCO	United Nations Educational, Scientific, and Cultural Organization
UNFCCC	United Nations Framework Convention on Climate Change
UNICEF	United Nations International Children's Emergency Fund
USA	United States of America
USSR	Union of Soviet Socialist Republics
UV	Ultraviolet
VOCs	Volatile organic compounds
WCED	World Commission on Environment and Development
WFP	World Food Program
WHO	World Health Organization
yr	Year(s)
ΔN	Change in the number of individuals
ΔNn	Change in the number of new individuals
$\Delta Nn/N\Delta t$	Specific natality
$\Delta Nn/\Delta t$	Absolute natality
Δt	Time interval

1 The Environment and Its Components

Everything that surrounds everything is the environment. The environment is something that means everything to everything. The environment is a medium everything is contained within. The environment is not just a physical entity; it is also a phenomenon that makes all living beings happen to be in existence. The environment is a culture in which living beings are nourished. The environment is composed of everything living beings are made up of. The environment can exist without life, but life cannot be independent of the environment.

Life is a phenomenon of the environment. Whatever is there in life is out there in the environment. To keep life and life processes intact and going on with the pace of evolution, the quality of the environment has to be maintained. With the going-on changes in the composition indicative of its deteriorating quality, the environment has become a global issue.

There are two aspects of the environment we tend to interrelate: one, used in the global context (the atmosphere, the water, and the land together with all living organisms), and the other in the context of human-environment (social environment, cultural environment, business environment, etc.). The two are interrelated. When the environmental changes being registered become a human concern, the message is clear – that the environment is deteriorating due to human activities. The environment is not just an inevitable condition for life to prosper and sustain, but also for human development and socioeconomic progress.

Environmental issues are not natural issues in themselves. These are the issues emanating from human activities antagonistic to the environment. Environmental conditions are like two-way traffic. The environment influences living organisms and gets influenced by them. Among the organisms, it is the human species "credited," almost exclusively, to deteriorating environmental conditions. Thus, the environment is often defined with humankind central to it, such as the following:

- Physical components of the Earth where man is the key factor influencing his environment
- Total conditions surrounding man at a given point in space and time
- Sum of all physical, chemical, biological, social, and economic factors that constitute the surroundings of man, who is both creator and molder of his environment

1.1 Components of the Environment

The environment can be broadly classified into abiotic and biotic components (Fig. 1.1). These components are interchangeable. The phenomenon of interconnecting the two components lies in photosynthesis and chemosynthesis which convert the inorganics into organics vibrant with life, and DNA is the biotic entity governing the process and determining this conversion capacity.

V. Singh, *Textbook of Environment and Ecology*, https://doi.org/10.1007/978-981-99-8846-4_1

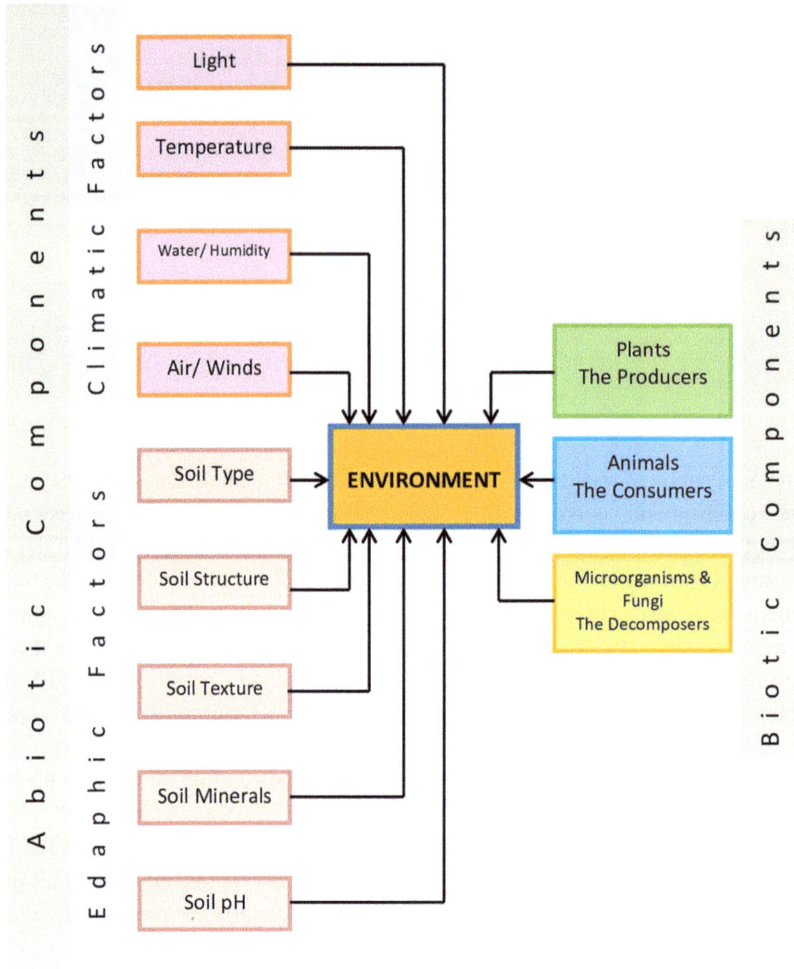

Fig. 1.1 The abiotic and biotic components of the environment

1.1.1 Abiotic Components

Abiotic components comprise the nonliving factors. This component is much larger and extends from the ocean floor to the lithosphere to the exosphere. The abiotic components can be divided into edaphic factors and climatic factors.

1.1.1.1 Edaphic Factors

The edaphic factors are the soil-relating factors and include the soil type (e.g., clay, sandy, loam, etc.), soil texture and structure, physical and chemical composition, pH, water/moisture, air, minerals, and organic matter. The soil forms a very complex medium comprising mineral matter, air, water, and organic matter in proportions varying spatially and temporally.

- **Soil type:** Plants do not grow well in heavy/clay soil because of oxygen deficiency in such soils. Plants do not grow well in sandy soil due to the leaching of minerals.
- **Soil texture:** Field capacity is reached faster in coarser-textured soil, such as loamy soil, than in the fine-textured soil, like clay soil.
- **Soil structure:** The amount of aggregation determines the extent of pores in the soil necessary to hold water.
- **Soil pH:** Many a plant species prefer soils with low pH, and many can grow even in alka-

line soils, while many in soils with neutral pH. Soil pH also determines the species, populations, and activities of soil flora and fauna.

- **Soil water:** Plants need soil water for photosynthesis, transpiration, and metabolism.
- **Soil minerals:** The deficiency of an essential mineral in the soil affects plant growth.

1.1.1.2 Climatic Factors

Climatic or atmospheric factors are responsible for creating conditions for life to prosper with all its potencies. These include light, temperature, water, humidity, and winds.

Light Light is the most essential and the most critical factor and a primary source of life on Earth. Light drives photosynthesis, a process through which chlorophyllous plants manufacture food that nourishes most of life in the biosphere. Many other vital attributes of light for plants include photoperiodism, seed germination, growth, stomatal opening, transpiration, phototropic movements, and circadian rhythms. Some distinctive attributes of light for animals are the eye size of marine animals, color vision, pigmentation, photoperiodism, metabolism, periodicity, and locomotion (for example, photokinesis in blind larvae of mussel crab; phototaxis or orientation of animals during locomotion, as in euglena and earthworm; and beeline in honey bees in which flight of the bees is determined by the angle of the sun).

Temperature Temperature is vital for all metabolic activities that go on in a narrow range of temperatures. Living organisms have evolved physiological and behavioral adaptations to withstand temperature extremes. To avoid extreme temperatures, many birds migrate to more favorable areas according to seasonal changes and cold-blooded animals undergo hibernation in winter and aestivation in hot summer to save themselves from extreme temperatures.

Water Most of the Earth is water and so are the organisms. Water is necessary for photosynthesis and all metabolic activities of organisms. The distribution of organisms depends on their water requirement. Water itself serves as a habitat for numerous organisms – plants, animals, and microorganisms.

Humidity It relates to the proportion of water vapor in the atmosphere. Humidity determines rates of evaporation from water bodies, transpiration in plants, perspiration in terrestrial animals, and precipitation. Various plants and animals have evolved various adaptations to withstand dry conditions.

Atmospheric Gases The atmosphere as a whole builds up a pressure – the atmospheric pressure necessary for living organisms to exist. The atmospheric pressure maintains balance with organisms' internal pressure. Some of the atmospheric gases incorporate in living processes; for example, CO_2 incorporates in photosynthesis, oxygen incorporates in generating energy essential for metabolism, and nitrogen fixes into soil and water and then incorporates in plant proteins and from there passes on to animal proteins.

Wind It determines the weather conditions and helps in the dispersal of seeds for the propagation of plants.

1.1.2 Biotic Components

Biotic components of the environment comprise all the living organisms – all plants, animals, and microorganisms, broadly classified based on their functions as producers, consumers, and decomposers, respectively. Fungi also serve as decomposers.

1.2 The Planet Earth

The planet Earth, a member of the Solar Family, is the only so-far-known living planet in the cosmos. Human space explorations and pictures of the planet captured from space reveal that our Earth is the most beautiful planet in the universe and it is often called Blue Beauty. When we imagine how our Earth looks like floating in space, an iconic image of the Blue Beauty hangs in front of our eyes which we cannot help marvel at. Some crucial features of the planet are presented in Table 1.1.

The physical environment of planet Earth has three realms, lithosphere, hydrosphere, and atmosphere, intricately interacting with each other. Parts of the lithosphere, hydrosphere, and atmosphere make up the biosphere – the living component of the Earth's environment (Fig. 1.2).

Table 1.1 Some dimensions of the planet Earth

1. Age	About 4.5 billion years
2. Distance from the sun	149,598,262 km
3. Surface area	510,072,000 km^2
4. Land area	148,940,000 km^2 (29% of the surface area)
5. Water	361,132,000 km^2 (71% of the total surface area)
6. Mass	5.97237×10^{24} kg
7. Volume	1.08321×10^{12} km^3
8. Circumference	40075.017 km (equatorial), 40007.86 km (meridional)
9. Mean radius	6371.0 km
10. Orbital speed	29.78 km per second
11. The highest place	Mount Everest, 8848 m
12. Lowest place under the ocean	Mariana Trench, 10,971 m
13. Lowest place on land	Dead Sea, 420 m
14. Mean temperature	14 °C
15. Hottest place	Furnace Creek, California, USA (56.7 °C)
16. Coldest place	Soviet Vostok Station in Antarctica (−89.2 °C)

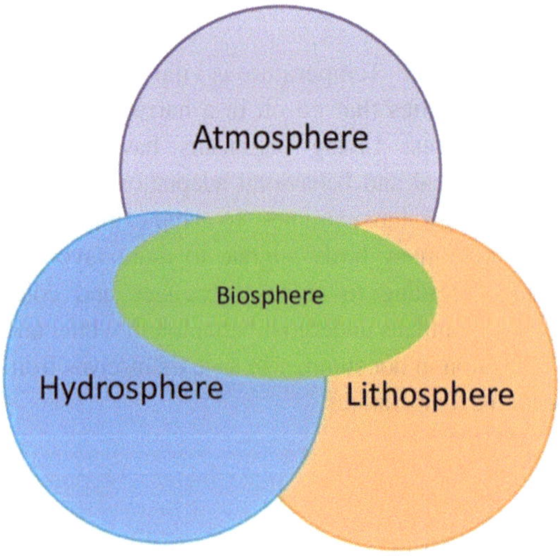

Fig. 1.2 The interconnected physical domains engaging and interacting within the biosphere

1.3 The Lithosphere

The lithosphere is the outer crust of the Earth and the base for supporting continents and basins. It is made up of the crust and upper mantle. It is heterogeneous comprising a variety of land masses and landforms with a spectacular diversity of ecosystems. The lithosphere consists of plains, hills, highlands, and mountains and covers about 30% of the Earth's surface. In the continental region, the lithosphere is the thickest with an average thickness of about 40 km. With a maximum thickness of about 10–12 km, the lithosphere is the thinnest below the oceans. The oceanic lithosphere is denser than the continental lithosphere.

The lithosphere encompasses horizontally several kilometers long 15 slowly drifting plates that constantly move at a rate of 1.3–20 cm in a year on the asthenosphere, a weaker, hotter, and deeper part of the upper mantle about 100–700 km below the Earth's surface lying beneath the lithosphere. The movement of the tectonic plates causes a variety of geological events, such as earthquakes and volcanoes.

The uppermost part of the lithosphere, the soil, is the most active part of the Earth constantly interacting with the hydrosphere (water, moisture, humidity) and atmosphere and supporting terrestrial life.

The crust and the upper mantle are distinguishable based on their mineralogy and chemistry. Several elements which constitute the bulk of the lithosphere include oxygen, silicon, aluminum, iron, calcium, sodium, potassium, magnesium, hydrogen, titanium, chlorine, carbon, and many others. Oxygen contributes as much as 93.8% of the composition of the Earth's crust.

1.4 The Hydrosphere

All the water on the surface of the Earth – the rivers, the streams, the lakes, the seas, the oceans, and the groundwater – constitutes a hydrosphere. It is the largest component of the planet's environment. Boundaries of the hydrosphere are more flexible than those of the lithosphere. From the groundwater reserves, it extends to the lithosphere (soil water/moisture) to the lower part of the atmosphere (water vapors/humidity/fog/clouds) and prevails in the three states of matter – solid ice and snow on the poles and high mountain peaks, liquid in major surface water bodies, and gaseous form in the air. Some dimensions of the hydrosphere are presented in Table 1.2.

The interchange of water between the hydrosphere, lithosphere, and atmosphere is governed by the water cycle or hydrological cycle. Ever since the planet evolved into a water planet, not a single drop of water exists that has not been recycled countless times. Being incessantly in cyclic mode is one of the key characteristics of the water component of the Earth's environment. It is thanks to the water cycle that the biosphere's ecological integrity is maintained.

Table 1.2 Key facts about the planet's hydrosphere

1. Total volume	1386 million km^3
2. Earth's area covered by oceans	361 million km^2
3. Total mass A. In the form of water vapor at a given time	1.4×10^{18} tons 20×10^{12} tons
4. Salt water	97.5%
5. Freshwater A. Ice and permanent snow B. Groundwater C. Lakes, reservoirs, and river systems	2.5% 68.9% 30.8% 0.3%
6. Average salinity of seawaters	3.5%
7. Turnover of water per year	577,000 km^3
A. Water evaporation from oceanic surface	502,800 km^3
B. Water evaporation from land surface	74,200 km^3
(a) Atmospheric precipitation on the oceans	458,000 km^3
(b) Atmospheric precipitation on land	119,000 km^3
(i) Total runoff of the Earth's rivers (119,000–74,200 km^3)	42,700 km^3
(ii) Direct groundwater runoff to the oceans	2100 km^3

1.5 The Atmosphere

The atmosphere is a thick transparent layer of a mixture of gases called air. Though the atmosphere generally refers to the layer of gases, it also contains water vapor, aerosols, and fine particulate matter. Wrapping up the globe, the atmosphere extends up to about 500 kilometers above Earth's surface. The atmosphere as a whole is colorless, odorless, tasteless, and expansible as well as compressible. The lower portion of the atmosphere is a mixture of nitrogen, oxygen, and carbon dioxide along with many trace gases (Fig. 1.3).

The total mass of the atmosphere is about 5.15 x 1018 kg. About 75% of the atmosphere's mass is concentrated only up to 11 km from Earth's surface. The density of the air goes on decreasing with increasing altitude. There is no clear-cut demarcating line between the atmosphere and the area where it merges into outer space. Kármán Line at 100 km – or 1.57% of Earth's radius – is often used as the "border" between the atmosphere and outer space.

The major constituents of dry air are N_2 (78.084%), O_2 (20.946%), Ar (0.934%), CO_2 (0.0415%), Ne (0.001818%), He (0.000524%), CH_4 (0.000187%), and Kr (0.000114%). There might be many other gases in just traces. Water vapor varies from place to place, from season to season, and from time to time over a short period and may range from 0 to 3% and constitutes about 0.25% of the atmospheric mass. There might be several kinds of pollutants including dust particles, organic pollutants like pollen grains, and natural pollutants like volcanic ash, sea sprays, etc. The concentration of these pollutants – man-made as well as natural – might be critical in some places.

The Earth's atmosphere is classified into five distinguishable layers, segments, or zones, especially based on their temperature regime and gaseous composition. These are the troposphere, stratosphere, mesosphere, thermosphere, and exosphere (Fig. 1.4).

1.5.1 Troposphere

The troposphere is the lowest atmospheric layer with an average thickness of 11 km above the Earth's surface, 8 km from the Earth's poles, and 14.5 km from the equator. This layer comprises nearly 75% of the atmospheric mass. It is the zone comprising life-supporting gases in the proportion vital for the sustenance of life – for example, N_2, CO_2, and O_2 that incorporate protein structure, photosynthesis, and energy production, respectively. The temperature in the zone goes on decreasing pretty uniformly with increasing altitude at an approximate rate of 6 °C fall per km until it decreases to −50 to −60 °C at the uppermost limit of the troposphere.

This layer is the most crucial and the most critical for life on Earth. Earth's weather system and

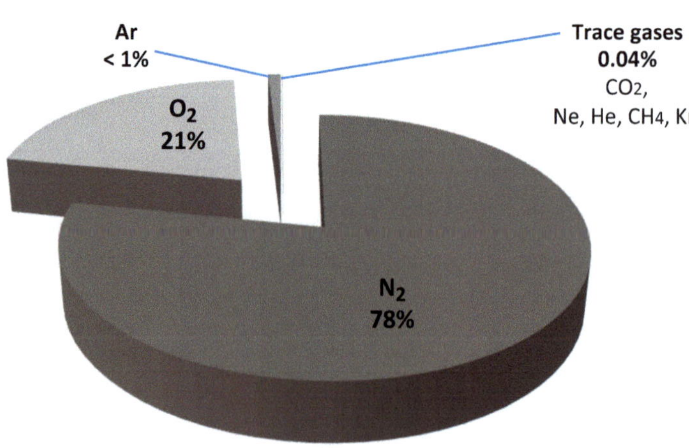

Fig. 1.3 Composition of the Earth's atmosphere

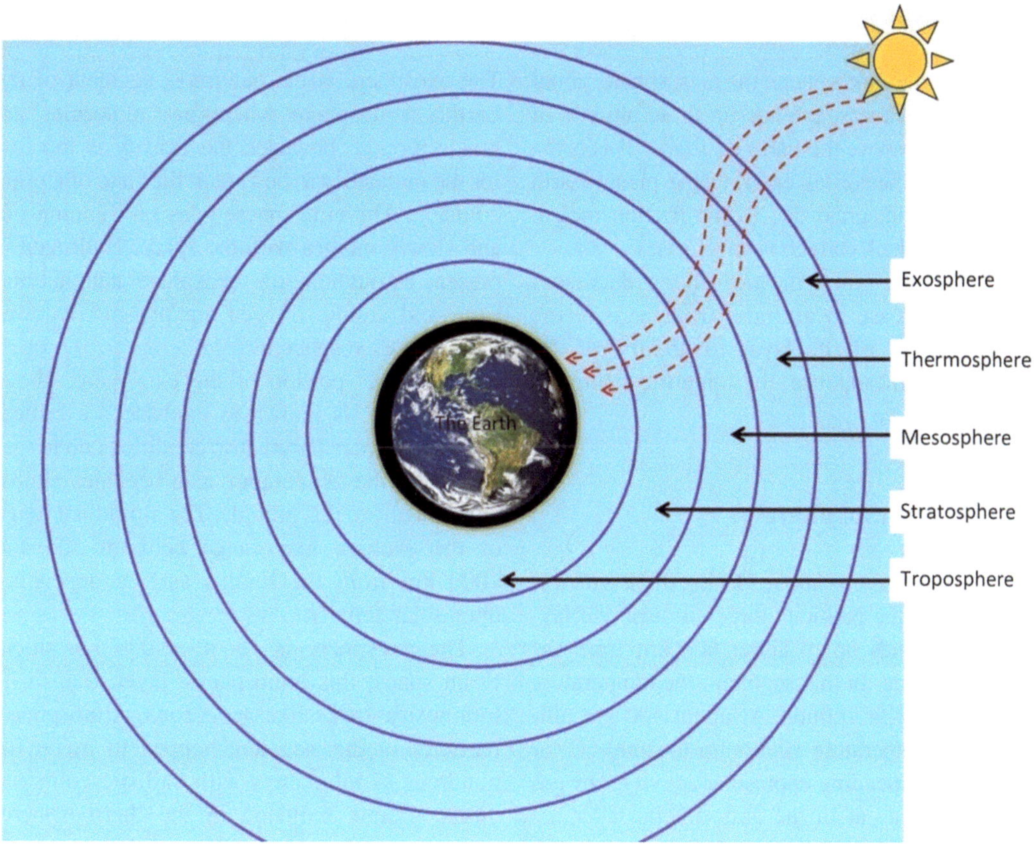

Fig. 1.4 The discernible atmospheric strata

water cycle are accommodated in this sphere. Strong winds, storms, and clouds are the other main features of this atmospheric zone. The troposphere is the meeting zone of the pedosphere of the lithosphere, hydrosphere, and biosphere. This is the atmospheric layer that has an enormous bearing on the biosphere which itself flourishes in this sphere.

The zone where the troposphere gradually ends after attaining minimum temperature is known as the tropopause.

1.5.2 Stratosphere

Above the troposphere lies the stratosphere about 39 km in thickness and extending up to an altitude of 50 km from the Earth's surface. One of the unique features associated with the stratosphere is the presence of an ozone layer known as the ozonosphere. The ozonosphere is of critical importance for life on Earth as it filters the harmful ultraviolet rays, thus serving as a protective umbrella for the living planet. It also contributes to the decline in the cooling rate of the planet. This atmospheric zone has a more stable atmosphere than that of the troposphere and, therefore, is often used for operating jet plains and long-distance airplanes.

The temperature in the stratosphere stays almost constant up to 20 km upwards and then temperature rise is recorded with an increase in the altitude up to a maximum of zero degrees Celsius. The uppermost segment of this atmospheric layer is known as stratopause.

1.5.3 Mesosphere

Just above the stratosphere, the mesosphere, about 35 km in thickness, extends up to an altitude of about 85 km above the Earth's surface. This is the zone where meteorites entering the planet Earth burn and disintegrate due to the friction met to them by the thick enough layer of gases.

The temperature in the mesosphere decreases with an increase in altitude, from zero in the lower portion of the layer to about −95 °C towards the mesopause, the uppermost limit of the mesosphere.

1.5.4 Thermosphere

The thermosphere is about 515 km thick which is thicker than the previous three atmospheric layers and extends up to about 600 km from the Earth's surface. In this segment, the temperature rises along with altitude. At about 500 km altitude, the temperature shoots up to 2000 °C, or even more depending upon solar activity. The gas molecules present in the atmospheric layer are laden with kinetic energy culminating in such high temperatures. However, their sparse presence makes them unable to transfer a significant amount of energy to a thermometer. A question often strikes our eardrum: What would happen if an astronaut stretches his/her arm outside a spacecraft? The answer is: There would be no effect as the sparsely distributed gas molecules cannot transfer enough energy to burn the cosmonaut's exposed arm.

The ultraviolet radiation in this zone causes photoionization of the gas molecules. The thermosphere also constitutes a larger part of the ionosphere. The atmospheric particles become electrically charged due to intense radiation in this atmospheric layer resulting in the refraction of radio waves that can be received on Earth's surface. This segment of the Earth's environment is of importance for the operation of artificial satellites. The uppermost portion of the thermosphere is called thermopause.

1.5.5 Exosphere

The exosphere is the outermost segment of the Earth's atmosphere where gas molecules are gravity-bound; however, the density is too low for the molecules to be kept in the state of atomic collision. The atmosphere goes on thinning out and slowly merges with the space. Hydrogen is present throughout the exosphere and helium, CO_2, and atomic oxygen are present in lower parts of the exosphere.

The lower portion of the exosphere, above thermopause, is known as exobase – a critical "altitude" where barometric conditions no longer apply and the temperature also becomes almost constant above this altitude. The altitudinal range of the exobase may range between 500 and 1000 km from the Earth's surface depending upon solar activity.

The exosphere is about 10,000 km thick. From space, this atmospheric layer, due to its luminosity, looks like geocorona. Atmospheric particles in this segment happen to travel for hundreds of kilometers with ballistic collision, unlike atomic collision in the thermosphere, before bumping into any other particles and can also escape into deep space. The upper boundary of the exosphere is referred to as the distance at which the influence of solar radiation on atomic hydrogen exceeds that of the Earth's gravitational pull and it happens at half the distance between the Earth and the moon, that is, about 192,200 km.

If the altitudes of different environmental layers are plotted against the temperature change trend, then we get a Σ-shaped graph that comes to the fore (Fig. 1.5), revealing a typical vertical stratification of the Earth's atmosphere.

1.6 The Biosphere

The biosphere is the sphere of life. In other words, the biosphere is the home to live. This sphere provides refuge to all organisms, the phys-

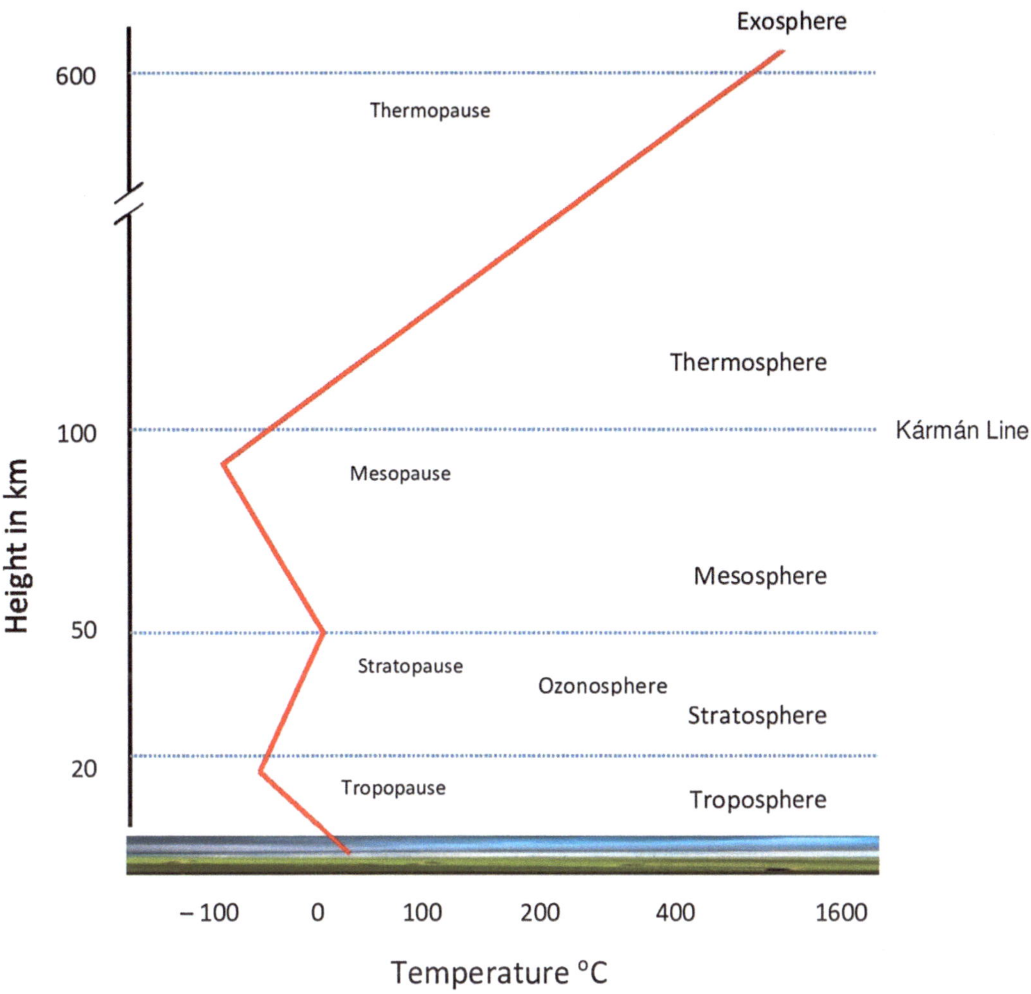

Fig. 1.5 Vertical layering of Earth's atmosphere: altitude versus temperature

ical environment supporting life, and the evolutionary processes of life. The biosphere is the fourth realm of the Earth's environment encompassing parts of the lithosphere, hydrosphere, and atmosphere and supports life in its entirety. It is the only sphere in the entire universe known so far where evolution of life is nurtured. The biosphere gives origin to, supports, nourishes, and sustains biodiversity, all ecosystems, and all biomes and holds life into biotic-abiotic integrity, that is, ecological integrity.

Biosphere's dimensions are limited by the absence of life-supporting elements of the environment, for example, light, oxygen, nitrogen, water, etc. In the lithosphere, the biosphere extends up to as deep as the trees' roots can reach, or the burrowing animals can make their access to, or the extent to which microorganisms can thrive. In the hydrosphere, the biosphere expands up to the ocean floor, up to 9 km, and even deeper. The deepest life in the hydrosphere is represented by the chemosynthesis-based hydrothermal community, of which the giant tube worm is one of the often talked about organisms, on the ocean floor. In the atmosphere, the biosphere is vibrant towards the lower portion of the troposphere. Then, with increasing altitude, it is limited by deficiency of oxygen, lack of suitable temperature, and low atmospheric pressure. Some birds have been seen flying into the lower zone of the stratosphere. Some bacteria can prevail even up to 15 km, or more.

1.7 The Environment-Organism Relationships

There is no boundary in the environment. All the spheres of the environment interact with each other. The biotic and abiotic components exist in reciprocal interactions. The abiotic environment influences the biotic environment and vice versa. There are relationships among the organism and their populations of various species living in a common environment of an ecosystem.

The organisms generally interact with their environment at several spatial and timescales. For example, a nitrogen-fixing bacterium interacts with air, water, and mineral elements within a space of a fraction of a cubic centimeter. A banyan tree, on the other hand, occupies a much larger space in the lithosphere and troposphere, interacting with soil environment, air, and light. The environment is not uniform throughout. It varies from space to space, from place to place due to variations in topological, edaphic, and climatic factors – light, temperature, humidity, and winds. Exchanges of matter and energy through various organisms within an environment phenomenally influence the upper layer of the lithosphere, the hydrosphere, and the lower part of the troposphere.

The organisms have to cope with their physical environment that tends to change over a certain timescale varying from a few minutes to a period stretching to a geological timescale. For example, populations of cyanobacteria in a nutrient-rich water body may change in a few minutes or hours due to light and nutrient concentrations, while variations in the lithosphere might take a very long timescale spanning over centuries, or even millennia. We shall discuss interactions occurring through the food chain and food web in ecosystems in the next chapter.

1.8 Ecology

Ecology is defined as the study of reciprocal relationships between organisms and their environment. Simply speaking, ecology is concerned with the interrelationships between organisms and the environment they are part of. Ecology reflects the totality of the pattern of organism-organism and organism-environment relationships.

The term ecology was first of all introduced by a German naturalist, zoologist, artist, and philosopher Ernst Haeckel (1834–1919) in 1866. Ernst Haeckel applied the term "oekologie" (Gr: oikos = home, logos = study or discourse) to the "relation of the animal both to its organic as well as its inorganic environment." The term for the first time was used in his notable work *Generelle Morphologie* (General Morphology) in 1866. "By ecology, we mean the whole science of the relations of the organism to the environment including in the broad sense, all the 'conditions of existence.'"

At an inaugural lecture to the philosophical faculty of the University of Jena in Germany (rechristened as Friedrich Schiller University, Jena), Ernst Haeckel further elaborated on his definition of ecology: "By ecology, we mean the body of knowledge concerning the economy of nature – the investigation of the total relations of the animal both to its inorganic and to its organic environment." The term "ecology" began sailing through the English language upon the translation of Ernst Haeckel's two-volume work, *The History of Creation*, into English.

Ecology as a distinctive discipline began capturing roots after the 1930s. After the 1970s, the scope of ecology began expanding as a subject. Henryk Skolimowski (1930–2018) integrated ecological principles into philosophy and created yet another branch of ecology in 1974 which is called ecological philosophy or eco-philosophy (Singh 2019). Ecology also got integrated with economics (ecological economics), agriculture (agroecology), plant sciences (plant ecology), animal sciences (animal ecology), forestry (forest ecology), aquatic ecosystems (aquatic ecology), soils (soil ecology), molecular biology (molecular ecology), anthropology (human ecology), social sciences (social ecology), and urban sociology (urban ecology). With the release of the Brundtland Commission Report, *Our Common Future*, in 1987, ecology emerged as a core of sustainable development. In the field of ecology, significant contributions have been made by various researchers, including Whittaker et al. (1973), Charles et al. (2001), Molles (2005), Wiens and Graham (2005), and Singh (2019), have advanced our understanding of ecological principles. And in our contemporary times, ecology offers the solution to all planetary problems. Our explorations of other planets and efforts to colonize the Moon and Mars have extended the scope of ecology to a cosmic scale.

1.9 Summary

The sum of all physical, chemical, and biological factors that constitute the surroundings of all living beings constitutes the environment. The environment can be broadly classified into abiotic and biotic components, interchangeable and inseparable. Abiotic components comprise nonliving factors, such as those relating to soil (edaphic factors) and the atmosphere or climate (climatic factors). Biotic components of the environment comprise all living organisms functionally classified as producers, consumers, and decomposers.

The physical environment of planet Earth has three realms: lithosphere, hydrosphere, and atmosphere, intricately interacting with each other. The biosphere – the living component of the Earth's environment – comprises part of the biotic components. The lithosphere is the outer crust of the Earth and the base for supporting continents and basins. It is made up of the crust and upper mantle and covers about 30% of the Earth's surface. All the water of the planet constitutes a hydrosphere. The atmosphere is a thick transparent layer of a mixture of gases called air. Major constituents of the dry air are N_2 (78.084%), O_2 (20.946%), Ar (0.934%), and CO_2 (0.0415%). Water vapor may range from 0 to 3% and constitutes about 0.25% of the atmospheric mass. The Earth's atmosphere is classified into five distinguishable layers, segments, or zones, especially based on their temperature regime and gaseous composition. These are the troposphere, stratosphere, mesosphere, thermosphere, and exosphere. The troposphere, nearly 75% of the atmospheric mass, is the lowest atmospheric layer with an average thickness of 11 km above the Earth's surface (8 km from the Earth's poles and 14.5 km from the equator). The temperature in the zone goes on decreasing pretty uniformly with increasing altitude until it decreases to −50 to −60 °C at the uppermost limit of the troposphere called the tropopause. The stratosphere is about 39 km in thickness and extends up to an altitude of 50 km from the Earth's surface. The temperature in this zone increases with an increase in altitude until it reaches about zero degrees towards the uppermost portion of this layer called stratopause. One of the unique features associated with the stratosphere is the presence of an ozone layer known as the ozonosphere that filters the harmful ultraviolet rays preventing them from reaching the Earth's surface. The mesosphere, about 35 km in thickness, extends up to an altitude of about 85 km above the Earth's surface. Meteorites entering the planet burn and disintegrate due to the friction met to them by the thick enough layer of gases in this zone. Temperature decreases with altitude, from zero in the lower portion of the layer to about −95 °C towards the mesopause, the uppermost limit of the mesosphere. The thermosphere is about 515 km thick and extends up to about 600 km from the Earth's surface. The temperature rises along with altitude, being 2000 °C at about 500 km altitude. The uppermost portion of the thermosphere is called thermopause. The atmosphere goes on thinning out and slowly merges with the space. The exosphere is about 10,000 km thick, looking like a geocorona. If the different vertical environmental layers are plotted against the temperature change trend, then a Σ-shaped graph comes to the fore suggesting the pattern of vertical stratification of the Earth's atmosphere. The biosphere is the sphere of life providing refuge to all organisms, the physical environment supporting life, and the evolutionary processes of life.

The biotic and the abiotic components exist in reciprocal interactions; the abiotic environment influences the biotic environment and vice versa. The organisms generally interact with their environment at several spatial and timescales. Ecology, first of all, introduced by Ernst Haeckel in 1866, is defined as the study of reciprocal relationships between organisms and their environment.

1.10 Exercises

1.10.1 Multiple-Choice Questions

1. Edaphic factors pertain to
   (a) soil
   (b) water
   (c) atmosphere
   (d) rainfall
2. The orbital speed of the Earth is about
   (a) 30 km per second
   (b) 30 km per minute
   (c) 30 km per hour
   (d) 30 km per day
3. Which on Earth is the lowest place on the land surface?
   (a) Mariana Trench
   (b) Dead Sea
   (c) Black Sea
   (d) Mediterranean Sea
4. Where does the hottest place on Earth lie?
   (a) Sahara Desert, Africa
   (b) Furnace Creek, California
   (c) Thar Desert, India
   (d) Gobi Desert, Mongolia/China
5. Asthenosphere lies
   (a) in the atmosphere
   (b) in the hydrosphere
   (c) below the lithosphere
   (d) on top of the Himalayas
6. How much of the Earth's surface is covered by the lithosphere?
   (a) 10%
   (b) 20%
   (c) 30%
   (d) 50%
7. Freshwater makes up only 2.5% of the total water on the planet. The largest proportion, about 69%, of the freshwater is in the form of
   (a) rivers
   (b) lakes
   (c) groundwater
   (d) ice and permanent snow
8. The minimum proportion of the freshwater on Earth is in the
   (a) ice and permanent snow
   (b) groundwater sources
   (c) seas
   (d) river systems
9. Most of the atmospheric mass is concentrated in
   (a) troposphere
   (b) stratosphere
   (c) mesosphere
   (d) thermosphere
10. Kármán Line at 100 km in the atmosphere from Earth's surface – or 1.57% of Earth's radius – is often used as the
   (a) the demarcation between the stratosphere and mesosphere
   (b) "border" between the atmosphere and outer space
   (c) "border" between mesosphere and thermosphere
   (d) the outermost boundary of the exosphere
11. Ozonosphere lies in the
   (a) troposphere
   (b) mesosphere
   (c) stratosphere
   (d) thermosphere
12. In which vertical atmospheric layer(s) does the temperature rise along increasing altitude?
   (a) Troposphere
   (b) Mesosphere
   (c) Stratosphere
   (d) Stratosphere and thermosphere
13. Which of the atmospheric layers records minimum temperature?
   (a) Troposphere
   (b) Stratosphere
   (c) Mesosphere
   (d) Exosphere
14. From space, this atmospheric layer, due to its luminosity, looks like geocorona. This is
   (a) Stratosphere
   (b) Mesosphere
   (c) Thermosphere
   (d) Exosphere
15. The term "oekologie" (ecology) was first of all used by Ernst Haeckel in his notable work titled
   (a) Natural Selection
   (b) Origin of Species
   (c) *Generelle Morphologie* (General Morphology)
   (d) The History of Creation

16. The phrase “economy of nature” in defining ecology was used by
    (a) Robert T. Paine
    (b) R. H. Whittaker
    (c) J. Wiens
    (d) Ernst Haeckel
17. Who is also often described as the Father of Modern Ecology?
    (a) G. Evelyn Hutchinson
    (b) Robert T. Paine
    (c) R. H. Whittaker
    (d) Raymond Lindeman
18. Match the items in column A with those in column B:

Column A	Column B
1. Ernst Haeckel	(a) Niche
2. Henryk Skolimowski	(b) Keystone species
3. Charles S. Elton	(c) Father of modern ecology
4. G. E. Hutchinson	(d) Ecology
5. Robert T. Paine	(e) Eco-philosophy

    (a) 1-a, 2-b, 3-c, 4-d, 5-e
    (b) 1-d, 2-e, 3-a, 4-c, 5-b
    (c) 1-b, 2-c, 3-d, 4-a, 5-e
    (d) 1-e, 2-d, 3-c, 4-b, 5-a
19. Which of the following statements is not true?
    (a) Predator species always reduce the biodiversity level in a community.
    (b) Predators might keep their prey populations below their carrying capacity.
    (c) The potential for competitive exclusion would be low in populations kept below carrying capacity.
    (d) If keystone species reduce the likelihood of competitive exclusion, their activities would increase the number of species that could coexist in communities.
20. What is Nile perch?
    (a) A tributary of the Nile River
    (b) A large predatory fish
    (c) A mutualist fish in coral reefs
    (d) The origin of the Nile River

Answers: 1-a, 2-a, 3-b, 4-b, 5-c- 6-c, 7-d, 8-d, 9-a, 10-b, 11-c, 12-d, 13-c, 14-d, 15-c, 16-d, 17-a, 18-b, 19-a, 20-b

1.10.2 Short-Answer Questions

1. What do you understand by environment?
2. Name the components of the environment with examples.
3. What are edaphic factors? Write examples.
4. Name the three categories of the biotic component of the environment.
5. Why do we call our Earth the Blue Beauty?
6. How many parts/spheres can the Earth's environment be divided into?
7. Name different sources of water on Earth.
8. Name the different layers of the atmosphere.
9. Define ecology.
10. What do you mean by a habitat?

1.10.3 Long-Answer Questions

1. Give an account of the lithosphere and hydrosphere of the Earth.
2. Describe the important features of all the atmospheric layers.
3. Explain the environment-organism relationships.
4. Discuss the importance and scope of environmental science in our contemporary world.
5. Write short notes on the following:
    (a) Biosphere
    (b) Edaphic Factors
    (c) Climatic Factors

References

Charles CS, Wootton JT, Leibold MA (2001) Animal ecology. University of Chicago Press, Chicago, p 296

Molles MC (2005) Ecology: concepts and applications. McGraw Hill, Boston. 622 pp

Singh V (2019) Fertilizing the universe: A new chapter of unfolding evolution. Cambridge Scholars Publishing, London. 285 pp

Whittaker RH, Levin SA, Root RB (1973) Niche, habitat and ecotope. Am Nat 107(955):321–338. https://doi.org/10.1086/282837

Wiens J, Graham C (2005) Niche conservatism: integrating evolution, ecology, and conservation biology. Annu Rev Ecol Evol Syst 36:519–539. https://doi.org/10.1146/annurev.ecolysis.36.102803.095431

Section I

Ecosystem Analysis

2 The Ecosystems

Ecosystem analysis – or ecosystem approach – is considered important for understanding complex interrelationships among different organisms and their interactions with their physical environment, structure, composition, and dynamics of the community inhabiting the common space, functions being performed, and services rendered. The ecosystem is also the basis of socioeconomic development and ecological security for mankind.

2.1 What Is an Ecosystem?

The term "ecosystem" was introduced by Sir Arthur George Tansley (1871–1955), an English botanist and a pioneer in ecology, in 1935. He defined an ecosystem as "a particular category of physical systems, consisting of organisms and inorganic components in a relatively stable equilibrium, open and various sizes and kinds." He regarded the ecosystem as 'the whole system not only the organism-complex but also the whole complex of physical factors forming what is called the environment' (Tansley 1935).

Many new advances in the field of ecology have taken place since the days of Tansley. The term "ecosystem" has assumed many more concepts. There is no single definition of what constitutes an ecosystem. In essence, an ecosystem is an area accommodating a community of living organisms in conjunction with its environment.

An ecosystem is an energy system – a solar-powered system. The energy enters through photosynthesis and is stored in plants. Plants also pick up nutrients from the soil, water, and atmosphere. Nutrients and energy contained in the plants flow into animals through the food chain. Decomposers finally release nutrients into their respective pools – soil, water, and atmosphere. Energy trapping (photosynthesis) and energy loss (respiration) and nutrient cycling are the continuous processes operating in an ecosystem. The community within an ecosystem is self-sufficient in terms of its nutrients and energy demands. In other words, an ecosystem is a self-reliant and self-regulating entity.

An ecosystem is a dynamic entity subject to evolutionary changes. Internal and external factors influence the structure, composition, and functioning of an ecosystem. Every factor operating in an ecosystem depends on every other factor, directly or indirectly.

2.2 Types of Ecosystems

There are two bases on the Earth for the distinctive communities to accommodate an interaction among populations of diverse species and their physical environment: land and water. On this basis, all the Earth's ecosystems can be categorized as terrestrial and aquatic ecosystems. Human intervention in all the planet's ecosystems is so intensive and widespread that not all the terrestrial and aquatic ecosystems stay natural. Many, rather most of them, are the transformed ones and, therefore, should be referred to as artificial or man-

V. Singh, *Textbook of Environment and Ecology*, https://doi.org/10.1007/978-981-99-8846-4_2

Table 2.1 Various types of ecosystems

Terrestrial ecosystems	Aquatic ecosystems
A. Natural ecosystems	**A. Natural ecosystems**
Mountain/hill/upland ecosystems	Marine ecosystems
Forest ecosystems	Ocean ecosystems
Grassland ecosystems	Estuarine/coastal ecosystems
Tundra ecosystems	Coral reef ecosystems
Desert ecosystems	Freshwater ecosystems
Warm desert ecosystems	Flowing water/lotic ecosystems
Cold desert ecosystems	River ecosystems
B. Anthropogenic/ man-made ecosystems	Stream ecosystems
Cropland/ agroecosystems	Still water/lentic ecosystems
Garden/orchard ecosystems	Lake ecosystems
Industry	Marsh/bog ecosystems
Laboratory	Pond/tank ecosystems
Spacecraft	**B. Anthropogenic/ man-made ecosystems**
	Aquarium ecosystems
	Sewage ecosystem

made, or anthropogenic ecosystems. A comprehensive account of various types of the Earth's ecosystems is presented in Table 2.1.

2.3 Ecosystem Components

An ecosystem accommodates everything and every process necessary for the organization and sustenance of life. Broadly speaking, there are two interrelated, inseparable, and integrated components: biotic and abiotic.

The biotic components comprise all living beings – plants, animals, and microorganisms. The green plants constitute autotrophs or producers forming food for all animals and microorganisms. The animals and the microorganisms are the consumers, the former being macro-consumers and the latter micro-consumers. The abiotic components include nonliving substances – organic and inorganic – and physical or climatic factors.

2.3.1 Biotic Components

2.3.1.1 Producers

The producers in an ecosystem are generally the autotrophic chlorophyllous plants which produce food for all other organisms, the heterotrophs, using carbon dioxide and water in the presence of solar radiation through the process known as photosynthesis. The autotrophs, the dominant biotic component in terrestrial ecosystems, comprise the trees, shrubs, and herbs. Green algae and blue-green algae or cyanobacteria, as two other photosynthesizers, also serve as producers in moisture-laden soils and aquatic ecosystems. Some photosynthetic protozoa, which have algae, such as chlorella, as endosymbionts, also function as producers. In deep aquatic ecosystems, floating plants are called phytoplankton and serve as the dominant producers.

The other conspicuous category of autotrophs or producers is the chemosynthetic bacteria that manufacture food using the energy derived from chemical compounds (chemical energy). These bacteria are active in soil and aquatic ecosystems.

Since the photosynthetic producers convert solar energy into the energy contained in the bonds of the organic matter, they are also called converters or transducers. They utilize the energy in their organic compounds in the building up and maintenance of their bodies and in liberating energy for various metabolic and productive functions. A fraction of their energy is passed on to the heterotrophs through food.

2.3.1.2 Consumers

Consumers or phagotrophs are heterotrophic organisms, mostly animals, which derive their food from the autotrophs (producers) or the heterotrophs. Thus, consumers are of two types: herbivores and carnivores.

The herbivores directly feed on the producers. They are also called primary consumers. As they are close to photosynthesis performed by the plants and thus relish the abundance of ecosystem energy, they, among the consumers, constitute relatively more stable populations. Examples of herbivorous animals are elephants, deer, cattle, monkeys, rabbits, grasshoppers, etc. in terrestrial ecosystems and protozoa, mollusks, crustaceans, etc. in aquatic ecosystems. As the primary consumers in an ecosystem convert the plant material into animal material, Elton (1927) called them the "key industry animals."

The carnivores in an ecosystem are the animals that feed on other animals. The carnivores feeding on the herbivores are primary carnivores or second-order consumers, for example, frogs, cats, foxes, etc. The animals dependent on primary carnivores are the secondary carnivores or third-order consumers, for example, snakes, owls, peacocks, wolves, etc. The larger carnivores that prey upon lower carnivores are the tertiary carnivores or fourth-order consumers. These large carnivores, such as lions, tigers, cheetahs, etc., are not further preyed upon by other animals and occupy the position of top consumers in an ecosystem called top carnivores. As the top carnivores are not preyed upon by other carnivores, they are also called apex predators.

2.3.1.3 Decomposers

Decomposers constitute the saprophytic microorganisms, such as bacteria and fungi, which reduce producers and animals, as well as their organic wastes, into the components they are made up of. They derive their food from all organisms and, because of their microscopic size, are also called micro-consumers. Through the decomposition processes, the decomposers play a crucial role in removing the dead bodies of the organisms and their wastes. As the decomposers reduce the biomass, the organic content "composed" during the process of photosynthesis and passed on to consumer animals, Singh (2019) also called them dephotosynthesizers and the process of decomposition as dephotosynthesis (reversal of photosynthesis).

Decomposers act upon the organic matter through what is known as external digestion. The enzymes they secrete in the surrounding medium digest organic matter externally. A part of the nutrients released is absorbed by the decomposers for their survival, maintenance, and growth, and the remaining are released into the environment with minerals added to the substratum. This process is known as mineralization. The minerals released into the environment are again utilized by the producers.

Decomposition helps the basic constituents of the organism's structures (carbon, hydrogen, oxygen, nitrogen, and all mineral elements) release into their respective pools. The decomposers, thus, set the conditions for recycling and reutilization of the nutrients in an ecosystem. This process is an essential step in the biogeochemical cycles. Since decomposition breaks down the bonds between different elements in organic matter, it is an energy-releasing process.

In the ecosystem analysis, sometimes two other categories of consumers come into the picture: parasites and detritivores or scavengers. Parasites are a category of various organisms, including bacteria, protozoa, fungi, worms, etc., that derive their food directly from other living organisms, the hosts. Detritivores or scavengers comprise small animals, such as beetles, termites, worms, etc., that feed on the dead bodies of other organisms.

2.3.2 Abiotic Components

The abiotic (or nonliving) components include all factors surrounding the organisms and may be divided into (i) inorganic substances, (ii) organic substances, and (iii) climate regime.

The inorganic substances include all the mineral elements involved in material or biogeochemical cycles, such as carbon, hydrogen, oxygen, nitrogen, phosphorus, sulfur, etc., occurring in the atmosphere (CO_2, O_2, N_2, etc.), hydrosphere (H_2O), and lithosphere (mineral elements).

The organic substances include carbohydrates, proteins, lipids, nucleic acids, etc. derived from dead organisms and the wastes produced by living organisms.

The climate regime is built up by factors like light, heat, temperature, humidity, wind, precipitation, etc. The interaction of various climatic factors determines the nature of an ecosystem.

The inorganic materials keep on exchanging between biotic and abiotic components and circulating within an ecosystem. The amount of inorganic nutrients found in an ecosystem at any time is called the standing state. This tends to vary from one ecosystem to the other and from season to season.

2.4 Ecosystem Structure

The physical organization of the ecosystem components – biotic and abiotic – characterizes the structure of an ecosystem. The main structural features of an ecosystem are (i) species composition, (ii) stratification, (iii) trophic organization, and (iv) nutrients.

2.4.1 Species Composition

Every ecosystem is home to a variety of species that have evolved in and are well adapted to the prevailing environment. Species composition varies from ecosystem to ecosystem. A tropical rainforest, for example, is likely to inhabit extreme diversity of species and so is a coral reef ecosystem. A desert ecosystem, on the other hand, would witness fewer numbers of species.

2.4.2 Stratification

An ecosystem is likely to have more than one layer or strata, each comprising a population of different species. The vertical distribution of different species occupying different strata or levels in an ecosystem is known as stratification. Some natural and undisturbed forests, such as a tropical rainforest, may have as many as five strata (Fig. 2.1):

(i) Emergent layer: A few trees grow over and above the general canopy layer of the forest.
(ii) Canopy layer: The crowns of the dominant trees in the forest which cast their shadows on the vegetation layer below.
(iii) Understory: Small trees adapted to thrive under shade; when a spot in the canopy opens, the samplings below grow rapidly.
(iv) Shrub layer: Shrubby species below the understory.
(v) Forest floor: Covered with herbaceous plants, saplings, seedlings, leaf litter, and decomposing plant material.

Some ecosystems, such as deserts, on the other hand, would witness fewer vertical strata and fewer species in each layer.

Fig. 2.1 Vertical stratification of a forest ecosystem

2.4.3 Trophic Organization

Organisms in an ecosystem organize following their feeding habits, thus forming different trophic levels. A trophic level relates to the feeding and nutrition of a group of organisms in an ecosystem that occupies the same level in a food chain. Thus, there might exist many trophic levels in an ecosystem. The more diverse and complex an ecosystem, the more trophic levels in it. The producers (autotrophs) make up the first trophic level. The rest of the trophics are made up of the consumers, the heterotrophs.

Among the consumers, the primary or first-order consumers, feeding directly on producers, make up the second trophic level. There may be two or three trophic levels occupied by carnivores, the animals that survive only by eating other animals. The primary, the secondary, and the top carnivores make up the third, fourth, and fifth trophic levels, respectively.

Decomposers or detritivores feed on the dead material of all trophic levels and make up the ultimate trophic level, the detritus trophic level. The decomposers have not been considered to make up the sixth trophic level by most ecologists. The fifth trophic level the top carnivores – the apex predators – make up is considered the last one. However, looking at the decomposers' role in an ecosystem, they should be assigned to the sixth trophic level.

Human species selectively consuming the organisms almost at all trophic levels, including the decomposers (for example, bacteria through fermented products, alcoholic drinks, etc.), has also not been ascribed any trophic level by ecologists. Humans by their original species-specific nature – morphologically, anatomically, and physiologically – are strictly herbivorous (vegetarian) organisms. However, due to circumstantial developments imposed by heinous individuals, groups, traditions, and cultures, the majority of the human population the world over is omnivorous. Human populations are exerting pressures on all trophic levels and have all other living species and almost everything in nature in their control. Singh (2019) considers the human species as the seventh trophic level, impacting the ecological processes of the entire biosphere.

The trophic structure is also characterized in terms of the measure of the living material present in different trophic levels at a time, known as a standing crop. The standing crop is generally expressed as organisms' number or biomass (generally on a dry-matter basis) per unit area.

2.4.4 Nutrients

Essential nutrients all the organisms in an ecosystem need are important constituents of ecosystem structures. The nutrients accumulate in the biomass of all organisms and abiotic components, such as soil. The amount of the nutrients varies from time to time as being in the cyclic pathways, they keep exiting the ecosystem (for example, carbon, nitrogen, and oxygen loss to the atmosphere) and entering the ecosystem through photosynthesis, nitrogen fixation, and absorption by roots. The amount of inorganic nutrients found in an ecosystem at a time is known as the standing state. It represents part of the nonliving matter in the ecosystem.

2.5 Ecosystem Functions

The functional attributes of an ecosystem determine the state of the environment and, in turn, are determined by ecosystem structure in interaction with operating factors. Both ecosystem structures and functions have a high degree of compatibility. The former is ameliorated; the latter too attains amelioration and vice versa. Structural improvement means the attainment of a higher degree of complexity. Every ecosystem is laden with the natural tendency of persistence. The various functions the structural components of an ecosystem perform contribute to and ensure ecosystem persistence. For instance, tree canopies absorb solar radiation and convert the same into biochemical energy and the root system absorbs nutrients from the soil. Herbivores consume parts of the plant products and, in turn, are preyed upon by carnivores, a proportion of their population thus ending up as food for the carnivores. The carnivores, especially the top ones, exert carnivory pressure on herbivores, thus helping release excess herbivory pressure on vegetation and contributing to maintaining ecological balance. Decomposers break down complex organic matter into simple inorganic ones which in turn get recycled into the physical environment to be available for utilization by producers.

The main functional attributes of an ecosystem we need to understand are productivity, decomposition, energy flow, nutrient cycling, and development and stabilization. Nutrient cycles and development and stabilization aspects of ecosystem functioning have been covered in separate chapters.

2.5.1 Productivity

Productivity denotes the rate of biomass production or the mass of carbon generated. It is generally expressed in the unit of mass per unit area (or volume) per unit of time (for instance, if the productivity is to be measured in grams in the one-meter area in a day, it is expressed as g $m^{-2}d^{-1}$). Ecosystem productivity is of two types: (i) primary productivity and (ii) secondary productivity.

Primary productivity is expressed through the rates at which solar radiation is captured by the green plants for the synthesis of organic compounds employing photosynthesis. Primary productivity, thus, is concerned with the producers, the photosynthesizers, of an ecosystem. It is expressed as g m^{-2} yr^{-1} for dry matter and kcal m^{-2} yr^{-1} for energy.

Primary productivity is of two types: gross primary productivity (GPP) and net primary productivity (NPP). GPP is the rate of the total capture of energy or the rate of the total production of organic matter or biomass by the producers per unit area and time. NPP is defined as the rate at which energy or organic matter is stored by the producers after respiration and maintenance per unit area and time. NPP is the balance of energy or biomass left after meeting the cost of respiration and maintenance of producers.

$$\text{NPP} = \begin{pmatrix} \text{GPP—loss due to respiration} \\ \text{and maintenance} \end{pmatrix}$$

Secondary productivity is the rate of increase in the biomass of consumers per unit area and time.

In a terrestrial ecosystem, primary productivity is generally limited by temperature and moisture. The same in an aquatic ecosystem is influenced by nutrient availability. Consumers, both in terrestrial and aquatic communities, exert an impact on primary production.

2.5.2 Decomposition

Decomposition pertains to organic matter and refers to the process in which complex organic matter is broken down into inorganic raw materials, particularly CO_2, H_2O, and various inorganic nutrients. In this process, the available inorganic raw materials become available for utilization by the plants.

2.5.3 Energy Flow

Energy tied with the matter (biomass/nutrients) flows through an ecosystem via various trophic levels – from producers to top carnivores. At the producers' level, solar energy is converted into chemical energy through photosynthesis. The same energy flows along with food from the producers through herbivores and various levels of carnivores. The two basic laws of thermodynamics govern the energy flow in an ecosystem:

1. Energy can neither be created nor destroyed, but can be transformed from one form to another, or can be transferred from one component to another.
2. Every transformation or transfer of energy is accompanied by its dispersion or loss of energy in the form of heat (thus making it impossible for a cent percent transformation or transfer of energy from one organism to another).

The energy flow through an ecosystem presents two basic features, viz.:

1. Energy flow is always unidirectional. The biochemical energy in the autotrophs or producers flows towards herbivores and carnivores. The reverse flow, that is, from top carnivores to herbivores and from herbivores to autotrophs, is impossible to happen.
2. The amount of energy flow decreases with every successive trophic level. Autotrophs or photosynthesizers are capable to utilize only a small fraction of solar energy to convert it into chemical energy through photosynthesis. Most of the solar energy is dissipated. A fraction of the energy assimilated by photosynthesizers for gross primary production (GPP) is utilized for the maintenance and growth of their standing crop and for providing food for herbivores. The unutilized net primary production, eventually, is converted into detritus. The detritus also holds a fraction of the energy which is utilized by decomposers in the soil in the case of the terrestrial ecosystems and water in the case of the aquatic ecosystems (Fig. 2.2).

The energy assimilated by the herbivorous animals is used in respiration; a fraction of the unassimilated energy is utilized by decomposers. About 10% of the energy in herbivores is transferred to carnivores, and in carnivores, energy utilization takes the same fate. The top carnivores invest their energy in respiration; it is not transferred to the next trophic order. Upon the death of the producers, the herbivores, and the carnivores, their energy is utilized by the decomposers.

The cost of respiration increases with successive trophic levels. The producers consume approximately 20% of their gross productivity in respiration. The herbivores consume about 30% of their assimilated energy in respiration. In carnivores, the proportion of the assimilated energy consumed in respiration is as high as 60%. It is due to this enormous loss of energy along successive trophic levels that not much residual energy is left in an ecosystem to support another higher trophic level.

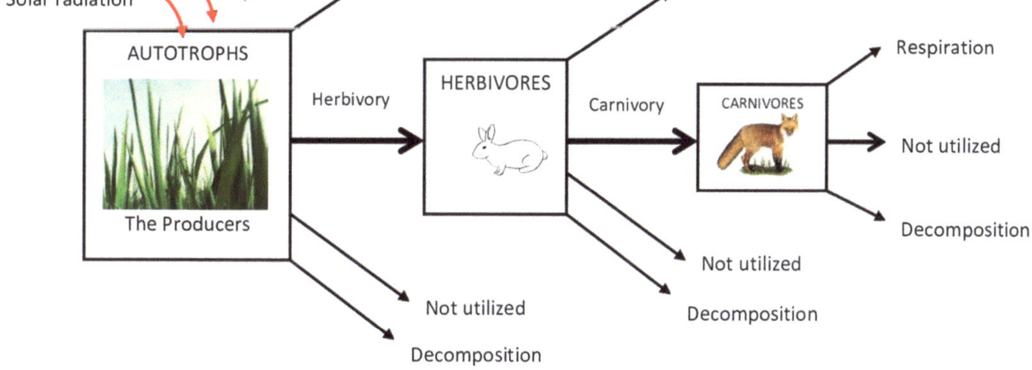

Fig. 2.2 Energy flow in an ecosystem

2.6 Food Chains

In an ecosystem, various trophic levels are related to each other through food chains. At the producers' level, the energy is converted into biochemical energy. Among the trophic levels, this energy is transferred through food for which each higher trophic level depends on the immediate lower trophic level. Some organisms among the carnivores depend on more than one trophic level. A food chain is defined as the transfer of food energy from the source to the producers, through a series of organisms, at each of the successive trophic levels. In other words, the transfer of energy from one to the next trophic level in an ecosystem is called a food chain. All types of ecosystems have two types of food chains, viz., (i) grazing food chain and (ii) detritus food chain.

2.6.1 Grazing Food Chain

The basis of a grazing food chain is constituted by the photosynthetic producers, the autotrophs. The energy flow in the grazing food chain, thus, occurs from the producers, through the herbivores, to the carnivores. It is the food synthesized by green plants that becomes the source of energy for all organisms at various trophic levels. The largest proportion of the energy occurs in the producers. Thus, the largest proportion of energy in the producers is shared by the organisms living closest to them, the herbivores, and the least is left for the top carnivores. With the natural principle of 10% energy transfer from a lower to the next higher trophic level, the abundance of energy goes on decreasing up to the highest trophic level constituted by top carnivores.

The grazing food chain is of two types: (i) terrestrial food chain and (ii) aquatic food chain. Some common grazing food chains prevailing in the terrestrial and aquatic ecosystems, as also shown in Fig. 2.3, are as follows.

2.6.1.1 Terrestrial Food Chains

Plant → grasshopper → frog → snake → hawk

Wheat grains → rabbit → fox → wolf → tiger

Rose plant → butterfly → frog → snake → peacock

2.6.1.2 Aquatic Food Chain

Phytoplanktons → zooplanktons → crustaceans → predator insect → small fish

Phytoplanktons → zooplanktons → small fish → seagull

Seaweed → herbivore fish → large fish → shark

2.6.2 Detritus Food Chain

The detritus food chain prevails in the environment of dead organic matter called detritus, such as in the soil and river and pond sediments. Detritus-feeding organisms consume a fraction of the energy in the detritus for their growth and maintenance. A fraction of the energy they consume is passed on to the predators of the detritus feeders.

The detritus food chain does not directly depend on photosynthesis. However, the detritus itself is a product of photosynthesis. The detritus food chain depends on the flux of the carbon produced in a different ecosystem. The detritus food chain is a sub-component of another ecosystem.

A detritus food chain is no less important as quite a large amount of energy in ecosystems flows through this vital food chain. A notable example is that of a mangrove ecosystem:

Leaf litter → detritus → microorganisms → crabs / shrimps → small fish → large fish

2.6.3 Y-Shaped Model of Energy Flow

All the participating organisms in the grazing food chain and the detritus food chain are so intimately associated with each other that sometimes their

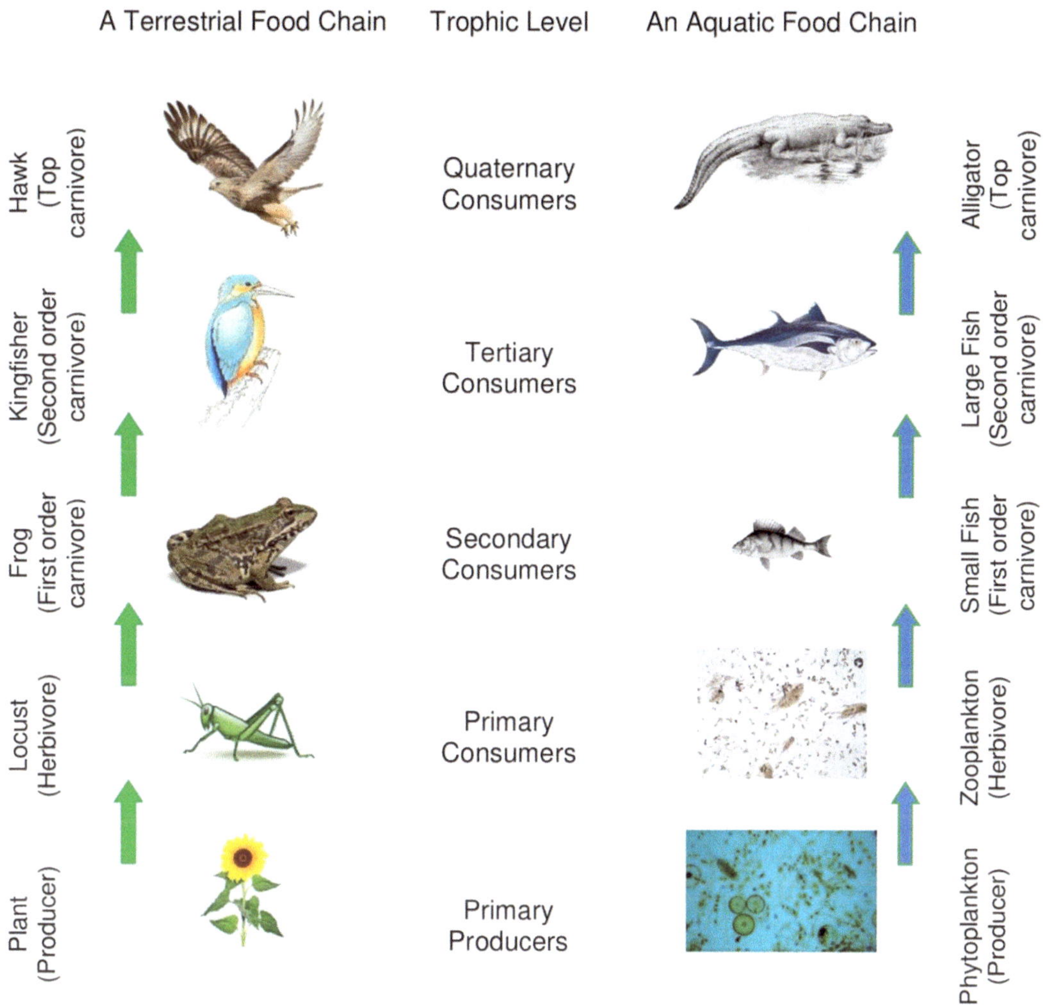

Fig. 2.3 Food chains and trophic levels in terrestrial and aquatic ecosystems

relative effect on the breakdown of original primary production is difficult to distinguish. As represented through Fig. 2.4, one arm of the model represents the grazing food chain and the other detritus food chain. Under natural conditions, the two types of food chains are inseparable. Small dead animals that once participated in the grazing food chain, in addition to animal wastes, add to the detritus. This interdependence of the two food chains when represented through a flow diagram comes out like the "Y" letter. It is why Odum (1983) called it a Y-shaped energy flow model.

The Y-shaped energy flow model is practically better compared to the single-channel models on account of the following outcomes:

(i) It represents the basic stratified structure of the ecosystems.
(ii) It distinguishes the grazing and the detritus food chains both in time and space.
(iii) Macro- and micro-consumers greatly differ in size metabolism relations.

2.6.4 The Ten Percent Law

The 10% law relates to the transfer of energy from one lower trophic level to the next higher trophic level. Proposed by Lindeman (1942), the 10% law states that from one trophic level to the other, only 10% of energy is passed on. In other

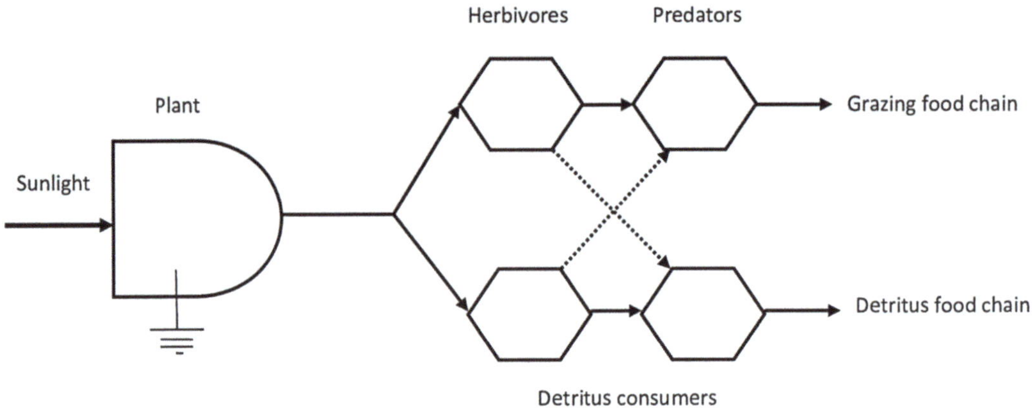

Fig. 2.4 The Y-shaped energy flow model showing linkages between the grazing and the detritus food chains in an ecosystem

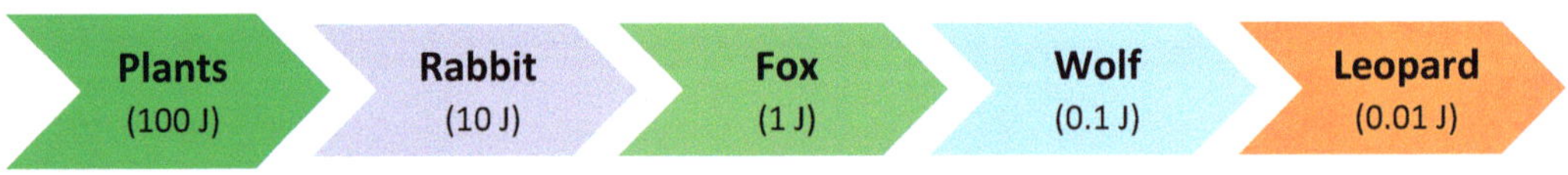

Fig. 2.5 Energy transfer efficiency in a food chain in compliance of the 10% law

words, the efficiency of energy transfer from one trophic level to the next is about 10%. Furthermore, this law indicates the inefficiency of energy capture at each successive trophic level.

If the producers receive 10,000 J of solar energy, they would capture through photosynthesis only 1% of it, i.e., 100 J. If the plants are consumed by a rabbit, it will take on 10% of energy, i.e., 10 J. If a fox eats the rabbit, it will capture only 1 J, and if the fox is eaten by a wolf, it will have only 0.1 J. Subsequently, if the wolf is preyed upon by a leopard, the latter will take only 0.01 J, according to the 10% law (Fig. 2.5).

Raymond Lindeman articulated a view of ecosystems centered on energy fixation, storage, and flows. He concluded that the ecosystem concept is fundamental to the study of trophic dynamics which, according to him, is the transfer of energy from one part of an ecosystem to another (Molles 2005).

The amount of energy transferred along the trophic hierarchy in an ecosystem in consonance with the 10% law can be calculated by using the following formulae:

$$\text{Energy at } n\text{th trophic level} = \frac{\text{Energy given by the sun}}{10^{n+1}}$$

$$\text{Energy at } n\text{th trophic level} = \frac{\text{Energy given by the plant}}{10^{n-1}}$$

Thus, if the energy given by the sun is 10,000 J, the energy at the 5th trophic level will be:

$$\text{Energy at 5th trophic level} = \frac{10{,}000}{1{,}000{,}000} = 0.01\,\text{J}$$

Similarly, if a plant has 100 J and the same is eaten by a rabbit, rabbit by a wolf, and the wolf by a leopard, the energy received by the leopard at the 5th trophic level will be:

$$\text{Energy at 5th trophic level} = \frac{100}{10{,}000} = 0.01\,\text{J}$$

2.7 Food Web

As we have seen, many food chains prevail in an ecosystem nourishing all living organisms. We have also seen how the fallen leaves and dead organisms make up detritus and thus an organic environment for several other food chains to prevail. All the food chains involving all the living organisms at all trophic levels are interconnected in an ecosystem as illustrated in Fig. 2.6. None of the food chains drifted from the rest. Thus, all the prevailing food chains make a complex web fuelling the entire ecosystem with energy.

A network of food chains interconnected at various trophic levels and forming several feeding connections among all the members of a living community is known as a food web. When we define a community as an association of interacting species, we come to realize that the food web is one of the most basic and revealing descriptions of community structure.

The food web creates a provision for an organism to derive its food from more than one type of organism in the lower trophic levels. Thus, an owl, the top carnivore, can feed on a rat and a snake. A wolf can eat a fox, a jackal, and a deer. A food web, thus, provides an opportunity for "food security" for all organisms in a biotic community.

Molles (2005) opines that a food web is essentially a community portrait. His reflections about a food web suggest that (i) feeding activities of a few keystone species may control the structure of communities, (ii) exotic predators can alter and collapse the structure of food webs, and (iii) mutualists can act as keystone species.

2.8 Ecological Pyramids

In an ecosystem, the trophic structure can be analyzed qualitatively and quantitatively by comparing the standing crop in terms of the number of individuals of all species, biomass, and energy at every trophic level.

When the three ecological parameters at all trophic levels in an ecosystem – the number of organisms, the biomass, and the energy – are plotted, they assume shape like a true pyramid. Their geometrical representation, therefore, is referred to as an ecological pyramid. An ecological pyramid, thus, is a graphic representation of

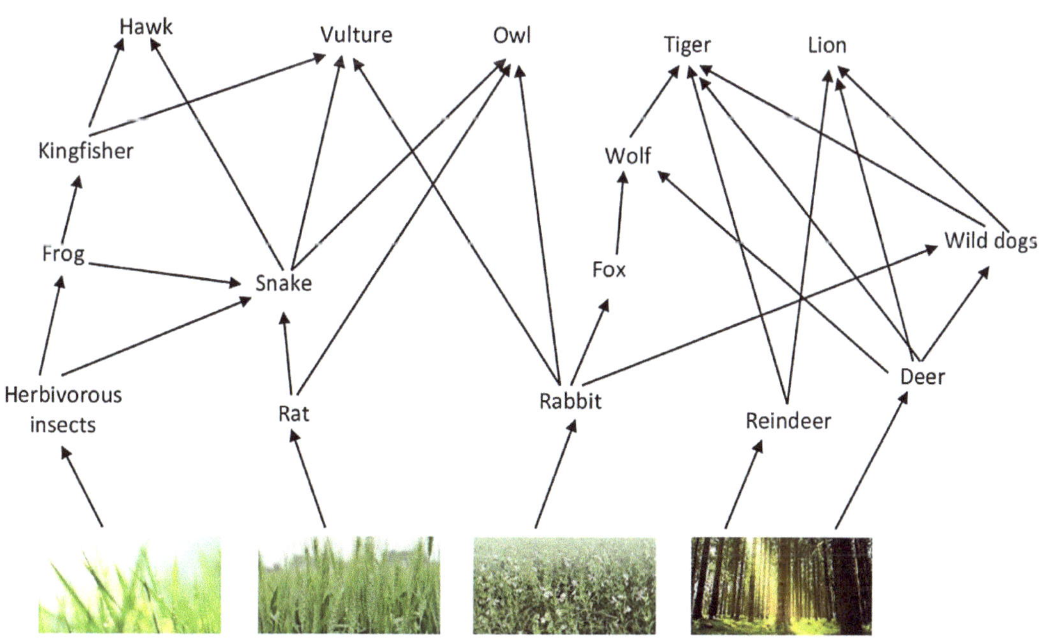

Fig. 2.6 A food web showing interconnections of several food chains in an ecosystem

an ecological parameter, viz., energy content, biomass, or the number of all organisms, at various trophic levels interconnected through food chains in an ecosystem.

An ecological pyramid is divided by bars making spaces for various trophic levels. The producers or autotrophs are accommodated in the bottom portion, the base, of the pyramid. The portion just above the autotrophs is composed of herbivores. The subsequent portions towards the peak of the pyramid contain various levels of carnivores with the top carnivores occupying the uppermost portion reaching up to the tip of the pyramid.

There are three types of ecological pyramids: (i) pyramid of numbers, (ii) pyramid of biomass, and (iii) pyramid of energy. The ecological pyramids can assume three shapes: straight or upright, inverted, and spindle shape.

2.8.1 Pyramid of Numbers

A graphic representation of the number of all the organisms of all species per unit area of various trophic levels is called a pyramid of numbers. The individuals of the producers lie at the bottom and the top carnivores are immediately below the tip of the pyramid of numbers. The pyramid of numbers describes the numerical relationships among different trophic levels of a food chain.

The producers at the bottom of the pyramid prevail in the largest numbers which support a much less number of herbivores, which, in turn, support still fewer number of the first-order carnivores, and, thus, the top carnivores in the tip portion represent the least number of individuals.

Take an example of a grassland community. The largest number of grass plants, the autotrophs, support a fewer number of insects, the herbivores. The birds still in lower number feed on the insects. The birds are fed by the predatory birds, the top carnivores with the least number of individuals (Fig. 2.7).

Another example to describe in terms of a pyramid of numbers is a pond ecosystem. Phytoplanktons, the autotrophs, support a smaller population of herbivores, the zooplanktons. Small fish feed on the zooplankton. The small fishes are fed by large fishes, which are fed by predatory birds. The number of organisms from the autotrophs to the top carnivores in the food chain goes on decreasing (Fig. 2.8).

The above-cited examples are of straight or upright pyramids. The inverted pyramid of numbers is encountered in the case of a tree ecosys-

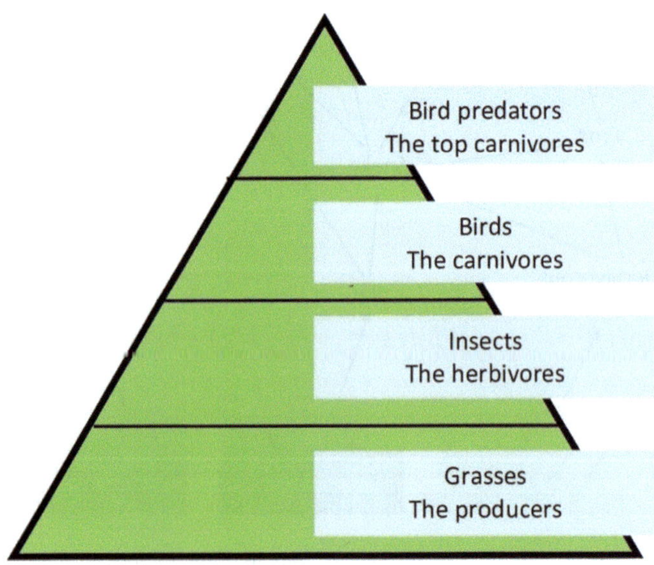

Fig. 2.7 A straight ecological pyramid of numbers representing a grassland ecosystem

tem. A single tree has several herbivorous birds which still have a very large number of ectoparasites that perpetuate on the birds for their nourishment. Again there are numerous hyperparasites, the parasites whose host is itself a parasite (Fig. 2.9). Such an inverted pyramid emerges in an individual tree laden with fruits for the birds to feed on.

There is another case of an individual tree serving like an ecosystem. Such a tree supports a large number of birds by providing them with the necessary fruits. The birds in turn are preyed upon by a very small number of carnivorous birds. In this case, a spindle-shaped ecological pyramid of numbers is carved out (Fig. 2.10).

2.8.2 Pyramid of Biomass

Biomass is the total content of living organisms or the organic matter available in an ecosystem measured on a fresh-weight or dry-matter basis. A pyramid of biomass denotes a graphic representation of the biomass available per unit area in different trophic levels in an ecosystem with producers in the base and top carnivores at the tip of the pyramid.

In a terrestrial ecosystem, the largest proportion of the biomass occurs in the producers at the base of the pyramid and goes on decreasing towards higher trophic levels being the minimum among the top carnivores at the tip of the pyramid. A pyramid of biomass in a terrestrial ecosystem, therefore, is upright or straight. It is assumed that about 10% of the biomass gets transferred from a lower to a higher trophic level. Thus, if the biomass of the producers is 10,000 kg, the amount of it transferred to the top carnivores via a straight food chain in an ecosystem will be only 1 kg (Fig. 2.11).

In an aquatic ecosystem, however, a pyramid of biomass is often of an inverted shape with the bottom of the pyramid upside and the peak downwards. The biomass of the organisms in different trophic levels in the aquatic environment depends on the reproductive potential and longevity of the organisms. Biomass is usually higher only in the case of long-lived organisms in the water. Thus, the biomass of the phytoplanktons would be less than the herbivores and it will gradually increase towards the trophic level occupied by the top consumers (Fig. 2.12).

2.8.3 Pyramid of Energy

The energy content of an ecosystem is the key source of all the functions it performs. Different trophic levels differ in their energy content. A

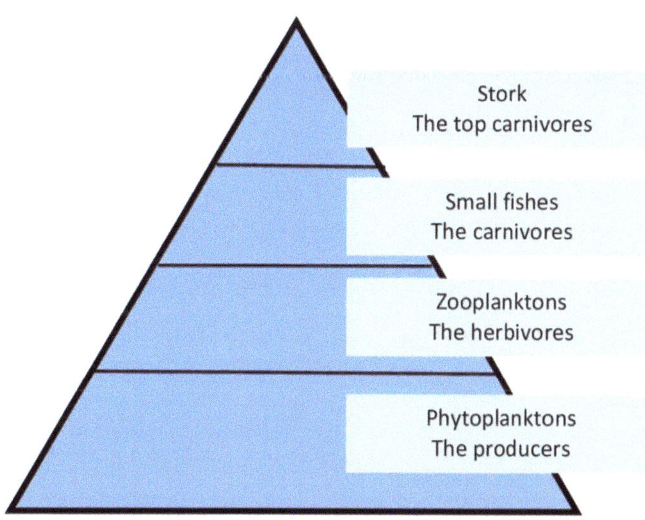

Fig. 2.8 Straight ecological pyramid of numbers of a pond ecosystem

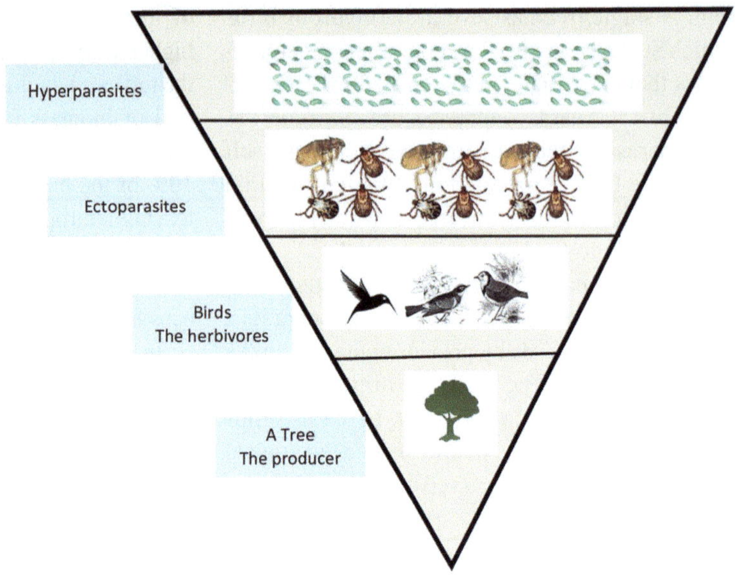

Fig. 2.9 An inverted pyramid of numbers

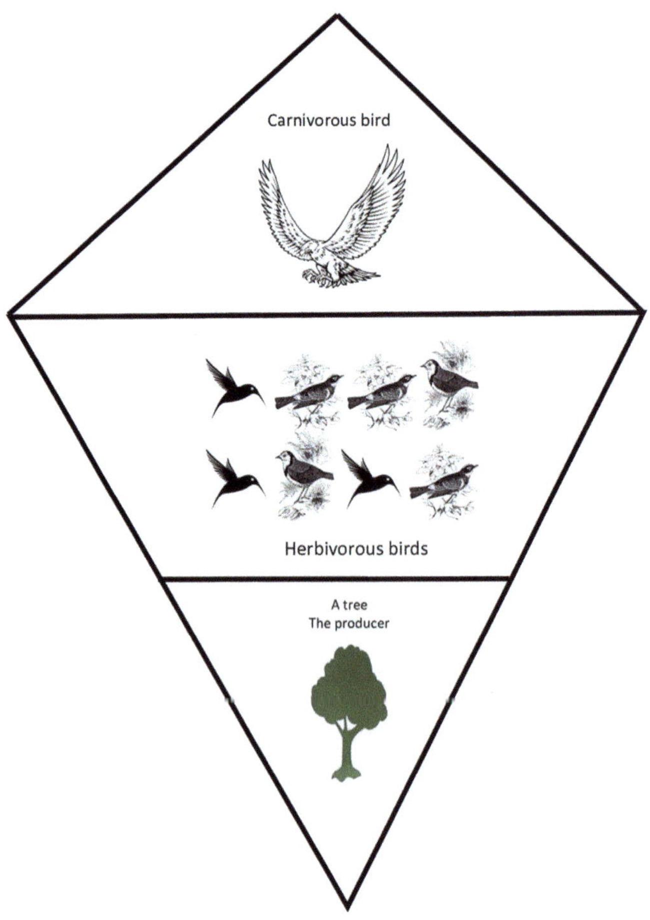

Fig. 2.10 A spindle-shaped pyramid of numbers sculpted out of a tree

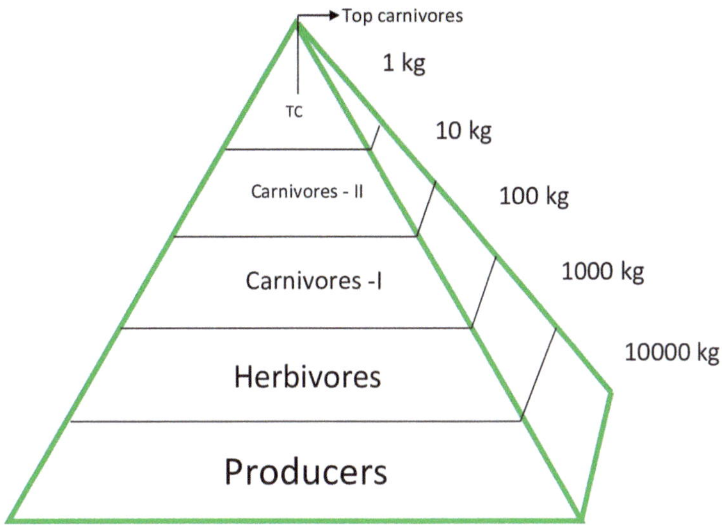

Fig. 2.11 A straight pyramid of biomass in a terrestrial ecosystem

graphic representation of the amount of energy trapped or utilized per unit area and time at different trophic levels in a food chain is called an energy pyramid. The energy content in an ecosystem is generally measured as kcal m^{-2} yr^{-1} or kJ m^{-2} yr^{-1}.

The producers or the autotrophs directly trap solar energy through photosynthesis and serve as the ultimate source of energy for all the organisms at all trophic levels. Therefore, the producers have the maximum amount of energy in an ecosystem. Herbivores, the primary consumers, derive some 10% of energy from the producers and supply 10% of their energy to the first-order carnivores and so on up till the 5th trophic level comprised of the apex predators comprising the least amount of energy in an ecosystem. Therefore, an energy pyramid is always straight or upright (Fig. 2.13).

Ecologists often document the energy pyramid based on a pond ecosystem constructed by Odum (1971). Phytoplanktons, the primary producers, trap 31,080 kJ m^{-2} yr^{-1} of the solar energy reaching the pond surface. Zooplanktons, the herbivores, contain as much as 7980 kJ m^{-2} yr^{-1} passed on to them from the phytoplanktons. They supply 2100 kJ m^{-2} yr^{-1} to the primary carnivores, such as larvae, insects, and small fish. The apex consumers feeding upon the secondary carnivores contain 126 kJ m^{-2} yr^{-1}.

2.8.4 Comparison of the Ecological Pyramids

Upon comparing all three types of ecological pyramids emerging in various ecosystems, it can be inferred that the pyramids of energy offer a true and best picture of the functional nature of communities. The following are the reasons:

- The energy pyramid is usually upright or straight pyramid in shape irrespective of variations in size and metabolic rates of the organisms.
- Whether an ecosystem is under biotic or abiotic stress, degraded or at the climax stage, anthropogenic (transformed), or natural, an energy pyramid does not assume a shape different from what it always stands in: straight.
- The number and biomass of the organisms at any trophic level depend on and vary according to the rate at which the producers are manufacturing food for the organisms.
- The number and biomass pyramids exclude the vital role of decomposers in community dynamics.

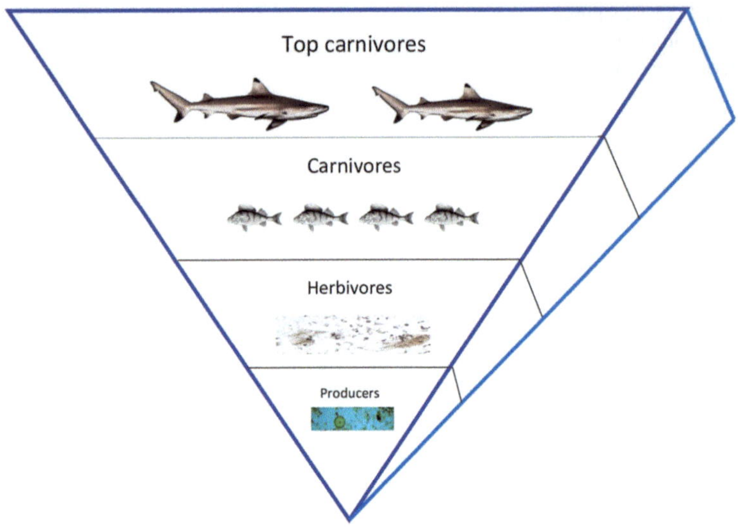

Fig. 2.12 An inverted pyramid of biomass in an aquatic ecosystem

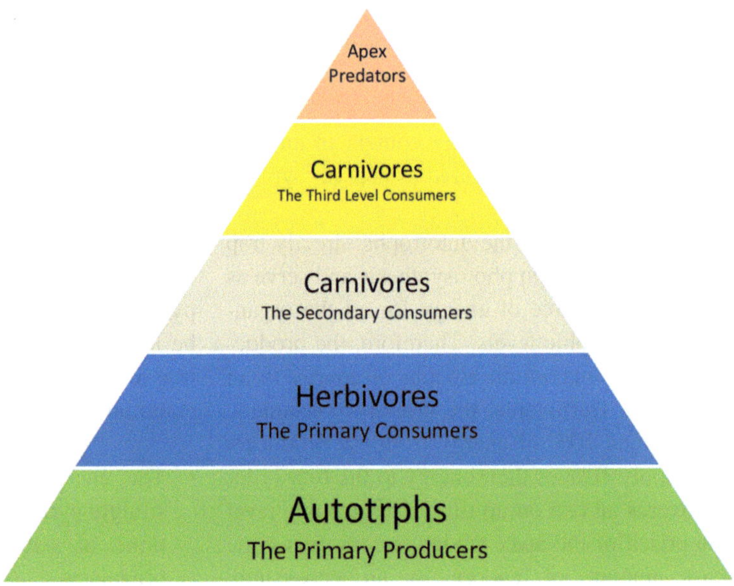

Fig. 2.13 An ecological pyramid of energy

- The pyramids of energy offer the means for determining the relative importance of the populations in a community.
- Energy flow through various populations shows that all the populations are functioning at almost the same trophic level
- A pyramid of energy – by plotting energy flow between different trophic levels – provides a suitable index for comparative evaluation of all ecosystem components.

2.8.5 Limitations of the Ecological Pyramids

Although ecological pyramids convey a crucial message about the structure, composition, functioning, and prospects (such as sustainability) of an ecosystem, there are certain limitations these graphic illustrations are faced with:

1. Ecological pyramids do not consider different species of a community. Only the organisms grouped based on their food sources are considered.
2. A species in a community may derive its food from more than two trophic levels which an ecological pyramid does not include in its count. Only the total numbers, biomass, and energy amount of the individuals based on their respective trophic levels make the criteria for their analysis. Positions of the organisms in more than one or two trophic levels remain uncertain.
3. An ecological pyramid assumes only a simple and linear food chain that hardly exists in an environment. It does not accommodate a food web that almost inevitably exists in a natural ecosystem.
4. An ecological pyramid does not take the saprophytes playing a crucial role in an ecosystem into account.
5. Detritus – for example, leaf litter and humus – is a very important source of energy, but does not find consideration in the analysis.
6. Ecosystem degradation, for example, overexploitation, fire incidence, simplification by agricultural practices, etc., remains outside the scope of an ecological pyramid.
7. The nature of an ecosystem, for example, its development as a result of primary or secondary succession, is not explained by an ecological pyramid.
8. A type of ecological pyramid – number, biomass, or energy – overlooks the aspects and significance of the other type of pyramid.
9. Seasonal and diurnal variations occurring in an ecosystem are not explained by an ecological pyramid.

2.9 Major Ecosystems

A variety of ecosystems is there in the biosphere. The diversity of the Earth's ecosystems becomes the basis of the species and intraspecies diversity. All the ecosystems would essentially differ in their structure, composition, and functional attributes. No two ecosystems would be exactly alike. Some of the biosphere's major terrestrial and aquatic ecosystems have been briefly discussed under the following subheads.

2.9.1 Terrestrial Ecosystems

2.9.1.1 Forest Ecosystem

The forests of the Earth constitute amazing diversity of ecosystems comprising a baffling variety of communities dominated by trees and inhabiting diverse species of animals. Forests have been discussed at length in a separate chapter. Forests constitute one of the most important ecosystems conserving most of the Earth's biodiversity and performing vital functions – productive, protective, and regulatory.

Inorganic and organic substances are present in the atmosphere and the soil. The proportion of the abiotic substances in the forest soils, e.g., minerals, differs from forest to forest. Forest floor litter is an important feature of most forests. Light, rainfall, temperature, microclimate, etc. also vary according to the geographical setting of a forest ecosystem.

Among the biotic components are the producers with the tree species being the dominant producers along with shrubs and herbaceous plants. The high degree of vertical stratification, especially in tropical rainforests, is one of the main features of the forests. Each layer or stratum of a forest comprises distinctive species. Natural forests are rich repositories of wild animals. Many forests embrace a very high degree of biodiversity of plants, animals, and microorganisms.

Consumers comprise a variety of insects including butterflies, locusts, leafhoppers, beetles, ants, etc., and small to large grazing animals (rats, squirrels, rabbits, deer, reindeer, wild buffaloes, elephants, etc.) are the major primary consumers or herbivores. Frogs, lizards, foxes, birds, etc. are the secondary consumers. Snakes, wolves, wild dogs, etc. are the tertiary consumers. Hawks, vultures, pythons, leopards, tigers, lions, etc. are the top consumers inhabiting a forest.

A wide variety of decomposers are active in playing their role in a forest ecosystem. They include a great many varieties of fungi, bacteria, and actinomycetes. Forests are also characterized by the overwhelming presence of photosynthesizing and nitrogen-fixing bacteria in the soil. Rates of decomposition also vary from forest to forest, being very high in the case of tropical rainforests and low in temperate forests.

2.9.1.2 Grassland Ecosystem

Grassland ecosystems are found in regions where the climate is cool to cold during the winter season and hot during the summer. These are important ecosystems from the viewpoint of their role in the conservation of a variety of herbaceous plants and economy, especially the livestock-based economy.

The abiotic components include the nutrients in grassland soils and the atmospheric/climatic factors. The soils are often rich in minerals and organic matter.

The biotic components include herbaceous plants dominated by the Gramineae or Poaceae family. Some of the main grasses include *Cynodon*, *Phragmites*, *Dichanthium*, *Cenchrus*, *Imperatus*, *Saccharum*, etc. Some grasslands in some regions may include shrubs and thin stands of trees.

Grazing animals such as cattle, buffaloes, horses, sheep, and goats among domesticated ones, and rats, rabbits, deer, etc. among the wild animals are the major primary consumers of grassland communities. A variety of insects as primary consumers also inhabit the grasslands.

Frogs, lizards, birds, foxes, jackals, birds, etc. are some of the secondary consumers found in grassland ecosystems. The top consumers include snakes, wolves, wild dogs, hawks, etc.

2.9.1.3 Desert Ecosystem

A desert ecosystem is characterized by extremely low rainfall. Vegetation is composed of trees, shrubs, and herbaceous plants and is generally sparse. In hot and arid areas, the plants performing CAM photosynthesis survive well. Most other plants are well adapted to high temperatures and water-stressed conditions. The root system of the desert plants is normally well developed and the shoot system is modified to suit extreme environmental conditions. In cold deserts, plants adapted to cold-stressed conditions are found. Lower plants, like lichens and xerophytic mosses, may also occur in desert ecosystems.

Wild animals among herbivores range from insects to camels. Domestic animals like sheep, cattle, and camels feed on desert vegetation. Reptiles, vultures, and many varieties of birds are the main carnivores. Relatively fewer animals, including blue sheep and snow leopards, are members of the food chain of the cold deserts.

2.9.2 Aquatic Ecosystems

Nearly 70% of the Earth's surface is occupied by aquatic ecosystems. These ecosystems range from small and seasonal puddles to the largest on the planet, the oceans.

2.9.2.1 River or Stream Ecosystem

A river or a stream, the form of a running or lotic ecosystem, is one of the unique ecosystems reverberating with life forms many of which do not occur in lentic (standing water) ecosystems, such as a lake or a pond. The running and freshwater of a river or a stream ecosystem are usually well oxygenated as it has a large surface to dissolve oxygen. Further, a running water body has a higher capacity to dissolve oxygen than a standing one. A river ecosystem is not uniform in its features throughout its journey from its origin to its merger into a sea.

Mineral content in a river or stream ecosystem is much less than in an ocean ecosystem. Due to being a freshwater body, sunlight penetrates the ecosystem up to deeper layers.

Freshwater algae, water grasses, weeds, amphibious plants, etc. are the main producers in a river. Floating plants and cyanobacteria proliferate serving as producers in river portions where water movement is slow.

A variety of consumers prevails in a river ecosystem. Water insects, flatworms, leeches, snails, many species of freshwater fishes, several kinds of birds, crocodiles, etc. are the main consumers inhabiting the rivers and streams.

Several species of protozoa, bacteria, and fungi are the decomposers active in the lotic ecosystems. The presence of bacteriophage is also quite common in some rivers.

2.9.2.2 Pond or Lake Ecosystem

A pond is a quiet water body. It is normally so shallow that hardly any temperature difference is noticeable from surface to bottom. A lake, on the other hand, is deep enough to demonstrate a temperature gradient from surface to bottom. However, both are almost indistinguishable from the viewpoint of their complexity in terms of the relationships between abiotic and biotic components and resulting nutrient and energy flows.

A lake ecosystem can be categorized into four major zones (Fig. 2.14) which present different characteristics of a biotic community. These zones are determined on the basis of the depth of the lake and distance from the shoreline.

1. **Littoral zone:** The near-shore area where solar radiation penetrates sediment, providing an appropriate environment for the plants, also called macrophytes, to grow well. The littoral zone can be narrow or wide. Generally young or oligotrophic ponds have a narrow littoral zone. Old or eutrophic ponds generally show a wide littoral zone due to gently sloping shorelines. The littoral zone is shallow in its depth and receives nutrients in abundance due to water runoff and nonpoint sources facilitating the plants and algae to grow profusely. The zone provides appropriate habitat for reeds, insects, crawfish, small fish, snails, etc.
2. **Limnetic zone:** The zone represents the open water area of a pond or a lake. It is quite an expansive area, larger in the young or oligotrophic lakes than in the old or eutrophic ones. The upper portion of the limnetic zone in which the sunlight can penetrate is known as the euphotic zone. This is also classified as epilimnion, the warm water region of the lake. This zone is rich in algae and other photosynthesizers. The fish population is also quite large and dense in this zone because of the higher oxygen content.
3. **Profundal zone:** This zone, also known as the hypolimnion or cold-water region, lies below the thermocline where solar radiation cannot

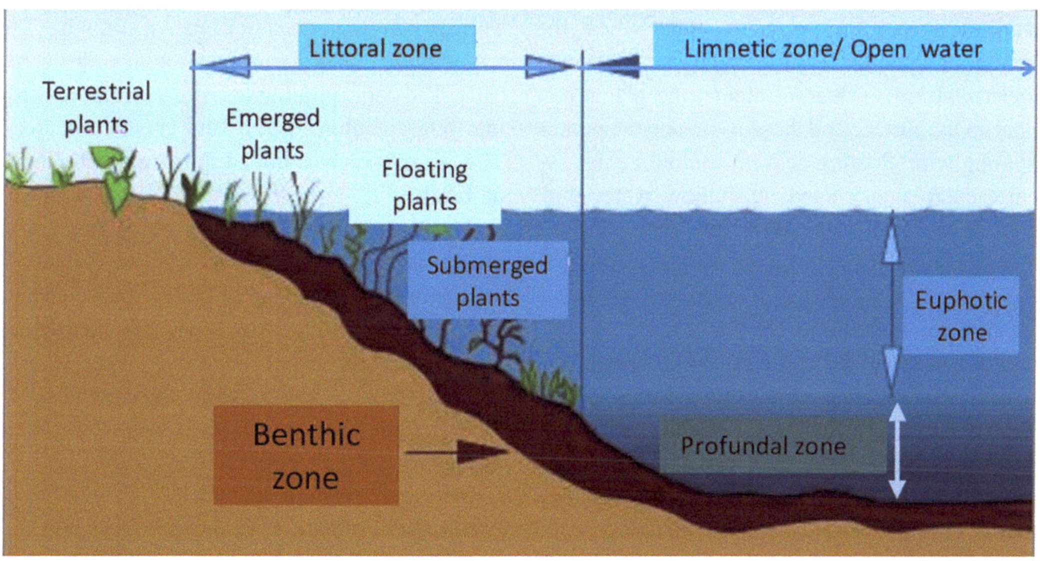

Fig. 2.14 Distinct zones within a pond or lake ecosystem

penetrate. An old or eutrophic lake would have a comparatively larger area in this zone. The fish population is also thin because of oxygen deficiency in this zone.

4. **Benthic zone:** This is the lowermost zone comprising the bottom of a lake or a pond. This zone consists of soil and sediments rich in organic matter that provides a rich culture for the decomposers to proliferate. The organic matter is sourced into dead algae, plants, animals, and their products and may also be added into the water body through water runoff. Fermentation, aerobic as well as anaerobic, of the organic matter in the benthic zone takes place. This zone functions as a digestive system of a water body and increases in thickness with the aging of a lake.

The ponds and the lakes receive energy from solar radiation. Green plants are the major primary producers. Nitrogen and phosphorus are the main nutrients for plants. Their presence promotes the growth of the producers. However, too much quantity of these nutrients in a pond or a lake ecosystem promotes an algal boom and leads to eutrophication.

The producers in a pond and a lake are the photosynthesizers that include (i) phytoplanktons, the microscopic algae or protists floating on the water; (ii) periphytic algae attached to substrates like rocks doing photosynthesis near the bottom of a pond; (iii) submerged plants thriving under water; (iv) floating plants, the plants that float on the surface and those rooted on the pond's bottom with floating leaves and/or stems; (v) emergent plants, rooted in shallow water with their leaves and stem above the water surface; and (vi) shore plants growing on the wet soil at the edge of the water body.

The major consumers in the aquatic ecosystem of a pond or a lake are (i) zooplanktons, the microscopic animals, some being single-celled, tiny crustaceans, or tiny small stages of large animals all floating on the water; (ii) invertebrates, including some macroinvertebrates; and (iii) vertebrates, like fish, frogs, salamanders, and turtles.

Decomposers in a pond or lake ecosystem depend on the detritus composed of dead and decaying plants and animals at the bottom of the water bodies. Upon decomposition, the mineral nutrients return for reuse by the primary producers. In this process, microorganisms produce water and CO_2. The decomposers also feed on the nutrients dissolved in water and thus contribute to water purification.

2.9.2.3 Ocean or Marine Ecosystem

An ocean forms the largest ecosystem on Earth. This ecosystem is ecologically the most stable one. High concentrations of salts and mineral ions in marine environments represent one of the unique characteristics of this ecosystem. Due to its largest size covering most of the Earth's surface, this ecosystem has bearing on all other ecosystems in the biosphere.

Compared to freshwater ecosystems, a marine ecosystem shows a more stable chemical composition of its saline water. An ocean creates tides on account of the gravitational influence of the moon and the sun, which is one of the unique physical phenomena associated only with this ecosystem on Earth. Sunlight penetrates the water column up to about 200 m depth. This zone is known as the photic zone. Below the photic zone lies the aphotic zone that extends up to the ocean floor, occasionally as deep as 10,000 m. The aphotic zone does not receive sunlight. The temperature of the marine ecosystem varies vertically. The concentration of oxygen varies according to temperature, depth, and chemical factors. The oceans serve as one of the largest CO_2 sinks on Earth.

Marine plants – seaweeds and algae – and phytoplanktons (diatoms and dinoflagellates) are the major photosynthetic producers of the oceans prevailing in the photic zone. The sea algae belong to Chlorophyceae, Phaeophyceae, and Rhodophyceae. The other photosynthesizers include some angiosperms and many other plants that serve as producers.

Primary consumers include zooplanktons, crustaceans, mollusks, and some fish species. Carnivorous fishes, such as mackerel, herring, etc., are the major secondary carnivores. Large carnivores, such as cod, haddock, and shark (the apex predator), etc., are the top carnivores.

Numerous types of bacteria and fungi are the major decomposers actively disposing of dead and decaying animals and plants and organic matter in the ocean ecosystems.

2.10 Summary

Introduced by Tansley in 1935, an ecosystem is an area accommodating a community of living organisms in conjunction with its environment. It is a self-reliant and self-regulating unit of nature. The Earth's ecosystems can be categorized as terrestrial and aquatic ecosystems, each comprising natural and anthropogenic ecosystems. There are two interrelated ecosystem components: biotic and abiotic. The biotic components comprise all the living beings – plants (the producers or autotrophs), animals (the consumers), and microorganisms (the decomposers). The abiotic (or nonliving) components include all factors surrounding the organisms and they are inorganic substances (mineral elements, such as carbon, hydrogen, oxygen, nitrogen, phosphorus, sulfur, etc.), organic substances (carbohydrates, proteins, lipids, nucleic acids, etc.), and climate regime (like light, heat, temperature, humidity, winds, precipitation, etc.). The inorganic materials keep on exchanging between biotic and abiotic components and circulating within an ecosystem. The amount of inorganic nutrients found in an ecosystem at any time is called the standing state. The physical organization of the ecosystem components characterizes the structure of an ecosystem. The main structural features of an ecosystem are species composition, stratification, trophic organization, and nutrients. Species composition varies from ecosystem to ecosystem. An ecosystem is likely to have more than one layer or strata, each comprising a population of different species. The vertical distribution of different species occupying different strata or levels (emergent layer, canopy layer, understory, shrub layer, and forest floor) in an ecosystem is known as stratification. Organisms in an ecosystem organize per their feeding habits, thus forming different trophic levels. There may be many trophic levels in an ecosystem. The producers (autotrophs) make up the first trophic level. The rest of the tropics are made up of the consumers, the heterotrophs. There may be two or three trophic levels occupied by carnivores. Decomposers or detritivores feed on the dead material of all trophic levels and make up the ultimate trophic level, the detritus trophic level. The trophic structure is also characterized in terms of the measure of the living material present in different trophic levels at a time, known as standing crop generally expressed as organisms' number or biomass on a dry-matter basis per unit area. Essential nutrients all the organisms in an ecosystem need are important constituents of ecosystem structures.

Functional attributes of an ecosystem determine the state of the environment and, in turn, are determined by ecosystem structure in interaction with operating factors. The main functional attributes of an ecosystem are productivity, decomposition, energy flow, nutrient cycling, and development and stabilization. Productivity denotes the rate of biomass production or the mass of carbon generated. It is generally expressed in the unit of mass per unit area (or volume) per unit of time. Ecosystem productivity is of two types: primary and secondary productivity. Primary productivity is expressed through the rates at which solar radiation is captured by the green plants for the synthesis of organic compounds through photosynthesis. Primary productivity is of two types: gross primary productivity (GPP) and net primary productivity (NPP). NPP = (GPP – loss due to respiration and maintenance). Secondary productivity is the rate of increase in the biomass of consumers per unit area and time. Decomposition pertains to organic matter and refers to the process in which complex organic matter is broken down into inorganic raw materials. In this process, the available inorganic raw materials become available for utilization by the plants. At the producers' level, solar energy is converted into chemical energy through photosynthesis. The same energy flows through food from the producers through various levels of carnivores. The cost of respiration increases with successive trophic levels. In an ecosystem, various trophic levels are related to each other through food chains representing the transfer of food energy from the source in the producers, through a series of organisms, to each of the successive trophic levels. All types of ecosystems have two types of food chains, viz., grazing food chain and detritus food chain. Proposed by Lindeman (1942), the 10% law states that from one trophic level to the other, only 10% of energy is passed on. All the prevailing food chains make a complex web known as the food web.

An ecological pyramid is a graphic representation of an ecological parameter, viz., energy content, biomass, or a number of all organisms, at various trophic levels interconnected through food chains in an ecosystem. There are three types of ecological pyramids: the pyramid of numbers, the pyramid of biomass, and the pyramid of energy. Pyramids of numbers can be straight, inverted, and spindle-shaped. A pyramid of biomass in a terrestrial ecosystem is upright or straight. In an aquatic ecosystem, a pyramid of biomass is often an inverted shape. An energy pyramid is always straight. The major terrestrial ecosystems include the forest ecosystem, grassland ecosystem, and desert ecosystem. The major aquatic ecosystems include river or stream ecosystems, pond or lake ecosystems, and ocean or marine ecosystems. A pond or lake ecosystem can be categorized vertically into littoral zones, limnetic zone, profundal zone, and benthic zone.

2.11 Exercises

2.11.1 Multiple-Choice Questions

1. Who introduced the term "ecosystem"?
   (a) A. G. Tansley
   (b) Hutchinson
   (c) E. P. Odum
   (d) Raymond Lindeman
2. Which of the following is an autotroph?
   (a) Algae
   (b) Seaweed
   (c) Cyanobacteria
   (d) All of the above
3. Which of the following are examples of a lotic and a lentic ecosystem, respectively?
   (a) A pond and a lake
   (b) A river and a pond
   (c) A pond and a river
   (d) A lake and a river
4. What trophic level do the herbivores constitute in an ecosystem?
   (a) First
   (b) Second
   (c) Third
   (d) Fourth
5. Which of the following is the primary consumer in a forest ecosystem?
   (a) A rabbit
   (b) A deer
   (c) An elephant
   (d) All of the above
6. What trophic level in an ecosystem do the tertiary carnivores represent?
   (a) Second
   (b) Third
   (c) Fourth
   (d) Fifth
7. What is not true about the standing crop?
   (a) It is the amount of biomass present in an ecosystem.
   (b) It represents the entire living matter of an ecosystem.
   (c) It gives an idea about the continuous synthesis and consumption of biomass.
   (d) It is the same as the standing state of an ecosystem.
8. What is true about the standing state of an ecosystem?
   (a) It is the amount of total inorganic nutrients occurring in an ecosystem.
   (b) It circulates between the biotic and abiotic components of an ecosystem.
   (c) It represents part of the nonliving matter in an ecosystem.
   (d) All of the above
9. The net primary productivity represents the amount of energy
   (a) available in primary producers after cell respiration

(b) available in primary producers before cell respiration
(c) the plants provide to herbivores
(d) synthesized during photosynthesis

10. Who proposed the 10% law?
(a) Ernst Haeckel
(b) A. G. Tansley
(c) Raymond Lindeman
(d) E. P. Odum

11. What percentage of energy produced at the trophic level composed of producers is available to secondary consumers?
(a) 1%
(b) 10%
(c) 50%
(d) 90%

12. Around how much percentage of available energy is lost through heat at each trophic level?
(a) 10%
(b) 20%
(c) 60%
(d) 90%

13. Suppose a grassland ecosystem receives solar radiation equal to 100,000 kJ and 1% of it is transformed into biochemical energy through photosynthesis. How much energy would pass on to the insectivorous lizards via the "grass → herbivore insects → lizard" food chain?
(a) 1000 kJ
(b) 100 kJ
(c) 10 kJ
(d) 1 kJ

14. A tree laden with fruits serves as an ecosystem. Some 50 birds feed on the fruits of the tree. Two predator birds, in turn, feed on the primary consumers. What shape the ecological pyramid of numbers would assume in such a scenario?
(a) Straight
(b) Inverted
(c) Either straight or inverted
(d) Spindle shape

15. Which of the following ecological pyramids never shows variable shapes?
(a) The pyramid of energy
(b) The pyramid of numbers
(c) The pyramid of biomass
(d) All of the above

16. An energy pyramid is a geographical representation of the
(a) Number of individuals at each trophic level
(b) The flow of energy through each trophic level
(c) The flow of biomass across trophic levels
(d) Animal populations in an ecosystem

17. CAM plants are most likely to occur in a
(a) Tropical rainforest ecosystem
(b) Desert ecosystem
(c) Marine ecosystem
(d) Mangrove ecosystem

18. An open landscape of spongy mosses, lichens, and dwarf willows would represent the vegetation type of the
(a) deserts
(b) tropical savanna
(c) tropical rainforest
(d) tundra

19. A boreal forest is generally dominated by
(a) evergreen conifers
(b) oak trees
(c) bushes
(d) herbaceous plants

20. Match the items in column A with those in column B

Column A	Column B
(a) Boreal forest	(i) Most of India
(b) Tundra	(ii) Africa
(c) Tropical dry forest	(iii) Taiga
(d) Tropical savanna	(iv) Southeast Asia
(e) Tropical rain forests	(v) Forests in the Arctic

(a) a-i, b-ii, c-iii, d-iv, e-v
(b) a-ii, b-v, c-iv, d-i, e-iii
(c) a-v, b-iv, c-ii, d-iii, e-i
(d) a-iii, b-v, c-i, d-ii, e-iv

Answers: 1-a, 2-d, 3-b, 4-b, 5-d, 6-d, 7-d, 8-d, 9-a, 10-c, 11-a, 12-d, 13-c, 14-d, 15-a, 16-b, 17-b, 18-d, 19-a, 20-d

2.11.2 Short-Answer Questions

1. Define an ecosystem.

2. Write examples of anthropogenic ecosystems among terrestrial and aquatic ecosystems.
3. Differentiate between lotic and lentic ecosystems.
4. What are the biotic components of an ecosystem?
5. What is the productivity of an ecosystem?
6. Write examples of the grazing food chain.
7. What do you know about the detritus food chain?
8. Why is the pyramid of energy always straight?
9. Name the layers/strata in a tropical rainforest.
10. Mention the main characteristics of a desert ecosystem.

2.11.3 Long-Answer Questions

1. Give an account of the ecosystem components and their interrelationships.
2. Explain ecosystem structure with an illustration of a forest's stratification.
3. What is an ecological pyramid? Discuss in detail the main features of all three ecological pyramids.
4. Write an essay on the biomes of the biosphere.
5. Write short notes on the following:
   (a) Trophic levels in an ecosystem
   (b) Food web
   (c) The Y-shaped energy flow model

References

Elton C (1927) Animal ecology. The Macmillan Co, New York. 207 pp

Lindeman RL (1942) The trophic-dynamic aspect of ecology. Ecology 23(4):399–418. https://doi.org/10.2307/1930126

Molles MC (2005) Ecology: concepts and applications. McGraw Hill, Boston. 622 pp

Odum EO (1971) Fundamentals of ecology. Saunders, Philadelphia. 574 pp

Odum HT (1983) Systems ecology: an introduction. Wiley, New Jersey. 644 pp

Singh V (2019) Fertilizing the universe: a new chapter of unfolding evolution. Cambridge Scholars Publishing, London. 285 pp

Tansley AG (1935) The use and abuse of vegetational concepts and terms. Ecology 16(3):284–307. https://doi.org/10.2307/1930070

3 Population Ecology

The population is defined as a group of individuals of a species occupying a particular area at a particular time. In an ecosystem population of a particular species is seldom isolated from that of other species. Therefore, ecologists categorize populations into monospecific populations and polyspecific or mixed populations. The former is the population of individuals of only one species, while the latter represents the population of individuals of more than one species. Polyspecific populations invariably denote a community involving heterogeneity of species' populations sharing a common habitat and interaction.

3.1 The Habitat

Habitat is an area or a place where a species lives, grows, and reproduces. A habitat provides a suitable environment for the population of a species (or several species). Clements and Shelford (1939) defined habitat as the physical conditions that surround a species, species population, or assemblage of species, or community. Habitats often vary from species to species or group or populations of the species. Immediate surroundings and physical factors of an individual plant and animal constitute a microhabitat.

Ecologists often use the term biotope for a habitat shared by many species and biome for the set of flora and fauna living in a habitat and occupying certain geography.

3.2 Defining Population Ecology

Population ecology is a branch of ecology that deals with the study of species populations, their dynamics, and their interactions with their physical environment. Population ecology can further be divided into autecology and synecology. Autecology means the study of individual species in relation to the environment, whereas synecology refers to the study of the organisms' groups in relation to the environment. Autecology, arguably, is species ecology. Since a population, not individual species, is a level of species organization making gene pool more coherent and stable, autecology is now merely an archaic term. Species ecology, a term used widely, is useful in understanding life history, behavior, and adaptation to the environment of individual species. Synecology is regarded as synonymous with community ecology underlining three components woven together, viz., population, community, and ecosystem. Odum (1953), a celebrated ecologist of the twentieth century, therefore, categorized synecology into population ecology, community ecology, and ecosystem ecology. Later works of Hutchinson (1978), Odum and Barrett (2005), Vandermeer and Goldberg (2013), and Rockwood (2015) led to a revolution in population ecology.

Population ecology is vital for the conservation of species populations in strategically managed areas, for example, in national parks,

V. Singh, *Textbook of Environment and Ecology*, https://doi.org/10.1007/978-981-99-8846-4_3

wildlife sanctuaries, and biosphere reserves. The study of population dynamics in ecosystems emanates from demography that involves the interests of biologists, ecologists, sociologists, mathematicians, and statisticians. Population ecology helps us understand and strike ecological balances.

The population of a species assumes a trend as a result of reproduction and environmental factors. While reproduction is a function of the fecundity of an individual species, including sex ratios, birthrates, and death rates, the population size, density, and distribution depend on the active and passive transport of the individuals. While active transport is a matter of individuals' instincts, passive transport is due to environmental factors, such as winds, water flows, etc. The environment is a major factor determining the population growth of a species. An appropriate environment with the provision of appropriate temperature range, moisture, nutrients, etc. is conducive to constant population growth. But the environment is never static or never in the "appropriate" state. It continuously changes and holds the population growth under check. Continuous growth in a species' population imposes constraints on the population of other species as well as on itself.

3.3 Population Characteristics

The population of a species occupying a particular area at a particular time has its distinguishable characteristics, especially in terms of size, density, dispersion patterns, age structure, natality, and mortality.

3.3.1 Population Size

Population size implies the number of individuals of a species in the assemblage in a particular place or area at a particular time.

3.3.2 Population Density

Population density is defined as the number of individuals of a species per unit area or volume of the environment. Population density measures might vary from organism to organism, for instance, the number of oak trees per hectare, the number of phytoplanktons per cubic meter of water, or the weight (biomass) of prawn per hectare of water surface.

Population density can also be measured in terms of crude density, i.e., number or biomass of individuals per unit of area, or ecological (specific) density, i.e., the number and biomass of individuals per unit area of the habitat they occupy.

3.3.3 Population Growth Rate

The population is a dynamic entity. Population size and density increase or decrease in tune with time. Any state of the population is time relative. There is also a possibility that the population size and density remain constant.

Such a condition might be a reality in human-managed systems, for example, the size of livestock population or the number of plants in a garden over time. Under natural conditions, however, a static population scenario is rare.

Population dynamics depends on or is determined by four factors, viz., (i) natality, (ii) immigration, (iii) mortality, and (iv) emigration. Should N be the number of organisms of a species and t the time, then the change (denoted by the symbol delta, Δ) will be as follows:

ΔN = change in the number of organisms

$\frac{\Delta N}{t}$ = growth rate, i.e., the average rate of change in the number of organisms per unit time (hours, days, months, or years)

$\frac{\Delta N}{N\Delta t}$ = specific growth rate

3.3.4 Dispersion

Organisms of a species occupy a specific location in relation to each other. In other words, the organisms in their population are dispersed in a specific fashion. Dispersion, thus, reveals the spatial pattern of the organisms in a population relative to one another. Dispersion patterns generally assume three dimensions: (i) random dispersion, (ii) clumped dispersion, and (iii) regular dispersion. Characteristics of each of the dispersion patterns have been described in Table 3.1.

Table 3.1 Features of population dispersion patterns

Characteristic	Random dispersion	Clumped dispersion	Regular dispersion
Position	Position of an organism in the population is not related with the positions of its neighbours.	Organisms in a population are aggregated into patches.	Organisms in a population occupy a place more or less equidistant from one another.
Frequency	Relatively rare in nature	Very common	Rare in nature, but common in human-managed systems
Example	Natural forests, grasslands	In most of the populations	Croplands, gardens, dairy farms
Illustration	** * *** *** ** ******** * * ** * ** *** * ** ** * *** ** ** *** *** ** *** * ** * *** * ** *** ** ** *	* *** **** * *** ** ** ***** ** * * ** * * *** * *** ** * * * ** **** * ****** *** ***** * ** *	* *

3.4 Ecological Age Structures and Age Pyramids

Every organism of a species in a population has its life span. Individuals of different ages, therefore, exist in a population. Organisms in each age group exist in a certain proportion constituting a specific age structure. Age structures prevailing in a population are major determinants of natality and mortality. The proportion of individuals of a certain age group will determine how stable and vibrant a population is. There are three types of ecological age groups characterizing a population:

1. Pre-reproductive
2. Reproductive
3. Post-reproductive

These three stages vary from species to species. Many tree species have pretty long pre-reproductive and exceptionally long reproductive periods. The human species enjoys almost equal length of each of the stages of life. There are yet some insect species that have long pre-reproductive phase, short reproductive phase, and no post-reproductive phase of their life span.

An age pyramid geometrically represents the proportions of different age groups (phases of life span) in a population of the organisms of a species. The bottom of an age pyramid comprises the pre-reproductive phase, the middle portion of the reproductive phase, and the uppermost tapering portion of the post-reproductive phase. There are three age pyramids, namely, (i) broad-base pyramid, (ii) bell-shaped pyramid, and (iii) urn-shaped pyramid. Each type of age pyramid represents the population size and growth pattern of the population of particular species. The main features of the age pyramids have been described in Table 3.2.

Density among human populations does not reveal a definite trend in population growth between various countries of the world, for example, the developed or developing countries. In human society, an age pyramid may also represent the age-sex structure of the population of an area or a country. A triangle-shaped or "expansive" pyramid reveals a very large population of young people and a high population growth rate. Such a pyramid is often a characteristic of developing countries with a much greater preponderance of young population, a high birthrate, and low life expectancy. This type of age-sex pyramid is typical of Mexico, Nigeria, and Bangladesh.

When the human population (age and sex) remains constant over time, the population pyramid emerges as what is called a bell-shaped or "stationary" pyramid. In stationary or stable populations, birthrates and death rates are more or less equal. This is typical of age-sex proportions of French and Czech Republic populations. Developed economies of the world often flourish within this kind of population pyramid. Most European countries and many other developed countries are glaring examples of stationary pyramids. Among them, Estonia records zero growth in its population.

When the human population comprises an overwhelming proportion of older people and an extremely low proportion of younger people and shows a declining trend, the pyramid emerging is urn-shaped or "constrictive" type. The life expectancy of the individuals is high and death rates, as well as birth rates, are low. Populations with decreasing growth rates generally preclude dependence on immigrants. Such a pyramid is typical of a country of highly educated and high-income people enjoying all amenities of life and exemplary health services and who are provided with incentives for birth control. A few adverse environmental factors might also be operating over the populations depicting a constrictive-type pyramid. Among the countries revealing the constrictive-type pyramid, the population of Puerto Rico registers a maximum decreasing rate (−1.73% per year).

The countries showing expansive-type pyramids are overwhelming in number, the countries with almost stable population growth are next in order, while those with negative growth rates and representing constrictive-type pyramids are fewer in number (UNO 2019).

Table 3.2 Key attributes of age pyramids

Pyramid features	Triangle-shaped or expansive pyramid	Bell-shaped or stationary pyramid	Urn-shaped or constrictive pyramid
	Broad base triangular gradually tapering towards the top; like ecological pyramid of energy	Rectangular base, tapering towards the top	Typically has inverted shape, tapering towards the bottom; looks like an urn or a beehive
Figure		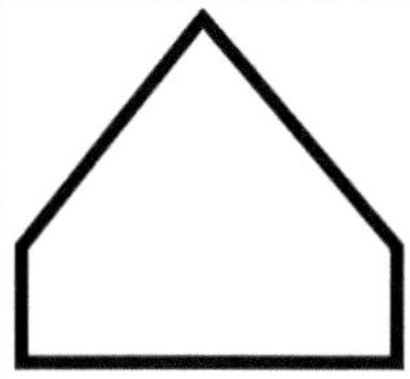	
Population type	Young population	Stable population	Declining population
Pre-reproductive population proportion	High	Moderate	Low
Reproductive population proportion	High	Moderate	Low
Post-reproductive population proportion	Low	Small	High
Birthrate and population growth	High; often exponential	Slow and stable; pre-reproductive and reproductive phases more or less equal in size; birth- and death rate almost equal	Declining; dwindling reproductive population
Examples	Yeast, paramecium, housefly, human population of some developing countries	Population of a species under control of environmental (or anthropogenic) factors	A population of a species constrained by environmental (or anthropogenic) factors

3.4.1 Natality

Coming into being of new individuals of an organism of a species adds to the population count. Natality covers this aspect of life. Individuals of a species come into existence through sexual or asexual reproduction. Cell division, germination, hatching, and birth are the various means by which individuals add to their species' population. Natality or birthrate especially counts for human beings (or other mammals) and is defined as the number of offspring produced per female per unit of time. Natality is of two types: (i) physiological and (ii) ecological.

Physiological natality, also called absolute, or potential natality, relates to the theoretically maximum production of offspring under ideal conditions, that is, the conditions not limited by ecological factors, but are regulated or limited only by physiological factors. This potential of offspring numbers per unit of time is also known as fecundity and varies from species to species.

Ecological natality is the growth in the population under an actual, existing condition. It takes into consideration all the operating or possible environmental factors and is also known as the fertility rate. It is expressed as follows:

$$\frac{\Delta Nn}{\Delta t} = \text{the absolute natality}(\mathrm{B})$$

$$\frac{\Delta Nn}{N\Delta t} = \text{the specific natality}(\mathrm{b}),$$

natality per unit of population

where N = initial number of organisms
n = new individuals in the population
t = time

The major characteristics determining the rate at which females of a species produce offspring are as follows:

(i) Age of attaining the reproduction stage
(ii) Clutch size, i.e., the number of offspring produced on each occasion
(iii) Length of time between successive reproduction events

The natality of an organism generally enhances during the maturity period and gradually falls as the organism becomes older. Breeding time is also different for different species and, in most cases, varies according to seasons of the annual cycle, with populations of some species showing a seasonal peak. In some species (human beings, for instance), breeding occurs throughout the year. Clutch size may also vary according to a climate type; for example, some plants, many birds, some small mammals, and some insects show a larger clutch size in temperate than in tropical climates.

3.4.2 Mortality

Mortality means the number of individuals in a population dying in a given period, or per unit of time. Mortality, often referred to as death rate, is of two types, akin to natality: (i) physiological and (ii) ecological.

Physiological mortality, also referred to as minimum, specific, or potential mortality, is determined by the physiological longevity of the individuals in a population of a species under non-limiting conditions. Every individual in a population is sure to die at a point in time according to the designs of apoptosis even if the most ideal environment is prevailing for its survival.

Ecological mortality is the actual or realized rate of death of individuals in a population as determined by ecological conditions or by operating environmental factors. Ecological mortality, unlike physiological mortality, is not a constant and varies following varying external factors.

A birth-death ratio, i.e., $\frac{\text{births}}{\text{deaths}} \times 100$, is known as vital index. The population as a whole at a certain point in time takes into consideration only the surviving members, not the dead ones. Therefore, survival rates, not the death rates, matter in the demographic analyses at a point in time.

3.4.3 Biotic Potential

The power of growth every population possesses is inherent. The population growth rate is subject to individual vigor and the environment. If the environmental conditions are favorable, i.e., when space, food, and environmental factors do not impose limitations, the population growth rate per individual or the specific growth rate is maximum as well as stable. This is characteristic of a particular age structure. Let it be expressed as symbol r, exponent in the differential equation for population growth in a favorable environment not flawed by constraining factors. This is an index of the inherent power of population growth.

$$r = b - d$$

where b = instantaneous specific natality (rate per individual per unit time)

d = instantaneous specific mortality

The overall growth of a population is conditioned by an appropriate environment, that is, under unlimited environmental conditions (r) depending upon (i) the age composition and (ii) specific growth rates. Thus, the value of r may vary for a species depending on its population structure. The value of r, thus, expresses together three parameters of a population, viz., natality, mortality, and age distribution.

The specific growth rate in a population with stationary age distribution is known as the intrinsic rate of natural increase (r *max*). The maximum value of r attained by a population of a species is known as biotic potential or reproductive potential. The values of r *max* range from 0.02 in large animals to 20,000 in bacteria.

At a certain point in time in some populations, biotic potential is readily realized. For example, when bacteria, say *Escherichia coli*, are grown in a Petri dish in a laboratory under ideal conditions, such as all essential nutrients and appropriate temperature range, bacteria reproduce at their biotic potential, and their population size doubles at a short interval of few (about 20) minutes. The other example is of blue-green algae in a nutrient-rich pond or a lake. Continuously fed by nutrients in a water body, blue-green algae grow at their potential and cover the entire water surface within a few days. If the graph of a population growing at its potential rate is plotted, it will be very steep.

Biotic potential in nature, however, is seldom realized due to limitations imposed by environmental resistance. Every population in a given environment is held into a kind of natural check as the operating environmental factors cannot practically allow unlimited conditions to persist sustainably. The most common factors imposing limitations on population growth are as follows.

3.4.3.1 Space

The space in which a given population prospers is never unlimited. Terrestrial populations have to grow within their ecosystems which have limited area. Aquatic plants and animals have to face the space limitations of their respective water bodies. A population of bacteria in a laboratory cannot cross the space limit of a Petri dish.

3.4.3.2 Food and Water

Supplies of food and water in an ecosystem might be adequate for small size and less dense populations. As a population grows in size and density beyond the limit of food and water supply rates, its individuals will usher in a state of competition for survival. As a result, the populations would resume their size to keep pace with the supplies of the basic needs of survival.

3.4.3.3 Light

Light is the most important resource for the photosynthesizers – green plants, algae, and cyanobacteria. Lack of light leads to the suppressed growth or death of the plants in an ecosystem. Some ecosystems with no access to adequate light or that get no light at all due to some obstructions would face suppressed growth rates and the elimination of many species of plants. For example, top-canopy trees and upper-story layer vegetation do not allow enough light to reach the ground flora in a forest ecosystem, thus imposing limitations on the growth of lower-level vegetation.

3.4.3.4 Predators

High-density populations in a natural ecosystem usually invite predators. The predators in a natural ecosystem serve as biological agents to maintain populations of their prey in an appropriate size. Insect pests inflict a barrier on the size and densities of the plants.

3.4.3.5 Parasitism

If an organism of a species feeds on the fluids or tissues of an organism of another species, the relationship between them is known as parasitism. The former organism is called a parasite and the latter a host. The parasite, usually much smaller in size than its host, goes on intensifying stress to the extent that the host's growth comes to cease, and in many cases, the host even dies. Population density and parasitism are generally closely related.

3.4.3.6 Diseases

Every organism in a population at a certain stage of life is prone to one or the other disease. High-density populations are generally affected more severely by diseases. The contagious diseases are generally density-dependent. Diseases are the natural barriers not allowing a population assume its potential growth.

3.5 Interactions Among Populations

Naturally evolved ecosystems on Earth generally harbor mixed populations of plants, animals, and microorganisms. These populations do not exist in isolation, but in interactions with each other. Interactions among populations are not simple, but complex ones. Interactions can be of two types: (i) intraspecific, i.e., interactions among the organisms of the same species, and (ii) interspecific, i.e., interactions among organisms of different species. Interactions can be among the populations of the species of the same kingdom and those belonging to different kingdoms.

Interactions among populations have been covered under an umbrella term "symbiosis" – a term introduced by De Bary in 1879. Symbiosis means togetherness or living together. As all the organisms of all populations in a natural ecosystem live together, i.e., in a symbiotic relationship, Odum (1971) endorsed the term but categorized the term into (a) positive and (b) negative interactions. However, contemporary ecologists use "symbiosis" for the interspecific interactions which are mutually beneficial for both species. Some ecologists also use the term "mutualism" for a relationship beneficial for the interacting organisms. On the other hand, a relationship in which one of the interacting species is harmed is often designated as antagonism.

Based on earlier classifications, such as those presented by McDougall (1918) and Haskell (1949), Burkholder (1952) presented another classification underlining co-actions between various species in an ecosystem (Table 3.3).

3.5.1 Competition

Competition is a natural and omnipresent phenomenon in life. Individuals in a population or individuals of different species often exist in the competition of various sorts. Space, resources, and time for an individual are not infinite to survive, perform, and accomplish their life cycle. An individual competes with others in a bid to acquire appropriate environmental conditions to survive out of the limited space, resources, and time (life expectancy) and performs and accomplishes their life cycle which, in essence, is the gist of life.

Competition among populations is of two types, viz., (i) intraspecific and (ii) interspecific. The former is the competition between the members of the same population and the latter is between the populations of different species.

Resources for which the populations compete can be of two types, viz., (i) space necessary for growth, movement, mating, preying, hiding from predators, and nesting for reproduction and (ii) resources like light, inorganic nutrients, and water for photo- and chemoautotrophs and organic foods and water for heterotrophs.

3.5.1.1 Intraspecific Competition

Also called scramble competition, intraspecific competition depends on population density and is a crucial factor to regulate population size. It often occurs in the populations of wild animals due to short supplies of food in the dry season. In some insects feeding on pulses (for example, adzuki bean weevil *Callosobruchus chinensis* found in legume stores), female fecundity declined due to the unsuccessful mating of individuals in the high-density population.

3.5.1.2 Interspecific Competition

In a natural forest, canopies of some tree species occupy the top story leaving little light available for the lower-story vegetation, and thus, competition between many species for the light is a natural outcome. In a forest, lions compete with leopards for their prey belonging to the same species. These two are examples of interspecific competition.

3.6 Summary

The population is defined as a group of individuals of a species occupying a particular area at a particular time. Habitat is an area or a place where a species lives, grows, and reproduces. Population ecology is a branch of ecology that deals with the study of species populations, their dynamics, and their interactions with their physical environment. Population ecology can further be divided into autecology and synecology. Autecology means the study of individual species in relation to the environment, whereas synecology refers to the study of the organisms' groups in relation to the environment. The population of a species has its distinguishable charac-

Table 3.3 Various types of interactions among different combinations of species within mixed populations

Combinations	Effects	Interaction type	Examples
0 0	Neither population affects the other	Neutralism	Birds perching on trees and sitting on the back of cows and buffaloes
– –	Direct inhibition of each species by the other	Competition (direct interference type)	Canopy competition in even-aged trees in a forest
– –	Indirect inhibition where common resource is in short supply	Competition (resource use type)	Weeds competing with food crops for soil nutrients
– 0	Population A inhibited, B not affected	Amensalism	Crop plants are affected by weeds
+ –	Population A, the parasite, generally smaller than B, the host	Parasitism	*Cuscuta* growing on other plants, root parasites *Orabanche* and *Epiphagus* on the roots of higher plants, *Rafflesia* on *Vitis* roots, viruses, bacteria, mycoplasmas, and rickettsias prospering on plants
+ –	Population A, the predator, generally larger than B, the prey	Predation	Insectivorous plants, *Zoophagus* (a fungus) capturing insects and nematodes
+ 0	Population A, the commensal, benefits while B, the host is not	Commensalism	Epiphytes, epizoans, lianas (vascular plants), etc.
+ +	Interaction favorable to both but not obligatory	Protocooperation	Sea anemone attached to hermit crab shells
+ +	Interaction favorable to both but not obligatory	Mutualism	Lichens, mycorrhizae, symbiotic nitrogen fixing bacteria, zoochlorellae and zooxanthellae, dispersal of fruits and seeds by animals, and pollination by insects

0 = no significant interaction; + = population growth and other attributes benefitted; – = population growth and other attributes inhibited

teristics, especially in terms of size, density, dispersion patterns, age structure, natality, and mortality. Population size implies the number of individuals of a species in the assemblage in a particular place or area at a particular time. Population density is defined as the number of individuals of a species per unit area or volume of the environment. Population dynamics depend on or are determined by four factors, viz., natality, immigration, mortality, and emigration. Dispersion reveals the spatial pattern of the organisms in a population relative to one another. Dispersion patterns generally assume three dimensions: random dispersion, clumped dispersion, and regular dispersion. Individuals of different ages exist in a population. Organisms in each age group are in a certain proportion constituting specific age structures determining natality and mortality. The proportion of individuals of a certain age group determines how stable and vibrant a population is. There are three types of ecological age groups characterizing a population: pre-reproductive, reproductive, and post-reproductive. An age pyramid geometrically represents the proportions of different age groups (phases of life span) in a population of the organisms of a species. There are three age pyramids: broad-base pyramid, bell-shaped pyramid, and urn-shaped pyramid. Each type of age pyramid represents the population size and growth pattern of the population of a particular species.

Coming into being of new individuals of an organism of a species adds to the population count. Natality covering this aspect of life is of two types: physiological and ecological. The natality of an organism generally enhances during the maturity period and gradually falls as the organism becomes older. Mortality means the number of individuals in a population dying in a given period, or per unit of time. The population growth rate is subject to individual vigor and the environment. The specific growth rate in a population with stationary age distribution is known as the intrinsic rate of natural increase. Individuals in a population or of different species often exist in the competition of various sorts. Space, resources, and time for an individual are not infinite to survive, perform, and accomplish their life cycle. An individual competes with others in a bid to acquire appropriate environmental conditions to survive within the limited space, resources, and time (life expectancy) and performs and accomplishes its life cycle.

3.7 Exercises

3.7.1 Multiple-Choice Questions

1. What can be equated with a community?
   (a) A population
   (b) Monospecific population
   (c) Polyspecific populations
   (d) All of the above
2. Autecology is the study of
   (a) individual species in relation to the environment
   (b) organisms' groups in relation to the environment
   (c) a plant in relation to the parasites prospering on it
   (d) food chains in an ecosystem
3. What is now merely an archaic term?
   (a) Autecology
   (b) Synecology
   (c) Snag Ecology
   (d) Evolution
4. Odum, a pioneer of ecology in the twentieth century, categorized synecology with
   (a) species ecology
   (b) population ecology
   (c) community ecology
   (d) ecosystem ecology
5. What is ecological (specific) density?
   (a) number of individuals of a species per unit area or volume of the environment
   (b) number or biomass of individuals per unit of area
   (c) biomass of individuals in a habitat they occupy
   (d) number and biomass of individuals per unit area of the habitat they occupy

6. If N is the number of organisms of a species, t the time, and Δ the change, then the specific growth rate of a population will be denoted by
   (a) ΔN
   (b) $\frac{\Delta N}{t}$
   (c) $\frac{\Delta N}{N\Delta t}$
   (d) $\frac{N\Delta t}{\Delta N}$
7. Which dispersion pattern is the most common among populations?
   (a) Random
   (b) Clumped
   (c) Regular
   (d) None of the above
8. Which of the following dispersion patterns is rare but common in human-managed ecosystems?
   (a) Random
   (b) Clumped
   (c) Regular
   (d) None of the above
9. A bell-shaped pyramid represents
   (a) stable population
   (b) rapidly growing population
   (c) rapidly declining population
   (d) either increasing or decreasing population
10. Which type of age pyramid denotes a high or exponential growth rate of the population?
    (a) Bell-shaped
    (b) Urn-shaped
    (c) Triangular
    (d) Both urn-shaped and bell-shaped
11. Among the countries, the human population of Puerto Rico registers a maximum decreasing rate (−1.73% per year) as per the population scenario in 2019. What type of the age-pyramid would the population of Puerto Rico reveal?
    (a) Constrictive type
    (b) Stationary type
    (c) Expansive type
    (d) Either Expansive or Stationary type
12. Many tree species have
    (a) long pre-reproductive and exceptionally long reproductive period
    (b) short pre-reproductive and long reproductive period
    (c) short pre-reproductive as well as short reproductive period
    (d) exceptionally long pre-reproductive and short reproductive period
13. A triangle-shaped or "expansive" pyramid reveals
    (a) a very large population of old people and a slow population growth rate
    (b) a very large population of young people but a slow population growth rate
    (c) a very small proportion of young people and a high population growth rate
    (d) a very large population of young people and a high population growth rate
14. An urn-shaped or a constrictive age pyramid represents
    (a) the low proportion of pre-reproductive and reproductive people in a population
    (b) the high proportion of post-reproductive people in a population
    (c) the declining birthrate and dwindling reproductive population
    (d) all of the above
15. Natality, or birthrate, is especially used for
    (a) human beings
    (b) birds
    (c) microorganisms
    (d) seeds of a plant
16. Which of the following is an example of commensalism?
    (a) Insectivorous plants
    (b) Epiphytes
    (c) A bird sitting on the back of a grazing cow
    (d) Lichens
17. What is the type of interaction in which population A is inhibited but population B is not affected?
    (a) Competition
    (b) Neutralism
    (c) Amensalism
    (d) Commensalism
18. If the environmental conditions are favorable, the specific growth rate is
    (a) maximum but unstable
    (b) maximum and stable

(c) minimum but stable
(d) minimum and unstable

19. The formulae $\frac{\Delta Nn}{\Delta t}$ and $\frac{\Delta Nn}{N\Delta t}$ depict, respectively, [N = initial number of organisms; n = new individuals in the population; t = time]
   (a) specific natality and absolute natality
   (b) absolute natality and specific natality
   (c) crude density and ecological density
   (d) ecological density and crude density
20. The major characteristics determining the rate at which females of a species produce offspring are:
   (a) age of attaining the reproduction stage
   (b) clutch size, i.e., the number of offspring produced on each occasion
   (c) length of time between successive reproduction events
   (d) all of the above

Answers: 1-c, 2a, 3-a, 4-b, 5-d, 6-c, 7-b, 8-c, 9-a, 10-c, 11-c, 12-a, 13-d, 14-d, 15-a, 16-b, 17-c, 18-b, 19-b, 20-d

3.7.2 Short-Answer Questions

1. Define a population.
2. What is the difference between monospecific and polyspecific populations?
3. What is a habitat?
4. Distinguish between autecology and synecology.
5. Differentiate between population size and population density.
6. How is the population growth rate measured?
7. What is dispersion, a characteristic of the populations?
8. Give some examples of the random, the clumped, and the regular dispersion patterns.
9. What are the characteristics of a triangular or expansive type age pyramid?
10. What is natality?

3.7.3 Long-Answer Questions

1. Discuss population characteristics citing appropriate examples.
2. What do you mean by age pyramids? Explain the three types of age pyramids giving examples.
3. Give an account of various interactions occurring among populations.
4. Discuss the factors determining natality and mortality.
5. Write short notes on the following:
   (a) Biotic potential
   (b) Parasitism
   (c) Competition among populations

References

Burkholder AR (1952) Co-actions in ecosystems: a classification of interactions among various species. J Ecol 23(4):456–467
Clements FE, Shelford VE (1939) Bio-ecology: with special reference to animals and man. John Wiley and Sons, New York
Haskell J (1949) An ingenious classification of co-actions on the basis of interactants. J Ecol Sci 15(3):212–225
Hutchinson GE (1978) An introduction to population ecology. Yale University Press, New Haven
McDougall W (1918) An introduction to social psychology. John W Luce & Co., Boston
Odum EP (1953) Fundamentals of ecology. Saunders, Philadelphia, 384 pp
Odum EP (1971) Fundamentals of ecology. W.B. Saunders Company, Philadelphia
Odum EP, Barrett GW (2005) Fundamentals of ecology. Thomson Brooks/Cole, Belmont, 598 pp
Rockwood LL (2015) Introduction to population ecology. Wiley, West Sussex, 378 pp
UNO (2019) World population prospects. United Nations, New York, 39 pp
Vandermeer JH, Goldberg DE (2013) Population ecology: first principles, 2nd edn. Princeton University Press, Princeton

4 Community Ecology

A variety of organisms – plants, animals, and microorganisms – and their populations occupying a common space and living in interaction and adjustment with each other constitute a community. A community shares a common environment and functions in a state of dynamism. As a community comprises only the living organisms of diverse species, it is also referred to as a biotic community. When a community is viewed in relation to its environment, it is regarded as an ecosystem.

4.1 The Community

The term community emanates from the days of Theophrastus (370–250 BC), the Father of Botany, who underlined the existence of plant communities, or association of species in varying environmental areas. Many scientists since ancient days have been describing various types of assemblages involving various organisms that resemble the term community in contemporary times. Carl August Möbius (1825–1908), a German zoologist, was the first to have described in detail the interactions among different organisms in an area. In 1880, he coined the term "biocoenose" which became quite popular among biologists.

A community is unique in itself. It has its evolutionary history, its structure, and functional attributes. A biotic community is never static or stationary. It exists with its dynamics and spells out its influence in various ways.

Seminal work of German zoologist Karl August Möbius (1825–1908) and later on aggressively executed experimental works of Clements (1916, 1936), Hutchinson (1961, 1978), Paine (1966, 1980), Odum (1971), Whittaker et al. (1973), and Odum (1983) phenomenally expanded the scope of community ecology.

4.2 Characteristics of a Community

Every community in the biosphere is characterized by certain unique features in terms of its origin, evolution, inter- and intraspecies diversity, species composition, and also its functional attributes, such as productivity, biomass, role in soil, water, biodiversity conservation, etc. A community cannot be similar to any other community. There is, thus, enormous community diversity in the biosphere: as much diversity as the total number of communities.

4.2.1 Species Diversity

A community is composed of a specific variety of plants, animals, and microorganisms. The number of species (species diversity) and population level in a community are different from that in others. While the number of species in a community is an evolutionary "content" and may not vary over time unless adverse environ-

V. Singh, *Textbook of Environment and Ecology*, https://doi.org/10.1007/978-981-99-8846-4_4

mental factors operate, their populations always stay in a dynamic state. Among all the species, a species, often specific to an ecological zone, prevails in a larger proportion. The species in a community organize themselves into trophic levels. A trophic group generally harbors many species. The dominant species in a trophic group as well as in the whole community is characterized in terms of a larger population (number of individuals), higher productivity, and greater biomass than all other species in the community. The other species exist in lesser proportion and those that make up the least proportion are often the rare ones. But all the species together count for what is known as species diversity. The plant species is dominant in community service as a source of nutrients and energy flows among all consumers within the community. Proportion, productivity, and biomass of the species go on decreasing with the successive higher trophic groups.

A measure to determine ratios between species number and importance values is known as the species diversity index. A natural forest, according to Odum, invests energy in structure and diversity by capturing the energy of light, recycling, and generating high gross energy. An increase in energy owing to structure and diversity may be estimated in terms of gross production. Gross productivity is a function of the energy used by an ecosystem. The energy needed to maintain structure and diversity comes through respiration. Species diversity, thus:

- is generally very high in undisturbed natural ecosystems (as in a rainforest) and often low in human-managed or anthropogenic ecosystems (as in cropland)
- in mountain areas (as in the Himalayas) generally decreases with an increase in altitude (decrease in a direction from higher stability to lower stability)
- is greatly influenced by the functional relationships between trophic levels (for example, the diversity of herbaceous plants in an overgrazed rangeland)
- generally increases with a decline in the ratio of antithetical maintenance to biomass – the R/B or ecological turnover ratio, or Schrodinger ratio (where, R = respiration, B = biomass)
- is directly correlated with ecological stability

Species diversity is phenomenally influenced by three major factors: geographical, developmental, and physical. The geographical factors are driven by natural evolution and have been dominant factors shaping and reshaping species diversity within spectacular diversity of geographical regions of the Earth. The developmental factors emanate from the anthropogenic management of natural species following socioeconomic and cultural demands varying from region to region and changing along the passage of time. The physical factors are often spontaneous, expected or unexpected, natural and/or human-induced that influence the state of species diversity.

Based on the abovementioned factors we come across what is known as species richness, expressed as simple ratios between total species, S, and total numbers, N.

4.2.2 Dominance

The nature and function of a community are not determined by all species. Out of numerous species, a community inhabits only relatively a few species (or species groups, not necessarily taxonomic ones) that generally exert significant influence through their proportion, size, numbers, productivity, or any other attribute inherent with them. These few species ruling over or controlling a community exist in synergistic, rather than competitive, relationships. Since the controlling species of groups do not belong to the same taxonomic group whether flora or fauna, they defy their classification based on taxonomy. The logical classification of the community-controlling species is based on the trophic groups comprising producers, consumers (primary, secondary, and tertiary), and decomposers. In these trophic groups some species phenomenally control energy flow and affect the environment of other species and are known as dominants or ecological dominants and the phenomenon is known as dominance or ecological dominance. The degree of dominance exercised by a species or many a species is expressed in terms of the index of dominance, i.e., the importance of each species in relation to the entire community.

In a Himalayan temperate oak-type forest, oak (*Quercus* spp.) species controls energy flow and positively influences the environment of so many associated species, mainly *Rhododendron*, *Myrica esculenta*, and *Lyonia* spp. The oak species, thus, has comparatively greater importance for the whole community. Dominance indices of the species in a forest community are also determined based on the contribution of each species to the total net primary production.

4.2.3 Keystone Species

Keystone species, a term introduced by Robert T. Paine (1966), are the species that play a crucial role in affecting many other species in an ecosystem. They also determine the ability of so many other species to perpetuate in their environment. The effect of a keystone species may be favorable for some species and detrimental for others. Thus, the keystone species affect the whole community by altering the structure, for example, species diversity, dominance, and multistory forest. They can even transform a forest into a savannah or grassland.

Some birds feed on the dominant species of insects and thus control the population of the specific insects. Such birds are the keystone species among carnivore predators. Some keystone herbivores, for example, elephants, are capable to transform a forest into open savanna. These two examples inflict detrimental effects on the community. Many pollinators, such as honeybees and butterflies, and seed dispersers, such as kangaroo, squirrels, birds, etc., serve as mutual keystone species imparting beneficial effects on a community.

4.2.4 Invasive Species

Invasive species, as the meaning of the term itself suggests, are those species that invade an ecosystem. These are often the introduced species not native to a region, area, or location. Introduction by human activity may be intentional or accidental. If it is intentional, the area invaded by the non-native species might be of any dimension. But if the introduction is accidental, its extension might be limited provided it is controlled by human actions. If the new location is favorable for the invasive species, it establishes disruptive effects. An invader establishes its strong breeding base in the new environment it is introduced in. An invasive species might alter ecological relationships among native species. It can suppress the native species by releasing allelochemicals from its roots, shoot, leaves, flowers, or the whole plant. It may alter the socioeconomic values of an ecosystem and even affect human health.

There are numerous examples of invasion by plants. One of the glaring examples of the invasive species introduced with good intentions is the *Lantana* bush in the subtropical Himalayan environment. Introduced as an ornamental plant, *Lantana* is proving to be havoc for the forests and native species therein. It flowers throughout the year and keeps on intensifying its invasions extending into larger areas and disrupting forest floor vegetation and forest diversity. With the ongoing climate change, this invasive species is expanding into high-altitude temperate Himalayan forests.

Pueraria lobata, popular as the kudzu vine, was planted to prevent soil erosion on roadsides. But later on, it covered extensive areas of forests and cultivated lands.

The other example of an invasive species intro duced in India accidentally is that of *Parthenium*. It is said that *Parthenium* seeds came from the USA mixed with wheat India imported. The *Parthenium*, known as one of the most noxious weeds, has invaded forests, grasslands, cultivated lands, and open spaces over extended areas.

Bromus tectorum, an invasive plant species also introduced accidentally, has displaced nutritive plant species in large areas of the pastures in Western North America.

4.2.5 Distribution Patterns of Structure

One of the ways to classify plant community is in terms of major growth forms: trees, shrubs, herbs, mosses, etc. Further, trees can be identified as broad-leaved trees, conifers, evergreen trees, etc. These vegetation groups form structures emanating from the distribution pattern of a forest community. The pattern, a term first used by Hutchinson in 1953, denotes the community structure resulting from the distribution of the organisms and their interactions with the environment. There exists a variety of such spatial structural arrangements in communities we call pattern diversity (Sharma 2013). The pattern diversity prevailing in a community can be of the following types:

(i) Stratification patterns or vertical layering
(ii) Zonation patterns or horizontal layering
(iii) Social patterns (in animal herds and flocks)
(iv) Activity patterns or periodicity
(v) Food web patterns
(vi) Coactive patterns (as a result of competition, mutualism, antibiosis, etc.)
(vii) Stochastic patterns (emanating from random forces)
(viii) Reproductive patterns (parent-offspring associations, plant clones, etc.)

On a vertical basis, a community represents its various strata, and the analysis of the changes according to the verticality of the community is stratification. A natural undisturbed forest community is often found comprising five strata or stories, viz.:

(i) Subterranean
(ii) Forest floor
(iii) Herbaceous vegetation
(iv) Shrubs
(v) Trees

Many tropic rainforest communities may be witnessed with as many as eight strata. It is because of the operating favorable factors (light, water, nutrients, microorganisms, and appropriate temperature range) that there might be more aboveground and belowground strata.

A grassland community, on the other hand, would be found with fewer strata: subterranean, floor vegetation, herbaceous layer, and characteristic fauna. Less number of strata is owing to a comparatively less complex grass community. Thus, the broader the base of biodiversity in a community, the greater the number of strata it possesses.

On a horizontal basis, the structure of a community represents subcommunities existing amidst homogeneous environmental conditions and ecological relations. The horizontal division is known as the zonation of a community. This kind of zonation of a community is very peculiar and recognizable in mountain areas.

The zonation of an aquatic community is different from that of a terrestrial one.

Generally, there are three zones in a deep pond or a lake, namely, (i) littoral zone, (ii) limnetic zone, and (ii) profundal zone. Each zone embraces different kinds of organisms. An open ocean is generally divided into three zones following its depth and amount of light it receives: (i) euphotic zone; (ii) bathyal zone, and (iii) abyssal zone. These have been compared in Table 4.1.

4.2.6 Trophic Structure

Trophic structure describes the organization of the organisms of different species in the form of trophic levels, each dependent on and serving as food for the other distinctive level. Trophic structure, thus, describes feeding relationships among organisms in an ecosystem over time. There can be many trophic levels a community depends upon (Fig. 4.1). Trophic structure is ultimately determined by the producers, the first trophic level, and the ensuing energy transfer efficiency between trophic levels (producers to consumers).

The trophic levels are explained explicitly by ecological pyramids, food chains, and food webs. Further, two basic processes regulate the trophic structure of a community:

(i) **Bottom-up process.** The amount of energy and nutrients at the producer level (or proportion of the producers) determine the state of the subsequent higher trophic levels.
(ii) **Top-down process:** Predators (carnivores) exert suppressive pressure on herbivores, thus favoring the producers.

Table 4.1 Zonation of an open ocean

Characteristics	Euphotic zone	Bathyal zone	Abyssal zone
Position	Upper most part	Middle part	Lower/bottom part up to a depth of 4000 to 6000 m
Presence of light	Enough	Twilight	No light/total darkness
Trophic state	Photosynthesis	No producers	Chemosynthesis
Producers	Phytoplanktons	No producers; only zooplanktons	Chemosynthetic bacteria
Biodiversity	High	Low	Medium

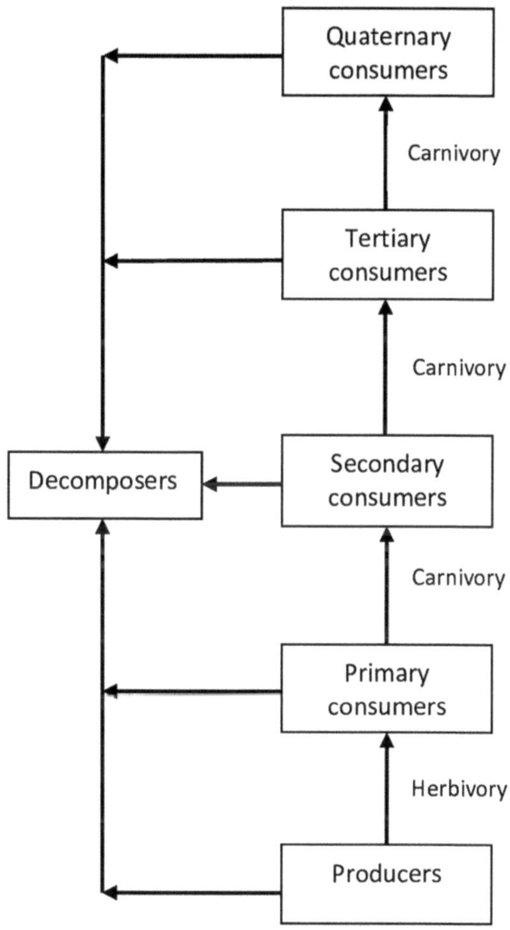

Fig. 4.1 Trophic levels in a community: positioning of organisms within a community's food chain levels

The trophic structure is also influenced when, subject to geographical features, herbivores easily hide in caves and places not easily accessible to predators. Some external factors, especially anthropogenic factors (e.g., hunting of some specific species or overexploitation of high-value plant species), lead to seminal influence on trophic structures.

4.3 Ecological Succession: Community Dynamics

A community of plants, animals, and microorganisms tends to be in a state of equilibrium. Prevailing environmental conditions, however, vary according to certain circumstances. The stability of a community is also not a rule of nature. Communities, in essence, are never stable or static. They are dynamic. They break the rule of stability. They are always changing in tune with space and time. Change is a rule of nature. Change is the gist of evolution. All the physical, chemical, and biological factors operating in the communities never stay in their "fixed" quantities. They are changing with space and time due to physiographic and climatic variations and so are the communities (Singh 2019, 2020). Interaction among species and organisms of a community also results in environmental variation.

Evolutionary phenomenon applying to living systems is what succession, a measure of ecological dynamics, is all about. Succession is often distinguished from evolution and both are discussed in somewhat different contexts. Evolution is a broad term amalgamated with a philosophy about living nature imbibing with distinctive theories, with telltales of the origin of life and species and its journey from the presumable ancient past till present times; it even attempts to look into the future. Thus, even the background of communities' succession is also a subject of evolution. Succession, on the other hand, involves only the patterns of changes among communities. Succession is tied to limits or a range, while evolution is an endless phenomenon. Succession occurs after disturbance, which evolution need not do so. Succession is divided into stages, while evolution is an indivisible phenomenon. Succession replaces one community with the other; evolution adds "content" to the community. Succession involves a successional trajectory, that is, a community that is predicted to occur after a particular type of disturbance on a site over a timescale. Evolution has no such trajectory and is unpredictable and there is no timescale for evolution to come out with something distinctively visible. Succession culminates into a climax, while the mysterious journey of evolution is endless and insatiable.

Ecologists have always been curious to understand how communities form and change over time. Out of this curiosity has come to the fore an understanding of community dynamics with a focus on ecological succession. The term ecological succession was first of all used by Ragnar Hult (1857–1899), a Finnish botanist and plant geographer. He published a comprehensive study of ecological succession in 1881. Ecological succession refers to a series of progressive and orderly changes in the composition of an ecological community over time. Ecological succession, in other words, is a natural process by which a community evolves. Frederic Edward Clements (1874–1945) was the first to carry out comprehensive studies on ecological succession. He published a theory of succession in 1916 and propagated it as a general ecological concept. His theory left a deep impact on subsequent ecological studies.

In 1916, Frederic Clements published a descriptive theory of succession and advanced it as a general ecological concept. His theory of succession had a powerful influence on ecological thought.

Changes in the structure and composition of a community are quite rapid in the earlier stages and then go slow until a more or less stable community comes into being. Odum (1971) referred to such an orderly progressive replacement of a community by another until the development of a stable climax community as ecosystem development.

4.4 Causes of Ecological Succession

Ecological succession is a complex process that – through successive changes in species structures – advances from a simple, less complex, and unstable community organization towards a more heterogeneous, complex, and stable community. This natural process is a prerequisite for the development of an ecosystem that provides refuge and self-sufficiency to a variety of populations of diverse species, that is, a complex and sustainable community. There may, therefore, not be the single cause of this process.

The main causes of succession are of three types, viz., initiating, ecesis, and stabilizing causes. The initiating causes pertain to climate and biotic interactions. Ecesis or continuing causes pertain to physicochemical, or edaphic, factors and their responses to species and vice versa. Stabilizing causes help the diverse species' populations prosper in appropriate environmental conditions (Fig. 4.2).

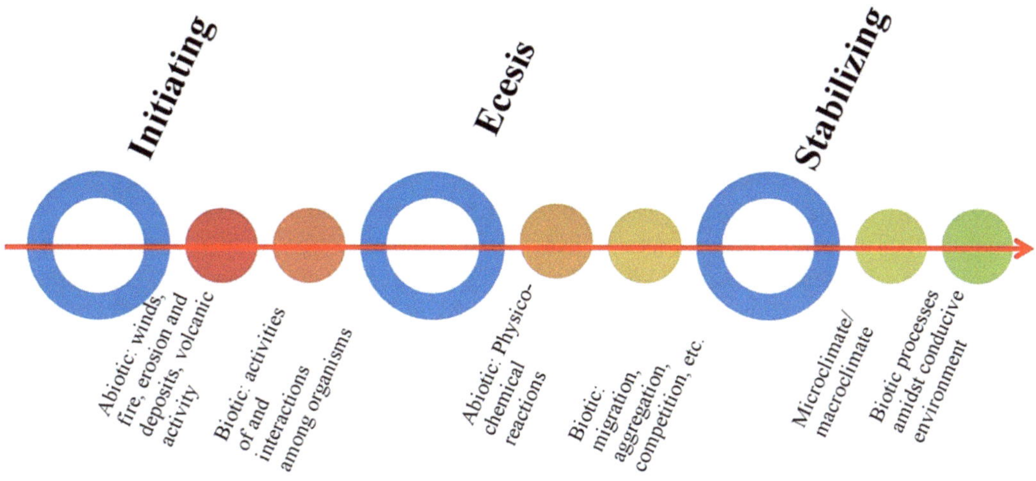

Fig. 4.2 The factors driving ecological succession at a site

4.5 Dimensions of Ecological Succession

Many types of succession are distinguishable based on different aspects. Most ecologists are confined to only two types, namely, primary and secondary succession. However, some other interesting dimensions of ecological succession are discussed under the following subheads.

4.5.1 Primary Succession

When planet Earth came into being, there was no soil, no environment, no organisms, and no habitat. In due course of geological times, soil and nutrient-rich water bodies formed and a typical environment also came into being. Beginning with prokaryotes and monerans, numerous types of fungi, plants, and animals and, subsequently, their communities evolved. This was the primary succession on planet Earth. Primary succession, thus, is the succession that starts from the primitive substratum (land or waterbody) which had no living organisms earlier. When a rock is exposed, a new island is formed. This is primary succession. When a waterbody comes into being, the type of ecological succession that would take place will be the primary succession.

The eruption of volcanoes followed by the mixing of lava with cold water forms rocks. First, weathering and other natural processes gradually break the rocks exposing them to be used as a habitat for the establishment of certain plants and the eventual development of a community. The Big Island of Hawaii, where approximately 32 acres of land are added each year, is an outstanding example of primary succession.

The first group of organisms/species established through the process of primary succession is known as pioneers, primary colonizers, or primary communities. The pioneers help improve the environment and set the tone for subsequent successional processes.

4.5.2 Secondary Succession

When an ecosystem is disturbed drastically due to natural hazards like fire, catastrophic landslides, floods, etc., or due to anthropogenic activities like logging, mining, blasting, bulldozing, etc., destroying entire communities, possibilities of its further ecological regeneration are never ruled out. The substratum of the once-destroyed ecosystem stays fertile enough for the succession to take place. In secondary succession, an area previously occupied by a community is recolonized following a partially or completely destroyed community.

Before a fire or anthropogenic factors engulfed it, the vegetation of a broad-leaf forest would have been dominated by tall trees. After their destruction, the tall trees do not reappear in the first place. Instead, the first plants to grow will be annual and then the perennial grasses. Over many years, shrubs will come into being. After a long period, about 150 years, if there is no further disturbance, a forest like one of the pre-fire days will flourish via secondary succession.

4.5.3 Autogenic Succession

In the course of its interaction with the environment, a community modifies its environment resulting in its replacement with new communities. Such a type of succession is known as autogenic succession.

4.5.4 Allogenic Succession

Unlike in autogenic succession, in some cases, the replacement of a community with a new one is caused by external, rather than internal, factors. This type of succession is known as allogenic succession.

4.5.5 Autotrophic Succession

When an inorganic environment prevails, autotrophic organisms – photoautotrophs and chemoautotrophs – lead the succession availing plentiful energy flow through the environment and continuing to be the dominant species throughout. Over time, the organic matter gets accumulated in the environment enabling the heterotrophs to become members of the community overwhelmingly dominated by autotrophs.

4.5.6 Heterotrophic Succession

This type of succession begins in a predominantly organic environment and involves the dominance of heterotrophs like fungi and animals in the beginning. As a result, energy content progressively decreases along the passage of succession.

4.5.7 Induced Succession

Ecological succession in our contemporary world is largely human-induced. The human species, owing to extraordinary capabilities attained through the course of natural evolution, has emerged as an all-pervading intervening factor. Most of the ecosystems of the biosphere are managed by human intervention by influencing natural succession. Croplands inhabiting cultivated plants and gardens are the main examples of induced succession. Natural parks, wildlife sanctuaries, and many natural lakes and artificial lakes are other examples of induced succession. Induced succession is harnessed for human comforts and desirable socioeconomic gains.

4.5.8 Cyclic Succession

Cyclic succession is a characteristic of a well-established community. A community during its course of normal succession keeps cyclically changing its structure too. There could be several species within a community that would thrive only during a particular season in a year and yet some emerge in their blooming and fruiting stage in more than a year. They would lie dormant for the rest of the period. And yet there could be certain species that vanish in a particular season and reappear out of their germplasm (e.g., seed or root) in another season of the year. Some animals of a community would migrate to other areas at a particular time and would cease to be part of the community until they return in the next season. And yet some other species from other communities might immigrate to a community in a particular season and become part of it until they emigrate in the next season. Thus, drastic structural changes are assumed by a community on a cyclical basis, a process known as cyclic succession.

4.5.9 Retrogressive Succession

Succession is usually successional or progressive, advancing from a simple organization to a more complex organization of a community. However, intensive biotic interference can reverse the process and the succession can turn in a backward direction, i.e., after reaching the climax stage, the succession can move backward towards bare land. For example, deforestation reduces a dense forest into shrubby vegetation and subsequently into a grassland. Such a reversible succession is called retrogressive succession.

4.6 The Process of Succession

Primary autotrophic succession, to accomplish its stages, involves many sequential steps from a lower level to a higher level of the hierarchy and from a state of instability to a state of ecological stability.

4.6.1 Nudation

The process of succession starts with the coming into existence of a bare area or a bare rock due to (i) natural topographic reasons, such as volcanic eruption, landslides, soil erosion, or formation of a new island, and climatic reasons, such as glacier melts and formation of a lake, cloud burst, hailstorm, floods, fires, diseases, etc., and/or (ii) anthropogenic or biotic activities, such as intensive mining and quarrying, digging, blasting, burning, logging, leveling of hills and mountains, inundation of large land areas by damming a river, overgrazing of grasslands and rangelands, etc.

4.6.2 Invasion

Invasion involves the arrival of new species employing their reproductive bodies (or propagules), such as flowers, fruits, seeds, vegetative organs, and spores, from other sites/communities/areas in the bare or new area.

Generally, there are three means of invasion, viz., (i) dispersal/migration, (ii) ecesis or establishment, and (iii) aggregation.

Dispersal or Migration Propagules of the species in the form of spores, flowers, fruits, and seeds disperse from other communities to the new site by natural agencies of winds, water, and animals. Small or large animals migrate to the new area adding to the growing community.

Ecesis It is a process of the successful establishment of new species at new sites. Many, rather most, of the propagules of the dispersed or migrated species in the new environment fail to establish themselves in the beginning. However, in the new environment at new sites, seeds and other propagules of many newly migrated species germinate and attain full growth. Adults of well-established species in the new environment also begin reproduction.

Aggregation In this final stage of invasion, organisms of the established species multiply their population through reproduction and aggregate to thrive better in the new environment.

4.6.3 Competition and Coactions

Aggregation of the invading species within a limited space causes their individuals to compete with each other for space and nutrients. The competition may be between species as well as among genotypes of a species. Organisms belonging to a species influence those of other species in various ways – physically as well as by releasing depressing allelochemicals. The action of two species leading to influence each other is co-action, which is vital for preparing a sound "ground" for the new competitively selected species to prosper in a new environment. If an invading species fails to compete with the other, it would vanish in due course of time or will stay at a subdominant position, and if it does, it will prevail in the new environment which would be largely determined by the reproductive capacity and ecological amplitude of the site-invading species.

4.6.4 Reactions

Living organisms prospering on a site continuously influence and eventually modify their environment. Subsequently, phenomenal changes occur in the light, soil, moisture, and temperature conditions of the site. The resulting modified environment would increasingly become unsuitable for the existing community, which, over time, is replaced by another community, known as the seral community. The sequences of the communities replacing one another at a site are referred to as sere. The various communities belonging to a sere are called seral communities, seral stages, or developmental stages. The changed vegetation amid a modified environment allures a variety of animals to join the community. Further, intensive interactions between organisms and their environment induce more changes in the environmental conditions conducive to the development of the final stage: the climax.

4.6.5 Climax or Stabilization

The new community with continuous competition among its species and organisms of the same species and constant interaction with the environment advances towards a state of its stabilization. Through successional advancement, the community eventually attains a state of stability and stays in equilibrium with the climate. This final community is known as a climax community and this state as a climax state. At the climax state, the community is not replaced by any other community.

A climax community is a diversity-laden, steady-state, healthy, resistant to external disturbances, vibrant, highly productive, resilient, and sustainable community. The soil of a climax forest community is rich in all the nutrients necessitated by a rich diversity of plants. Food chains and the food web operating in a climax community are very complex. A climax community strikes an equilibrium between gross primary production and total respiration and exists in perfect equilibrium with the environment. It builds its powerful microclimate and contributes to building up an appropriate macroclimate. Since an undisturbed climax community is overwhelmingly fed by photosynthesis, its carbon sequestration efficiency is very high. Therefore, an ecosystem vibrant with a climax community is vital for climate change mitigation our planet is in desperate need of.

A naturally evolved forest ecosystem constitutes a climax community. The primary succession takes about 1000 years to attain a climax state, while the secondary succession evolves into a climax community in about 200 years.

Successive communities are well presumed. In some cases, however, succession gets deflected from its presumed path and the final community is somewhat different from the presumed climatic climax community. Such a succession is called deflected succession. Thus, deflection occurs mainly on account of different site conditions, such as different soil compositions or microclimate.

A climax community is influenced by and best adapted to the climate it contributes to buildup. The climatic climax community is the one that maintains equilibrium with a zonal climate. When a climatic climax community resorts to anthropogenic intervention getting disturbed to a considerable extent, it is called disclimax or anthropogenic subclimax.

4.7 Successions in the Biosphere

We have already discussed two broad categories of succession, namely, primary and secondary succession. Based on the type of ecosystem, these are further categorized into:

1. Lithosere
2. Hydrosere
3. Psammosere
4. Halosere
5. Xerosere

Lithosere begins on a rock surface, hydrosere in an aquatic environment, psammosere

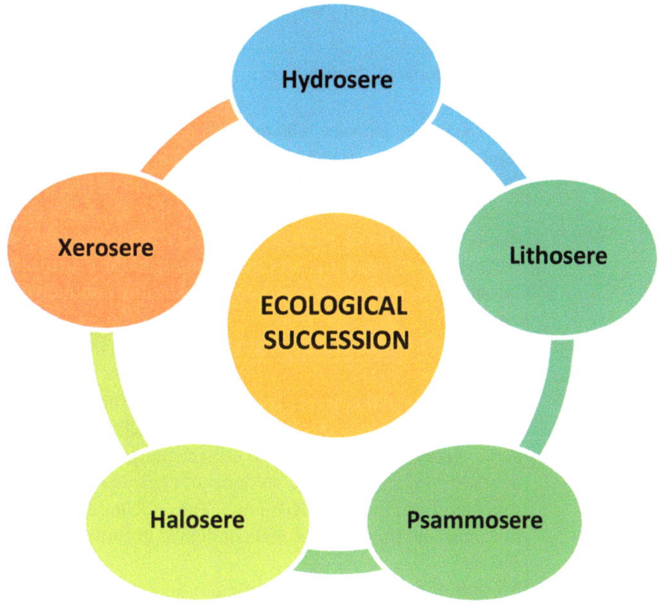

Fig. 4.3 Diverse aspects of succession as influenced by the specific site and its impact on community growth patterns

on a sandy surface, halosere on a salty surface, and xerosere in a xeric or dry environment. Wherever specific areas or ecosystem components prevail, the respective successions are a natural fact. Succession, however, is a single and indivisible process. It is just categorized based on the site specificities (Fig. 4.3). The psammosere, the halosere, and the xerosere are parts of the lithosere and these will be discussed together.

Community changes during succession include (i) an increase in species diversity, (ii) changes in species composition, and (iii) an increase in biomass, primary production, respiration, and nutrient retention at an ecosystem level.

4.8 Lithosere: Xerosere

Ecological succession called lithosere or xerosere (xerarch) begins with a bare rock surface. The original surface is deficient in water and mineral nutrients. Ecological succession on a bare rock involves several orderly steps each representing a unique plant community. The beginning of the succession occurs with the colonization of crustose-type pioneers. Step upon the step, from one lower seral community to the next higher one, the ecological succession culminates into a climax forest community as explained in Table 4.2 and also represented through Fig. 4.4.

4.9 Hydrosere or Hydrarch

Ecological succession taking place in a waterbody, notably a pond or a lake, is known as hydrosere or hydrarch. Like on a bare rock or in a dryland, succession in a waterbody begins with the colonization by autotrophs and ends up in a climax community. Compared to the lithosere, the hydrosere is more varied and interesting from the viewpoint that it involves both aquatic and terrestrial species. Various stages of a hydrarch are as in Table 4.3.

A habitat, in essence, tends to modify from extremely dry or aquatic conditions to mesic or moderately wet conditions through its long journey of ecological succession (Fig. 4.5).

Table 4.2 Stages of succession and main features of the lithosere

Successional stage	Main features of the lithosere
1. Crustose-lichen stage	***Flora*** Rock surfaces are very inhospitable as they are dry, moisture-deficient, and exposed to extreme temperatures. Crustose lichens being able to grow in extreme xeric conditions are the first to colonize the rock or dry land surfaces devoid of life. The common lichens of this species are *Lecanora, Rhizocarpon, Rinodina, Lecidia*, etc. Lichens produce lichenic acid, or carbonic acid, that reacts with rock surface. Slowly corroding the rock, colonization by crustose rock initiates rock weathering. Organic matter emanating from the crustose lichens mixes with small rock particles initiating the process to build up a substratum conducive to the growth of other lichens. ***Fauna*** Ants, mites, spiders in the beginning. Fauna is sparse.
2. Foliose-lichen stage	***Flora*** Foliose lichens, the higher forms of lichens, such as *Parmelia, Dermatocarpon, Umbilicaria*, etc., constitute the next seral stage. The large leaflike thalli overlap the crustose lichens of the first seral stage rendering them unable to utilize sunlight for photosynthesis. The crustose lichens meet slow death leaving behind large amounts of organic matter. The foliose lichens avail good amounts of moisture from the built-up humus. They accumulate dust particles enriching the substratum with nutrients and thin layer of soil conducive to further communities to develop. ***Fauna*** Species of ants, spiders, and mites.
3. Moss stage	***Flora*** Thin soil layer helps another seral stage, of the xerophytic mosses, such as *Polytrichum, Bryum, Barbula, Grimmia, Funaria, Hypnum*, etc., to occupy rock site gradually becoming suitable for advanced organisms. The mosses arrive to invade the site by means of their wind-blown spores. Now there is intensive competition between the mosses and the foliose lichens for moisture and nutrients. Gradually, the mosses outcompete the foliose lichens. Biomass of the dead and decaying foliose lichens further adds to the organic matter on the rock surface. ***Fauna*** Species of spiders, mites, springtails, etc.
4. Herbs stage	***Flora*** More of organic matter and humus and thicker layer of soil set in the conditions for herbaceous species, such as *Poa, Festuca, Sporobolus*, etc. , to grow and prosper on rock site. There are annual herbs in the beginning which are followed by perennial herbs, including some xeric ferns, like *Adiantum, Asplenium, Actiniopteris, Cheilanthes*, etc., to occupy the site. Organic matter and humus in the soil constantly increase on account of death and decay of short-lived herbaceous plants, a condition that further helps stabilize subsequent seral communities. ***Fauna*** Fauna begins stabilizing with the stabilizing community. Nematodes, larval insects, new species of insects, micro-anthropods, etc. begin proliferating. Depending on the areas, small micro-mammals can already appear.
5. Shrub stage	***Flora*** The habitat stabilized to a great extent by herbaceous community becomes favorable for the shrub species to grow and prosper. Invasion by shrubby species begins with the dispersal of their seeds or rhizomes from nearby habitats. Moderate climate is favorable for shrubs like *Zizyphus, Zygophyllum, Rhus, Capparis*, etc. to proliferate in their new habitat. Shrubs go on covering the herbaceous species. The latter begin getting vanished, and along the passage of time, a community dominated by shrubs comes into being. This seral community further enriches the soil and with organic matter and substantially improves environmental conditions for the next community to appear and proliferate. ***Fauna*** Fauna enrichment of the seral community continues to take place. Centipedes, lizards, birds, shrews, etc. add to the community under succession.

Table 4.2 (continued)

Successional stage	Main features of the lithosere
6. Climax forest stage	***Flora*** Tree species begin to replace shrubs along the passage of time. In the beginning, there may be xerophytic trees, like *Acacia*, *Prosopis*, *Boswellia*, etc., which are followed by replacement with deciduous tree species. Finally, with the passage of time, a community of climax forest comes into existence. ***Fauna*** A variety of animal species is a feature of the climax community. The animals in the early seral stages are joined by slugs, snails, frogs, salamanders, reptiles, birds, squirrels, shrews, foxes, etc.

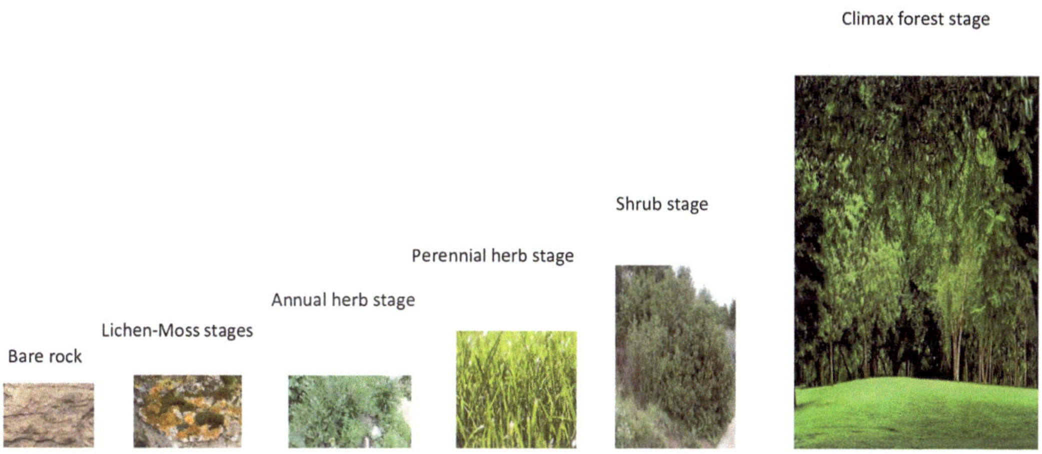

Fig. 4.4 Sequential stages in xerarch succession commencing from an initial bare rock

4.10 Ecosystem Changes During Succession

From the above discussion we have understood that succession brings about ecosystem changes in terms of increased biomass, primary production, respiration, and nutrient retention. Changes in ecosystem characteristics are the consequences of the hierarchical changes in biodiversity and composition of communities. Not only the biotic characteristics but also the edaphic properties register notable changes, such as an increase in organic matter and nutrient content in the soil. Some of the notable discoveries relating to ecosystem changes during succession as documented by Molles (2005) are as follows:

- A significant increase in soil depth and the depth of all major soil horizons from the pioneer community to the climax stage.
- Substantial increase in organic content, moisture, and nitrogen concentrations of the ecosystem soil.
- Soil bulk density, phosphorus contents, and pH record a decrease over the same successional sequence.

Thus, it comes to the fore that succession involves not only changes in the diversity and composition of the species but more than that. Succession pervading the terrestrial ecosystems, in essence, pulsates with ameliorative changes in key ecosystem properties. Changes in soil properties are inherent in succession occurring in terrestrial ecosystems. This is a very crucial

Table 4.3 The stages of hydrarch succession and the distinctive features of each

Seral stage	Characteristics of the hydrosere
1. Phytoplankton stage	Phytoplanktons comprising blue-green algae (Cyanophyceae), green algae (Chlorophyceae), diatoms (Bacillariophyceae) associated with protozoa and bacteria colonize the water surface as pioneers. These forms of life float on water and multiply profusely. In the process, they continue to add organic matter and nutrients to the waterbody they colonize.
2. Rooted submerged state	Phytoplanktons die and decay in the mud that was accumulated with the rainwater. The mud at the bottom of the waterbody, thus, becomes soft and rich in organic matter and conducive to support the plants wherever the light goes. The hydrophytes that establish under such conditions are the rooted submerged ones. *Hydrilla*, *Chara*, *Vallisneria*, *Elodea*, *Potamogeton*, etc. are some examples of the hydrophytes replacing the pioneer phytoplanktons. Propagating parts and seeds of these hydrophytes reach a waterbody by birds and animals that visit the site for drinking water and foods. Parts of the hydrophytes keep on decaying and organic matter keeps settling down in the bottom, enriching the bottom with organic matter, binding the loose bottom sediments into finer matrix, and making the waterbody shallow by decreasing the depth. Such changes in a waterbody pave way to a higher community to establish.
3. Rooted floating stage	With further decrease in the depth of the water column, rooted floating plants establish themselves at the bottom of the waterbody. Examples of rooted floating plants include *Aponogeton*, Trapa, *Nelumbium*, *Spirodela*, *Azolla*, *Salvinia*, *Pistia*, *Lemna*, etc. Changes in the vegetation type lead to changes in animal forms. The new animal species emerging at this stage are dragonflies, mayflies, crustaceans (*Daphnia*, *Gammarus*, *Cyclops*, *Asellas*, etc.). The decomposing biomass further enriches the bottom and makes the waterbody shallower. Such conditions favor another more complex seral community to establish.
4. Reed swamp stage or amphibious stage	Marshy rooted plants, such as *Sagittaria*, *Typha*, *Phragmites*, *Scirpus*, etc., invade the shallow waterbody when the conditions for the seral stage dominated by rooted floating plants become unfavorable. The plants are rooted in the bottom but their photosynthetic aerial parts remain exposed to the open atmosphere. These plants have well-developed rhizomes and form a very dense vegetation. Growth of the reed swamp stage leads to reduced level of water. Among the animals, birds like kingfisher, ducks, sparrows, etc. join the community. Some insect species, like giant water bug and scavenger beetle join the mayflies and dragonflies.
5. Sedge meadow stage (marginal mats)	When the water level comes to decrease enough as to just cover the surface, the swamp plant stage sets conditions for the succession of sedge meadow stage. Edges of the waterbody are occupied by plants like *Juncus, Carex, Cyperus, Mentha, Eleocharis,* etc. These plant species gradually spread over the central portion of the waterbody forming sod-like mats of the invading vegetation. This mat acts as a barrier against the wind-borne soil, depositing the same and building up a substratum. The earlier seral community adds to organic matter and nutrients in the substratum. Substantial changes in vegetation composition allure terrestrial animals to feed upon the nutritive plants.
6. Woodland stage	Water of the waterbody completely dries up and so does the surface. Terrestrial herbs, shrubs, and small trees invade the sedge meadow stage. The new vegetation rapidly transpires and the remaining water on the site gets exhausted. *Acacia*, *Alnus*, *Populus*, *Cornus*, *Salix*, etc. are the main plants of the woodland stage. These plants cast their shadows, gradually preparing ground for new vegetation. The soil is enriched with organic matter and humus. Soil is inhabited by a variety of bacteria that feed on organic matter and engaged in mineralization, ensuring plentiful nutrient supplies to the plants. Large herbivores, like rabbits, deer, reindeer, etc., migrate to the site.
7. Climax forest stage	As the site conditions become more appropriate, woodlands are invaded by a variety of other trees, shrubs, herbs, creepers, etc. Soils are rich in organic matter, humus, and nutrients, and sedimentary cycles go on rapidly. A variety of insects, non-vertebrates and vertebrates, and herbivores and carnivores inhabits the sites. A healthy, vibrant, and stable community – the climax forest – finally evolves as the final stage of the succession.

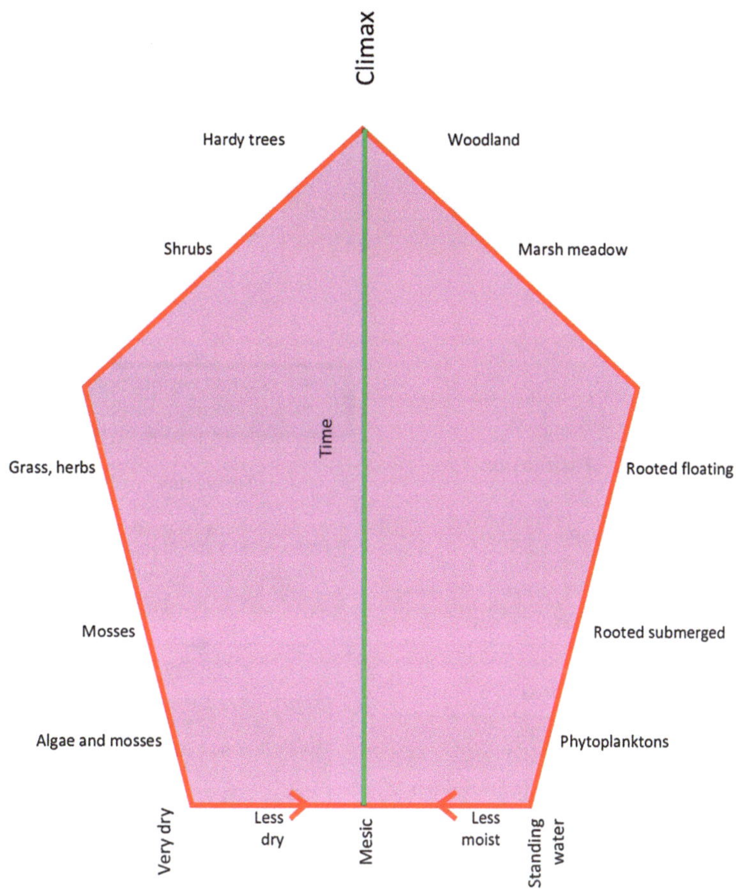

Fig. 4.5 Convergence of succession: transitioning from xeric and aquatic to mesic plant communities

successional attribute as the soils serve as a foundation on which terrestrial ecosystems are built.

4.11 Biomass Accumulation Model of Ecosystem Recovery

Disturbance to a forest ecosystem induces a series of recovery phases. Bormann and Likens proposed a model of ecosystems' recovery from disturbance known as the biomass accumulation model. This model emphasizes four phases of recovery of a forest ecosystem, viz.:

1. Reorganization phase (10–20 years): despite biomass accumulation, a forest loses biomass and nutrients.
2. Aggradation phase (>100 years): a forest ecosystem accumulates biomass unless attaining peak biomass.
3. Transition phase: biomass in a forest ecosystem declines from the peak to some extent.
4. Steady-state phase: forest biomass keeps fluctuating around a mean level.

The biomass accumulation model as tested by Bormann and Likens and as further elaborated by Molles (2005) is shown in Fig. 4.6.

4.12 Succession Mechanisms: Alternative Models

In the early quarter of the twentieth century, Clements (1916) underlined the crucial role of what he called facilitation in driving succession.

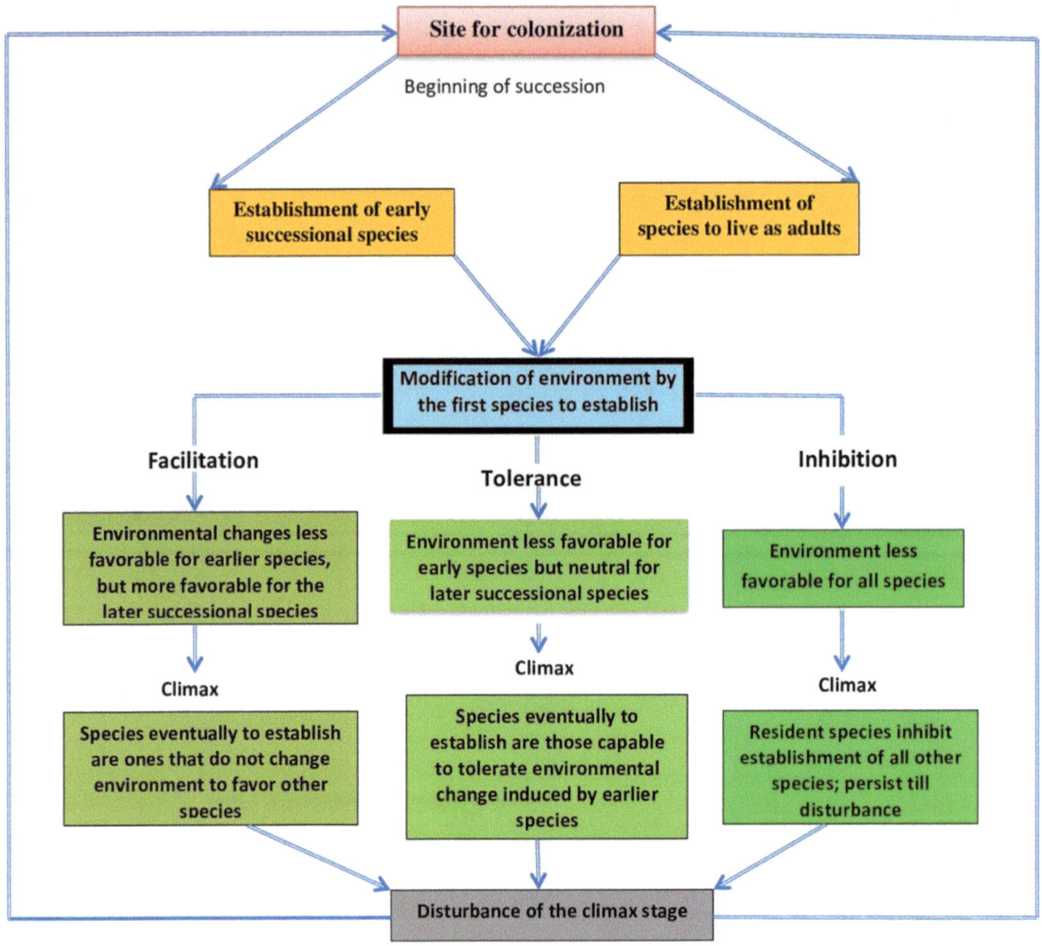

Fig. 4.6 Alternative mechanisms of succession (based on Molles 2005)

In the last quarter of the twentieth century, Connell and Slatyer (1977) proposed three alternative models, viz.:

1. Facilitation
2. Tolerance
3. Inhibition

4.12.1 Facilitation

According to this, perhaps a widely acknowledged model, several species occupy a space devoid of living organisms, such as a bare rock or a desert, where succession is to inevitably begin. However, only a few with specific characteristics would establish themselves. These fewer species with the traits of colonizing a new space are known as pioneer species. The pioneer species in the process of colonizing the site appropriate for them modify the environment to an extent that it becomes less favorable to them and more favorable for the species of the later successional stage. Putting it differently, the earlier successional species, that is, the pioneers, provide "facilitation" for the later successional species to colonize the site. The early colonizers, thus, facilitate their disappearance from the site to facilitate the establishment of the successional species until, after several stages of facilitations and replacements, a final stage, the climax community, emerges on the site.

4.12.2 Tolerance

There are fundamental differences between the tolerance and facilitation models. These are as follows:

1. The initial colonization stages do not involve only a few pioneer species, and juveniles of species can be present from the earliest stages of succession that finally dominate at the stable climax stage.
2. The species colonizing a site during an earlier stage of succession do not provide facilitation for the species to emerge during later successional stages. The earlier colonizers (the pioneers) also do not modify the environment to give way to the subsequent successional species to occupy the space. The successional species to emerge at the later stages are those tolerant to environmental conditions created during earlier successional stages.
3. When species tolerant to the environmental conditions created by earlier successional species are exhausted, then the stable climax community gets established.

4.12.3 Inhibition

According to the inhibition model, a species that goes on existing till its adult stage can colonize the site during earlier successional stages. According to the model's proposition, the initial species in a succession process alter the environment, leading to conditions less suitable for both early and late successional species. The later successional species can only establish themselves in the areas if space becomes available through the disturbance caused by the early colonizers. The species that culminate in a final stable community are those that can resist damages caused by physical and biological factors and, thus, are capable to live for a long time.

Among these three models (Fig. 4.6), as most studies focused on succession reveal, the facilitation model and the inhibition model, or the combination of both, gain larger support (Molles 2005).

4.13 Stability, Resistance, and Resilience

Succession leads a community or an ecosystem to a state of stability. Stability literally means "no change." In the context of a community, however, there is nothing like a state of "no change." Ecosystem stability implies an absence of damaging factors which could considerably affect the succession attaining climax. A climax stage of the succession attains ecological stability. Ecological stability is not something static. This is dynamic as vital processes, such as speciation, and new scope for the natural evolution of new life forms might be hidden in it. One of the main reasons for ecosystem stability is physical constancy. It is due to this factor that the benthic communities of the deep sea keep constantly stable on account of constant physical conditions.

Terrestrial ecosystems, for example, forests, however, cannot remain in a state of physical constancy. Many natural physical, biological, and/or anthropogenic factors keep operating on them. These ecosystems maintain their stability by employing what is known as resistance. How an ecosystem or a community maintains its stability persistently in the event of a disturbance of any kind is the question underlining the ability of the community to exercise its resistance. A community or an ecosystem bearing a high degree of resistance against potential disturbances is naturally the more stable one.

One of the critical attributes of ecological stability is resilience, that is, an ecosystem's capability to regain its original structure despite disturbances. Every community or ecosystem has the capability of a certain degree to recoup losses due to a certain degree of disturbance. Resilience is one of the traits of ecological sustainability. Biodiversity is one of the basic attributes of ecological stability, resistance, and resilience imparting a high degree of sustainability to an ecosystem. Since in an ecosystem, the relationships are reciprocal, biodiversity and sustainability are influenced by ecological stability, species resistance, and the ecosystem's resilience (Fig. 4.7). The greater the measure of biodiversity, the higher the

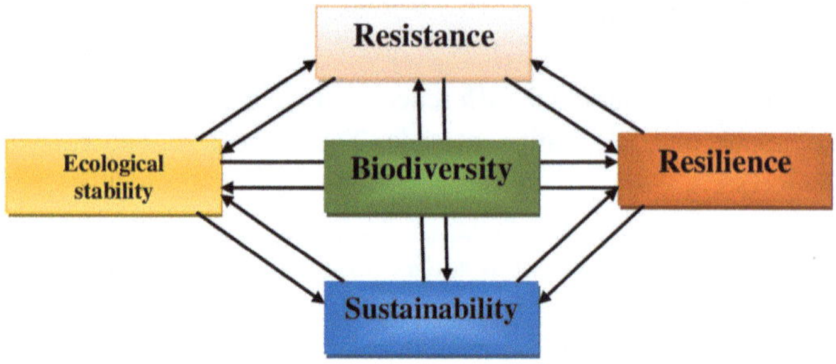

Fig. 4.7 The interactions existing within a community

degree of ecological sustainability (Singh 2023). Monocultures in anthropogenic ecosystems, such as in gardens and croplands, exhibit a lesser degree of resilience and, hence, are prone to damage by physical and biological factors.

What is inherent to resilience is succession. When an ecosystem is damaged due to physical and biological shocks, it is the succession that carries a community towards higher stages of the organization through gradual changes in its structures and composition. Ecological succession, in essence, is the very soul of resistance, resilience, and stability of an ecosystem or a community.

4.14 Summary

A variety of organisms and their populations occupying a common space and living in interaction and adjustment with each other constitute a community. A community, unique in itself, has its evolutionary history, its structure, and functional attributes. A biotic community, never static or stationary, exists with its dynamics and spells out its influence in various ways. Among all the species, a species, often specific to an ecological zone, prevails in a larger proportion. A measure to determine ratios between species number and importance values is known as the species diversity index. Species diversity in a community is phenomenally influenced by three major factors: geographical, developmental, and physical. In the trophic groups some species phenomenally control energy flow and affect the environment of other species and are known as dominants or ecological dominants and the phenomenon is known as dominance or ecological dominance. The degree of dominance exercised by a species or many a species is expressed in terms of the index of dominance, i.e., the importance of each species in relation to the entire community. Keystone species are the species that play a crucial role in affecting many other species in an ecosystem. They also determine the ability of so many other species to perpetuate in their environment. Invasive species are those species that invade an ecosystem. These are often the introduced species not native to a region, area, or location. The trees can be identified as broad-leaved trees, conifers, evergreen trees, etc. and form structures emanating from the distribution pattern of a forest community. There exists a variety of spatial structural arrangements in communities we call pattern diversity. On a vertical basis, a community represents its various strata, and the analysis of the changes according to the verticality of the community is stratification. Many tropical rainforest communities may be witnessed with as many as eight strata which are on account of the favorable factors (light, water, nutrients, microorganisms, and appropriate temperature range) that there might be more aboveground and belowground strata. The horizontal division is known as the zonation of a community. The zonation of an aquatic community is different from that of a terrestrial one. Generally, there are three zones in a deep pond or a lake, namely, littoral, limnetic, and profundal zones, each embracing different kinds of organisms. An open ocean is generally divided into three zones per its depth and amount of light it receives: the euphotic zone, the bathyal zone, and the abyssal zone. Trophic structure describes feeding relationships among organisms in an ecosystem over time.

Ecological succession is a complex process that – through successive changes in species structures – advances from a simple, less complex, and unstable community organization towards a more heterogeneous, complex, and stable community. The main causes of succession include initiating, ecesis, and stabilizing. There are many types of succession distinguishable based on different aspects. While most ecologists are confined to only two types, namely, primary and secondary succession, there could be many dimensions of ecological successions, such as autogenic succession, allogenic succession, autotrophic succession, heterotrophic succession, induced succession, cyclic succession, and retrogressive succession. Primary autotrophic succession involves many sequential steps from a lower level to a higher level of the hierarchy and from a state of instability to a state of ecological stability. Successive communities are well presumed. In some cases, succession gets deflected from its presumed path and the final community is somewhat different from the presumed climatic climax community. Such a succession occurring mainly on account of different site conditions, such as different soil compositions or microclimate, is called deflected succession. A climatic climax community maintains equilibrium with a zonal climate. When a climatic climax community resorts to anthropogenic intervention getting disturbed to a considerable extent, it is called dis climax or anthropogenic subclimax. Lithosere, hydrosere, psammosere, halosere, and xerosere are the major types of ecological succession. The lithosere or xerosere (xerarch) begins with a bare rock surface and its successional stages are the crustose-lichen stage, foliose-lichen stage, moss stage, herb stage, shrub stage, and climax forest stage. Ecological succession taking place in a waterbody, notably a pond or a lake, is known as hydrosere or hydrarch and its various seral stages are phytoplankton stage, rooted submerged stage, rooted floating stage, rooted swamp stage or amphibious stage, sedge meadow stage, woodland stage, and climax forest stage. A habitat, in essence, tends to modify from extremely dry or aquatic conditions to mesic or moderately wet conditions through its long journey of ecological succession. Changes in ecosystem characteristics are the consequences of the hierarchical changes in biodiversity and composition of communities. Not only the biotic characteristics but also the edaphic properties register notable changes, such as an increase in organic matter and nutrient content in the soil. Disturbance to a forest ecosystem initiates a sequence of recovery phases. Bormann and Likens presented a model of an ecosystem's recovery from disturbances known as the biomass accumulation model emphasizing four phases of recovery of a forest ecosystem, viz., reorganization phase (10–20 years), aggradation phase (> 100 years), transition phase, and steady-state phase.

4.15 Exercises

4.15.1 Multiple-Choice Questions

1. Who coined the term "biocoenose" that became quite popular among biologists in the beginning?
   (a) Carl August Möbius
   (b) Theophrastus
   (c) Ernst Haeckel
   (d) Charles Darwin
2. A natural forest, according to Odum, invests energy in
   (a) structure and diversity by capturing the energy of light
   (b) expanding greenery
   (c) phenological processes
   (d) controlling wild animals' populations
3. What is not true about species diversity?
   (a) It is generally very high in undisturbed natural ecosystems (as in a rainforest).
   (b) It is often low in human-managed or anthropogenic ecosystems (as in a cropland).
   (c) In mountain areas (as in the Himalayas), it generally increases with an increase in altitude.
   (d) It is greatly influenced by the functional relationships between trophic levels.
4. The "dominance" characteristic of a community is determined by

(a) all the species in the community
(b) only by the species in taxonomic relationships
(c) the species with larger proportion, size, number, and productivity
(d) the trophic level farthest away from the autotrophic species

5. Who introduced the term keystone species?
(a) Robert T. Paine
(b) E.P. Odum
(c) Charles Elton
(d) Raymond Lindemann

6. What is true about the keystone species?
(a) They do not affect the whole community by altering the structure.
(b) These species provide a favorable environment to all the species.
(c) They may not be detrimental to any species.
(d) None of the above.

7. What may be the role of an invasive species?
(a) It may release allelochemicals.
(b) It may affect the ecological relationships of the species.
(c) It may replace native species.
(d) All of the above.

8. Vertical analysis of a community provides information about its various
(a) strata
(b) taxonomic groups
(c) trophic levels
(d) food webs

9. Bottom-up and top-down processes in a forest community determine its
(a) productivity
(b) strata
(c) trophic structure
(d) none of the above

10. What is the difference between an ecosystem and a community?
(a) An ecosystem considers only environmental factors into consideration, while a community involves both biotic components and the physical environment.
(b) A community encompasses only the population of either plants or animals, while an ecosystem also comprises populations of decomposers.
(c) A community encompasses populations of varied species, while an ecosystem takes into account a community as well as its physical environment.
(d) A community comprises all the animals in interaction with each other, while an ecosystem comprises animals, plants, and microorganisms irrespective of their relationship with the environment.

11. The trophic structure is ultimately determined by the __________, the first trophic level, and the ensuing energy transfer efficiency between __________ levels.
(a) producers, trophic
(b) top carnivores, producers
(c) photosynthesis, biodynamic
(d) ecological pyramids, food chain

12. The process of successful establishment of a species at a new site is known as
(a) aggregation
(b) ecesis
(c) invasion
(d) nudation

13. There are two cases of succession, viz.:
A. A natural forest at the climax stage of succession was destroyed due to anthropogenic activities, namely, logging and mining.
B. The eruption of a volcano followed by the mixing of lava with cold water led to the formation of rocks on a large area.

What major process of succession would take place in these cases?
(a) Primary succession in case A, secondary in Case B
(b) Secondary succession in case A, primary in Case B
(c) Primary succession in both the cases
(d) Secondary succession in both the cases

14. The first group of organisms that help improve the environment and set the tone for subsequent successional processes is known as
(a) keystone species
(b) pioneer species

(c) invasive species
(d) climax species

15. In the course of its interaction with the environment, a community modifies its own environment resulting in its own replacement with new communities. Such a type of succession is known as
    (a) allogenic succession
    (b) autogenic succession
    (c) autotrophic succession
    (d) heterotrophic succession
16. Which of the following would represent an induced succession?
    (a) Croplands/cultivated lands
    (b) Artificial lakes
    (c) Wildlife sanctuaries
    (d) All of the above
17. Match the terms of column A with the appropriate terms of column B and select the most appropriate answer:

Column A	Column B
A. Psammosere	(i) The second stage of the hydrosere
B. Crustose-lichen stage	(ii) Autotrophic succession
C. Rooted-submerged stage	(iii) Succession on the salty surface
D. Halosere	(iv) Succession on a sandy surface
E. Prevalence of inorganic environment	(v) Lithosere
F. Ragnar Hult	(vi) Ecological succession

    (a) A-vi, B-i, C-v, D-ii, E-iv, F-iii
    (b) A-iv, B-v, C-i, D-iii, E-ii, F-vi
    (c) A-i, B-vi, C-ii, D-v, E-iii, F-iv
    (d) A-vi, B-v, C-iv, D-iii, E-ii, F-i
18. Reed swamp stage in hydrarch is also known as
    (a) amphibious stage
    (b) sedge meadow stage
    (c) woodland stage
    (d) climax stage
19. The climax community is a forest stage in
    (a) xerosere
    (b) hydrosere
    (c) psammosere
    (d) all of the above
20. In ecological succession, the final stable community in equilibrium with the climate is referred to as
    (a) seral community
    (b) climax community
    (c) pioneer community
    (d) mixed community

Answers: 1-a, 2-a, 3-c, 4-c, 5-a, 6-d, 7-d, 8-a, 9-c, 10-c, 11-a, 12-b, 13-b, 14-b, 15-b, 16-d, 17-b, 18-a, 19-d, 20-b

4.15.2 Short-Answer Questions

1. Define community.
2. What do you mean by a habitat?
3. What is ecological succession?
4. Trace the various stages of succession in a bare rock.
5. Differentiate between primary and secondary succession.
6. What is the difference between seral and climax communities?
7. What is lithosere?
8. Define hydrarch.
9. In the context of ecological succession, what do you mean by pioneers?
10. What is the amphibious stage in hydrarch?

4.15.3 Long-Answer Questions

1. Discuss various causes and processes of ecological succession.
2. What are the different types of succession?
3. Explain various stages found in xerosere.
4. Describe with examples the various stages occurring in hydrosere.
5. Write short notes on
    (a) Climax community
    (b) Biomass accumulation model of ecosystem recovery
    (c) Stability, resistance, and resilience

References

Clements FE (1916) Plant succession: an analysis of the development of vegetation. Carnegie Institution of Washington, Washington, D.C. Publication 242

Clements FE (1936) Nature and structure of the climax. J Ecol 24:252–284

Connell JH, Slatyer RO (1977) Mechanisms of succession in natural communities and their role in community stability and organization. Am Nat 111(982):1119–1144

Hutchinson GE (1961) The paradox of the plankton. Am Nat 95:137–145

Hutchinson GE (1978) An introduction to population ecology. Yale University Press, New Haven

Molles MC (2005) Ecology: concepts and applications. McGraw Hill, Boston. 622 pp

Odum EO (1971) Fundamentals of ecology. Saunders, Philadelphia. 574 pp

Odum HT (1983) Systems ecology: an introduction. Wiley, New Jersey. 644 pp

Paine RT (1966) Food web complexity and species diversity. Amer Naturalist 100:65–75

Paine RT (1980) Food webs: linkages, interaction strength and community structure. Jour Animal Ecol 49:667–685

Sharma PD (2013) Environmental biology and toxicology. Rastogi Publications, Meerut. 569 pp

Singh V (2019) Fertilizing the universe: a new chapter of unfolding evolution. Cambridge Scholars Publishing, London. 284 pp

Singh V (2020) Environmental plant physiology: botanical strategies for a climate smart planet. Taylor and Francis (CRC Press), Boston. 230 pp

Singh V (2023) Biodiversity: concept, crises, and conservation. NIPA, New Delhi, 190 pp

Whittaker RH, Levin SA, Root RB (1973) Niche, habitat and ecotope. Am Nat 107(955):321–338. https://doi.org/10.1086/282837

5 Biogeochemical Cycles

The biosphere exists in integrity thanks to the continuous flows of energy and materials. Within the biosphere energy flows in a unidirectional manner, the matter flows in a cyclic manner. There are three sources or components or pools in the biosphere the matter (elements and chemical compounds) flows from and into, viz., lithosphere, hydrosphere, and atmosphere. The matter in each of these pools comprises the nutrients necessitated for structure and functioning of living beings. These nutrients selectively flow from the (abiotic) pool to the biotic components of the biosphere and, after playing their metabolic role for a defined length of time, return back to the original pool. There is no beginning or end of the nutrient cycles.

While the energy flow does not remain confined to ecosystems (solar energy makes exit to the space in the form of heat), nutrients' flows are contained within ecosystem(s). Energy flow in an ecosystem is "outsourced" into inexhaustible solar energy; therefore, it is a profligate in nature. Nutrient flows, on the other hand, are drawn from finite pools of abiotic components of the environment and, therefore, are "conservative" in nature.

Nutrient cycles involve use, transformation, movement, and reuse of all nutrients that are essential for plants, animals, and microorganisms in an ecosystem. It is through the nutrient cycles that:

(i) Abiotic and biotic components exist in reciprocal relationships.
(ii) A balance in the environment is struck.
(iii) An ecosystem exists in a dynamic state.
(iv) An ecological equilibrium is maintained.
(v) Ecosystems (and hence the whole biosphere) exist in ecological integrity.

Nutrient cycles are critical for the growth, reproduction, population size, and ethology of all the species flourishing in an ecosystem. Thus, they are also vital for ecosystem functions. The nutrient cycles themselves represent ecosystem functions. If the nutrient cycles are obstructed or get deviated from their normal paths due to certain factors (e.g., anthropogenic interventions), the structure and functions of an ecosystem are bound to be affected.

Nutrient cycles and distribution of nutrients are phenomenally influenced by the community structure (composition of living species, trophic structures, and strata). Decomposition rates largely determine the pace at which the nutrient cycles in an ecosystem go on. Disturbances occurring due to environmental and/or anthropogenic factors lead to increased nutrient losses from the ecosystems.

An understanding of nutrient cycles helps us understand growth patterns and productive performance of the species, functional relationships among species of various trophic levels (food chains), decomposition rates, and dynamics of an ecosystem.

V. Singh, *Textbook of Environment and Ecology*, https://doi.org/10.1007/978-981-99-8846-4_5

5.1 Sedimentary and Gaseous Cycles

Based on their respective reservoirs, the nutrient cycles can be broadly categorized into sedimentary and gaseous cycles. Those with nutrient reservoir in the rocks are the sedimentary cycles and with reservoir in the air (and water) are the gaseous cycles. Phosphorus, sulfur, calcium, and magnesium cycles are the examples of sedimentary cycles, while nitrogen, carbon, and oxygen cycles are the examples of the gaseous cycles.

Gaseous cycles take place at a faster pace compared to sedimentary ones and therefore are also completed at faster rates. Sedimentary cycles are generally imperfect cycles because of the higher immobility of the nutrients from the Earth's crust. Gaseous cycles, on the other hand, are the perfect cycles in which nutrients, owing to large reservoirs and high degree of nutrient mobility, quickly attain self-adjustment.

5.2 Hydrological Cycle

Water, through its cycles, exchanges itself in different environmental components (oceans, atmosphere, land, and living organisms) within biosphere. An understanding of water cycle helps us to understand this unique nature of the water. A large proportion of the solar energy on Earth (about one-third) is utilized in driving hydrological cycle, i.e., melting of ice to water, raising water temperature and its evaporation from the oceans and seas, and cloud formation and rainfall. Water on Earth truly acts as a temperature buffer.

Water, actually, is not categorized as a nutrient. However, nutrition is possible only because of water. Water is fundamental to photosynthesis, as also to the structure and functions of all ecosystems and living organisms. All the enzymes responsible for the synthesis of biomolecules and for metabolism act only in aqueous medium. Movement of nutrients from the soil to the plants and within the bodies of all organisms takes place only in aqueous medium. Plants also retain large quantities of water in their bodies as the same is vital for their metabolism. An understanding of water cycle, therefore, is also essential for understanding the overall structure and functioning of all ecosystems as well as of all living beings.

About 97% of the total global water is contained in oceans. Distribution of water across biosphere, however, is not uniform and/or static at a given time. Be an ocean, a sea, a lake, a pond, a river, a glacier, or a biological reservoir (an organism, for example), water in every reservoir keeps on replenishing itself. The water resources receive freshwater supplies from precipitation. The precipitation itself is on account of the continuous evaporation from these resources.

Evaporation from all water sources occurs at all temperatures and slowly gets lost to the atmosphere. Ice and snow melt into liquid form while a proportion of the solid form of water directly sublimes into vapor. Water in the atmosphere also comes from living organisms. From soil and vegetation it enters into the atmosphere through evapotranspiration. Of the total evaporated water entering into atmosphere, a large chunk is contributed by the oceans alone. Major components of the air, i.e., N_2 (molecular mass 28) and O_2 (molecular mass 32), have higher density than that of H_2O vapor (molecular mass 18). Therefore, as a result of buoyancy, water vapor goes higher in the atmosphere. Water cycle is accommodated in the troposphere of the atmosphere where temperature and pressure decrease with an increase in altitude. In the upper troposphere, water vapor condenses into small droplets the huge mass of which assumes the form of clouds. With strong air currents in the upper atmosphere, cloud particles collide, grow, and precipitate and water returns back in the form of rain, snow, hail, sleet, etc. When moist air and cool air collide near Earth's surface, fog is formed.

Most of the precipitation in the form of rains occurs over oceans of the globe. A proportion of precipitation occurs in the form of snow, hail, and sleet that forms ice caps and glaciers. A proportion that falls as rain on land flows into streams, rivers, rivulets, etc. ultimately getting drained into oceans and seas. A proportion of the runoff water collects into lakes, ponds, and other small water bodies. A proportion of the same soaks into ground through infiltration. The water that goes deeper into ground replenishes aquifers.

The precipitation enables the water to collect in various water bodies on the planet. The hydrological cycle, thus, comprises seven major stages, viz., evaporation, condensation, sublimation, precipitation, transpiration, runoff, and infiltration.

The evaporative phase of the water cycle purifies water with which the water sources are replenished with fresh water. Melting of ice and snow and flow of water on land surface carry mineral resources and distribute them throughout the globe. Water flows also reshape geological features through erosion, landslides, and sedimentation (Fig. 5.1).

The biosphere is held in ecological integrity. It is the ecological integrity that ensures intactness of and helps the biosphere reverberate with life. The water cycle links all the three major components of the biosphere – the lithosphere, the hydrosphere, and the atmosphere – that impart essential ecological integrity to the biosphere. Water is critical for life as it is the basic source of photosynthesis. The water cycle helps circulate water among all ecosystems of the earth, thus nourishing the phenomenon of photosynthesis and helping all the photosynthesizers and animals and microorganisms acquire water for their sustenance. The water cycle, in other words, is vital for the maintenance of Earth's ecosystems.

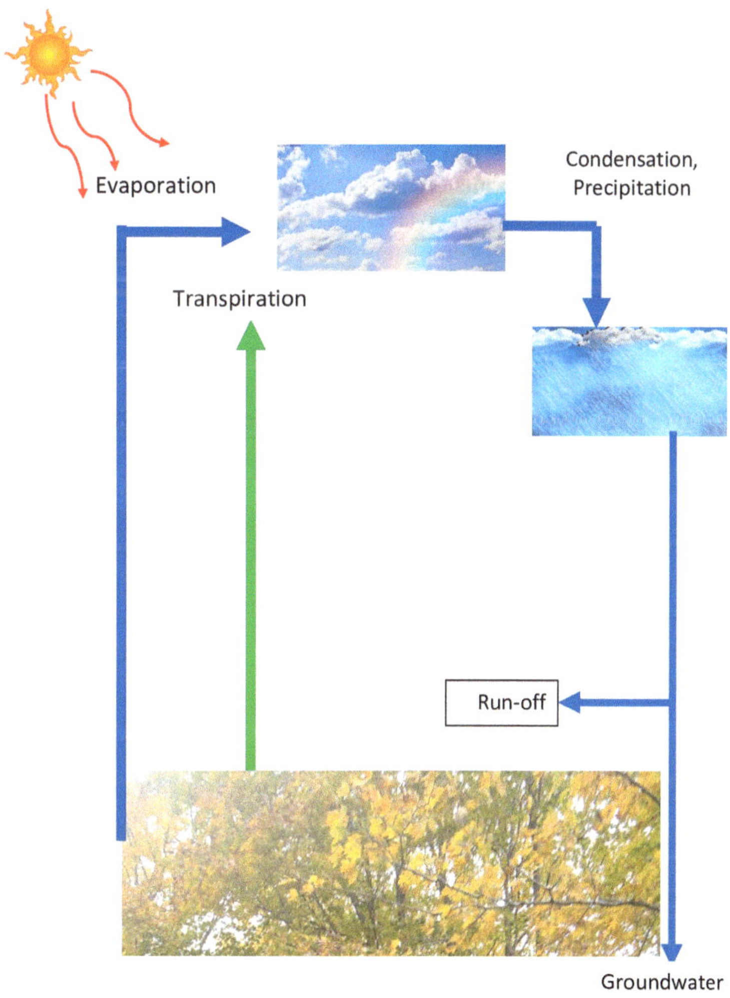

Fig. 5.1 The hydrological cycle

The water cycle involves energy exchange, and therefore, it contributes to influence the energy budget of the biosphere and determines weather cycle and climate patterns. Evaporation of water uses atmospheric energy leading to the cooling of the surroundings. Condensation, on the other hand, releases energy to the warm up the atmosphere.

5.2.1 Residence Time of Water

Not a single molecule of water on Earth exists in its original form. It has renewed itself undergoing recycle a numerous times ever since the water planet came into being. The average time a water molecule spends in a particular reservoir is what is known as the residence time of the reservoir within the hydrological cycle. In other words, the residence time of a reservoir is the average age of water that resides in that reservoir.

It is the Antarctica that records the longest residence time (approximately 20,000 years) followed by groundwater in deep layers of the planet (10,000 years). Oceans completely change their water in a cycle of about 3200 years. Soil moisture stays for a brief period of about 2 to 6 months only as it remains exposed to the atmosphere. Water residence time in the rivers is also very short (2 to 6 months). The shortest residence time of water is in the atmosphere (only 9 days) (Table 5.1).

Table 5.1 Average reservoir residence time

Reservoir	Average residence time
Antarctica	20,000 years
Groundwater: deep	10,000 years
Oceans	3200 years
Groundwater: shallow	100–200 years
Lakes	50–100 years
Glaciers	20–100 years
Seasonal snow cover	2–6 months
Rivers	2–6 months
Soil moisture	1–2 months
Atmosphere	9 days

Source: Pidwirny (2006)

5.3 Nitrogen Cycle

Nitrogen cycle, a gaseous type of the biogeochemical cycles, converts molecular nitrogen into various compounds that incorporate in structural, functional, and genetic biomolecules in living organisms. Most of the atmosphere (78%) is composed of nitrogen. However, plants and animals cannot utilize this nitrogen as such. Nitrogen comes into usable form where it can enter into plants (and through them into consumer animals) both physically and biologically. While nitrogen in the atmosphere exists as molecular nitrogen (N_2) almost exclusively (NH_4^+ and NO_3^- might also be present in the atmosphere which get deposited through precipitation), in other components of the environment it exists in various other forms, such as ammonia (NH_3), ammonium (NH_4^+), nitrite (NO_2^-), nitrate (NO_3^-), nitric oxide (NO), nitrous oxide (N_2O), nitrogen dioxide (NO_2), etc., as inorganic compounds. In chlorophyll, amino acids, peptides, proteins, nucleic acids among living organisms, and decomposed and decomposing compounds in the hydrosphere and the lithosphere, nitrogen prevails in the form of organic compounds. Nitrogen cycle involves transformation of nitrogen into many compounds. Many a processes of nitrogen cycle are mediated by a variety of microorganisms.

5.3.1 Nitrogen Fixation

Nitrogen fixation involves conversion of molecular nitrogen into nitrites and nitrates. In nature, this conversion takes place physically as well as biologically. In the Industrial Age, this is also carried out industrially. Since two atoms of nitrogen in N_2 are bonded by a triple bond ($N \equiv N$), breaking up of the molecule into atoms requires input of energy. Investment of substantial energy in converting molecular nitrogen into usable form imposes limits on nitrogen fixation into organic matter. It is why despite a pretty large proportion of N_2 in the atmosphere, it is often rendered deficient for living organisms in ecosystems.

5.3.1.1 Physical or Atmospheric Nitrogen Fixation

Physical fixation, or atmospheric fixation or non-biological fixation of the atmospheric nitrogen takes place due to lightning. Huge amount of energy generated during lightning leads to break up of the inert nitrogen molecules into atoms which react with oxygen in the air forming nitrogen oxides (NOx). During rainfall NOx dissolve in water and form nitrite acid which upon falling on the soils is converted into nitrate, a kind of nitrogen compound usable by plants. Of the total nitrogen fixed, only about 5%–8% is fixed through non-biological or physical means.

5.3.1.2 Biological Nitrogen Fixation

Hermann Hellriege, a German agronomist, and Martinus Beijerinck, a Dutch microbiologist, have the credit to have discovered biological nitrogen fixation that takes place when atmospheric N_2 gets converted to NH_3 by means of an enzyme complex nitrogenase (iron-molybdenum cofactor, FeMoco) that contains two proteins, viz., an iron protein and a molybdenum-iron protein. In this process, the role is played exclusively by prokaryotes. The reaction is as follows:

$$N_2 + 8H^- + 8e^- \rightarrow 2NH_3 + H_2$$

In the above reaction, 16 moles of ATPs are hydrolyzed with co-formation of a molecule of hydrogen. Biological nitrogen fixation, thus, is an energy-demanding process which the nitrogen-fixing prokaryotes invest at 16 moles of ATPs per molecule of N_2. In the process, N_2 molecule is first bound with the nitrogenase enzyme complex, FeMoco. First, iron protein is reduced by means of electrons ferredoxin donates. The reduced iron protein then binds ATP, reducing the molybdenum-iron protein, which, in turn, donates electrons to N_2 resulting in the production of HN=NH. This, as a result of further donation of electrons by ferredoxin in the two cycles, gets reduced to $H_2N\text{-}NH_2$ that subsequently is reduced to $2NH_3$. As the whole process involves investment of a large number of ATP moles, from the point of view of energy usage, it is relatively expensive for the plants. The ultimate source of energy, of course, is sunlight used by the plants by means of photosynthesis. Ammonia formed is converted into amino acids which serve as building blocks for subsequent protein synthesis.

A nitrogen molecule, in order to split into its two N atoms, invests 160 kcal. Reduction of a nitrogen molecule (two N atoms) into ammonia releases 13 kcal. Thus, the net energy input for nitrogen fixation is 147 kcal/mole. Biological nitrogen fixation (BNF) contributes to about 63% of the total nitrogen fixed in Earth's ecosystems.

BNF by Free-Living Bacteria The free-living bacteria are the asymbiotic organisms and BNF by these organisms is called asymbiotic nitrogen fixation. The organisms involving in asymbiotic N_2 fixation are (i) aerobic bacteria and anaerobic bacteria and (ii) blue-green algae (cyanobacteria).

Aerobic and Anaerobic Bacteria These are the saprophytic bacteria. *Clostridium* and *Azotobacter* are the examples of the anaerobic and aerobic bacteria, respectively, and inhabit soils where they play their role in N_2 fixation.

Blue-Green Algae (Cyanobacteria) These are photosynthetic autotrophs inhabiting water-logged areas and wet soil of rice fields. *Nostocales* and *Stigonematales* are important in playing their role in N_2 fixation. Some examples are:

(i) Anabaena with branched filaments made up of the green photosynthetic cells and the colorless and non-photosynthetic heterocysts involve in fixing N_2 aerobically. Some blue-green algae not having heterocysts also fix N_2.
(ii) Some free-living blue-green algae in the soil enter into plant tissues and inhabit there permanently, e.g., Nostoc in Anthoceras produce heterocysts in large numbers and fix N_2 with increased efficiency. In this symbiotically regulated process, nitrogenous substances in the cyanobacterial cells then pass onto the higher plant (host), and the N_2-fixing cells, in turn, take up carbohydrates

from the host plant for energy for N_2 fixation.
(iii) Nostoc and Scytonema algae establish symbiotic association with fungi forming lichens. The alga in the lichen fixes N_2 for fungi and fungi provide energy for algae.

5.3.1.3 Symbiotic Nitrogen Fixation

It provides very important pathways to nitrogen fixation. Several species of fungi and nodule-forming bacteria are the two major sources of N_2 fixation.

1. **Fungi (Actinomycetes):** Many species of Actinomycetes with ability to fix atmospheric N_2 establish mycorrhizal association – ectotropic and endotropic – with the roots of certain plants, such as *Pinus* and *Casuarina*.
2. **Nodule-Forming Rhizobium Species:** It is the most frequently discussed BNF. Rhizobium species infect roots of certain plant species leading to the formation of root nodules. Several leguminous plants, such as soybeans, gram, pea, etc., develop root nodules. All plants of family Papilionaceae develop nodules. About 30% of the species belonging to family Caesalpinaceae develop root nodules.

The two symbionts, the bacteria and the host plant, recognize each other by means of a chemical substance called phytoagglutinins. The nitrogen fixing (nif gene) in the rhizobium controls the synthesis of nitrogenase. The nodules on the secondary roots of the legume plants the symbiotic bacteria form contain leghemoglobin (LHb), a red-colored pigment. LHb acts as an oxygen scavenger contributing to maintain low O_2 levels inside the nodules, a condition helping the nitrogenase to activate and reduce N_2 to ammonia.

Amino acids are the main molecules the bacteria fix the nitrogen into. These amino acids following transpiration stream are transported to the host plant. The N_2-fixing bacteria, in turn, get carbohydrates as food from the host plants. This symbiotic relationship is vital for the fertility and consequent productivity of the soil in a natural way.

5.3.1.4 Industrial Nitrogen Fixation

In this process nitrogen and hydrogen (from petroleum or natural gas) are combined into ammonia at very high temperature (400 °C) and high pressure (200 atmospheres). This industrial process called Haber-Bosch process (Smith et al. 2004) contributes nearly 30% of the total nitrogen fixed.

5.3.2 Assimilation

Ammonium and nitrates in the soil are taken up by the plants through their root hairs. Upon absorption NO_3^- are reduced to NO_2^- and then into NH_4^-. The NH_4^- are incorporated into amino acids, nucleic acids, and chlorophyll. The legume-rhizobia symbiosis allows some N_2 to assimilate in NH_4^- directly from the nodules. A complex cycling of amino acid is known to occur between plants and *rhizobia* bacteroids through an interdependent relationship. In this unique relationship the host plant provides amino acids to the bacteroids which, reciprocally, provide amino acids synthesized with newly fixed nitrogen to the plants. Thus, ammonia assimilation in plants is not required.

5.3.3 Ammonification

Plants and animals in nature die and decompose. Wastes of the plants and animals also decompose. Proteins and nitrogenous wastes, like urea and uric acid, decompose. The decomposition process involves a variety of saprotrophic microorganisms – the ammonifying bacteria, e.g., *Bacillus ramosus*, *B. vulgaris*, and *B. mycoides*. During the decomposition process these saprotrophs convert organic nitrogen into NH_3 and NH_4^+. This process is exothermic.

$$\text{Glycine}\left(\text{Amino acid}\right) + O_2 \rightarrow NH_3 + CO_2 + H_2O + \text{Energy}$$

$$\Delta G^\circ = 176\,\text{kcal}/\text{mole}$$

A proportion of NH_3 exits into atmosphere while the NH_3 trapped into soil is absorbed by the plants (as NH_4^+) and converted into organic nitrogen. This process is also known as mineralization.

5.3.4 Nitrification

It involves the oxidation of NH_3 to nitrates via nitrites. Oxidation of NH_3 is necessary as it is toxic to the plant. The process occurs in the presence of nitrifying bacteria. *Nitrosomonas* convert NH_3 to nitrites and *Nitrobacter* oxidizes nitrites to nitrates:

$$2NH_3 + O_2 \rightarrow 2NO_2^- + 2H^+ + 2H_2O$$

$$\Delta G^\circ = +65\,\text{kcal/mole}$$

$$2NO_2^- + O_2 \rightarrow 2NO_3^- + \text{Energy}$$

$$\Delta G^\circ = +17\,\text{kcal/mole}$$

5.3.5 Denitrification

This final step of nitrogen cycle recycles the nitrogen in its molecular form into its pool, the atmosphere. This is the biological process that occurs in the presence of denitrifying bacteria, like *Thiobacillus denitrificans, Pseudomonas stutzeri, Bacillus subtilis, Micrococcus denitrificans,* etc. In denitrification, as opposed to nitrification, ammonium compounds, nitrates and nitrites, are reduced to the molecular nitrogen (N_2) that enters into the atmosphere, its own pool. The conversion reactions are as follows:

$$NO_3^- \rightarrow NO_2^-$$

$$NO_2^- \rightarrow NO$$

$$NO \rightarrow N_2O$$

$$N_2O \rightarrow N_2$$

Reduction of NO_3^- to N_2O takes place under anaerobic conditions and in the presence of glucose. It is an exothermic reaction releasing 545 kcal/mole. Complete denitrification of NO_3^- to N_2 releases 570 kcal/mole. When the soil ecosystem is exposed to conditions such as lack of aeration, waterlogging, poor drainage, etc., the process of denitrification is accelerated. The schematic representation of the nitrogen cycle in Fig. 5.2 attempts to show linkages among all the steps involved in the complex process discussed above.

5.3.6 Anthropogenic Effects on Nitrogen Cycle

The anthropogenic impact on the nitrogen cycle in the Post-Industrial Age is phenomenal. Since the advent of the Green Revolution in the 1960s, pathways of the nitrogen cycle have been altered significantly thanks to human activities. Nowadays, as much as 23% of the total nitrogen fixation takes place through the industrial process to be used mainly in food production-related operations. Cultivation of monocultures of cereal food crops deprives soils of nitrogen content. Indiscriminate use of pesticides wipes out nitrogen-fixing microflora in the soil. Thus, the soil capacity of biological nitrogen fixation is suppressed significantly. In order to fulfill nitrogen demands of the food crops, high doses of industrially synthesized nitrogenous fertilizers – especially urea – are applied repeatedly. Balanced soil fertilization requires supplementation of nitrogenous fertilizers with phosphorus and potassium. Further, adoption of intensive farming practices in the Green Revolution system of food production have also deprived soil of several micronutrients. To meet the micronutrient requirement, mined fertilizers are also applied in the soil. Excessive use of nitrogenous fertilizers has triggered increased nitrogen flow to groundwater and surface water sources aggravating the problem of eutrophication. Rice farming accelerates ammonification and generates enormous amounts of nitrogenous gases (NOx). Accumulation of nitrates in water sources causes human health hazards.

Industrial nitrogen fixation and application of nitrogenous fertilizers have also increased NOx gases that serve as potent greenhouse gas, thus

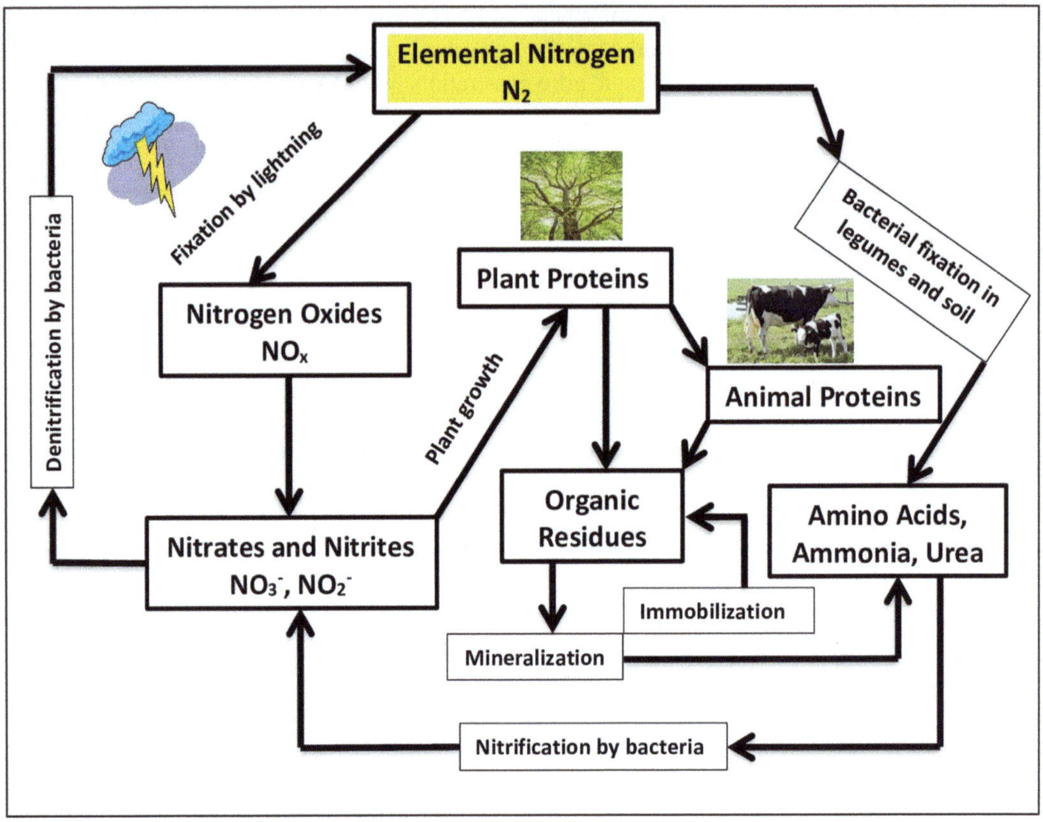

Fig. 5.2 Schematic representation of the nitrogen cycle

contributing to global warming. A spurt in the combustion of fossil fuels in our times is contributing copious amounts of NOx gases to deposit in the atmosphere, solid nitrogen compounds in the soil, and acid rain due to conversion of NOx into nitric acid.

Industrial livestock farming in our contemporary times has turned ruminant animals as one of the largest sources of ammonia and greenhouse gases. The nitrogenous compounds, including copious amounts of ammonia, in animal wastes enter into the soil, water, and atmosphere contributing to alter nitrogen pathways (Singh et al. 2017).

5.4 Carbon Cycle

In the universe carbon is the fourth most abundant element, the first three being hydrogen, helium, and oxygen. The organic nature of the biosphere is owing to carbon. Carbon is the key constituent of life. Carbon, in essence, defines life. Life, in essence, is organic: a miracle of perhaps the most incredible element – the carbon. Though only in trace amounts in the atmosphere, the inorganic carbon plays a key role to create and sustain life on Earth. The key element is not only the carbon itself but the process through which carbon moves in its various forms among the atmosphere, hydrosphere, pedosphere, geosphere, and biosphere of the Earth – called the carbon cycle. It is the carbon cycle, in addition to nitrogen and water cycles, that is phenomenal for the Earth to sustain as a living planet, perhaps the only living planet in the universe as far as we know. Discovered by Joseph Priestley and Antonie Lavoisier and popularized by Humphry Davy (Holmes 2008), the carbon cycle helps us understand the very nature, structure, and functioning of the biosphere. It also helps us understand one of the most serious problems the living planet is faced with in our contemporary times: climate change.

5.4.1 Carbon Reservoirs

There are several carbon reservoirs on Earth, all of them interconnected with each other. These are:

1. The atmosphere (in inorganic form)
2. Terrestrial biota (in organic form – in living biomass and soils)
3. Oceans (in dissolved inorganic form)
4. Aquatic biota (in organic form)
5. Sediments (in organic and inorganic forms)
6. Fossil fuels (in dead organic form)
7. Earth's interior, mantle and crust (organic and inorganic forms)

Amounts of carbon in various reservoirs on Earth are presented in Table 5.2.

Carbon in the atmosphere prevails largely in the form of CO_2 and to a lesser proportion as CH_4. Both gases are the greenhouse gases contributing to absorb and retain heat of the Earth. However, the greenhouse effect of CH_4 is much larger per volume than that of CO_2. Since CH_4 is much less in concentration and further is short-lived compared to CO_2, most of the greenhouse gas effect is due to CO_2. CO_2 is transparent for the incoming radiation but opaque for the infrared radiation from the Earth. The chemical bond between carbon and oxygen does not absorb the incoming visible radiation but does absorb the infrared (heat) radiation going back from the Earth to the space.

Table 5.2 Major carbon reserves and carbon pools on Earth

Pool	Quantity (in gigatons)
Atmosphere	720
Oceans (total)	38,400
Inorganic	37,400
Organic	1000
Surface layer	670
Deep layer	36,730
Lithosphere	
Sedimentary carbonates	>60,000,000
Kerogens	15,000,000
Terrestrial biosphere (total)	2000
Living biomass	600–1000
Dead biomass	1200
Aquatic biosphere	1–2
Fossil fuels (total)	4130
Coal	3510
Oil	230
Gas	140
Other (peat)	250

Source: Falkowski et al. (2000)

The terrestrial biosphere contains carbon in the biomass of all living (and dead) organisms (plants, animals, and microorganisms) as well as in the soils. The amount of the carbon in the soils is about three times that in the aboveground biomass in vegetation and other living organisms. Most of the carbon in the terrestrial biosphere is in organic form. However, in the soils about one-third proportion of the total carbon is in the inorganic form, such as calcium carbonate.

After the lithosphere, most of the carbon on Earth is in the oceans, which is the world's largest store of the actively cycled carbon. In the aquatic medium the carbon is stored as bicarbonates, limestone, and marble rocks. Oceans continuously absorb atmospheric CO_2 and serve as the largest carbon sink on Earth. The amount of carbon in the oceans is about 50 times more than that in the atmosphere. The dissolved carbon is in the form of carbonic acid and the reaction is reversible. Thus, the oceans are capable to regulate CO_2 in the atmosphere.

$$H_2O + CO_2 \leftrightarrow H_2CO_3$$

$$H_2CO_3 \leftrightarrow H^+ + HCO_3^-$$

$$HCO_3 \leftrightarrow H^+ + CO_3^-$$

The lithosphere of the Earth stores most of the carbon mostly in its inert form (Falkowski et al. 2000). Most of the carbon deposit in the mantle of the Earth is believed to be stored ever since the planet came into being. A proportion of the lithospheric organic carbon has come from the biosphere. As high as 80% of the total geospheric carbon is in inorganic form, stored as limestone which has built up thanks to the continuous sedimentation of $CaCO_3$ of the shells of marine organisms. The rest 20% of the carbon is deposited as kerogens which come into existence due to burial and sedimentation of the organisms under high temperature and pressure (see Fig. 5.3).

5.4.2 Carbon Dioxide Utilization

Photosynthesis fixes carbon into life. CO_2 is imperative for green plants – along with water in the presence of sunlight – for photosynthesis. It is the atmospheric CO_2 that is drawn by the green plants for photosynthesis. Carbon in the lithosphere is not available for getting fixed into plants through photosynthesis unless it gets released into the atmosphere though burning and/or chemical change. CO_2 sequestration from the atmosphere and H_2O synthesize carbohydrates in the presence of sunlight. Glucose in carbohydrates is also used to synthesize proteins, fats, and nucleic acids. Thus, within organisms carbon moves from one biomolecule to another.

From the autotrophs (the photosynthesizers or producers), carbon moves to herbivores and from there to carnivores, up to the top carnivores, through foods. Energy is also transferred to the organisms of all the trophic levels, from herbivores to top carnivores. Thus, carbon becomes a key source of energy and nutrients for the organisms through food chains. From producers and

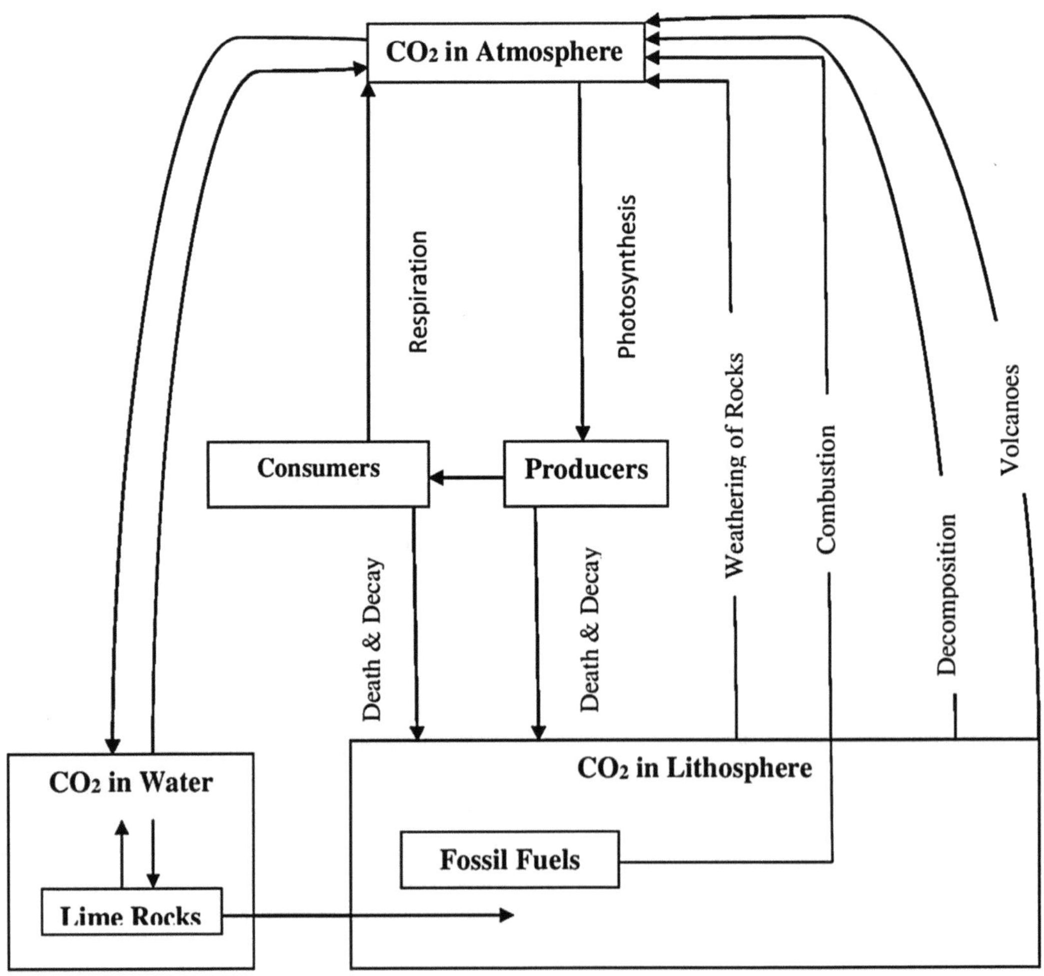

Fig. 5.3 Carbon cycle

macro-consumers (herbivores and carnivores), carbon moves to the micro-consumers, the decomposers.

Sequestered from the atmosphere through photosynthesis, carbon, via plant biomass, steadily keeps becoming deposited in the soil where it becomes a culture for soil organisms, a medium of detritus food chain and is also stored in humus and other carbon compounds and mixtures. Soil organic carbon is the basis of ecosystem functions with primary ecosystem productivity being one of its attributes.

In aquatic ecosystems inorganic carbon enters through dissolution. Oceans are very rich carbon sinks on Earth. Organic carbon reaches the oceans through rivers. CO_2 dissolved in water converts to organic carbon through photosynthesis by green aquatic plants and blue-green algae. The organic carbon is exchanged among aquatic animals through aquatic food chains directly by herbivores and, from them, to carnivores. A proportion of the carbon precipitates into deeper oceanic layers in the form of soft tissues, and another is deposited in the shells of the aquatic organisms. Carbon in deeper layers remains stored for a pretty long period of time.

5.4.3 Carbon Dioxide Production

5.4.3.1 Respiration

Respiration by all organisms – producers as well as consumers – releases CO_2 into the atmosphere. Respiration is just reversal of photosynthesis. What is an input of photosynthesis is an output of respiration and vice versa.

5.4.3.2 Decomposition

Decomposition of organic matter by decomposers – bacteria and fungi – releases CO_2 back into the atmosphere. Decomposition is also reversal of photosynthesis. What is synthesized through photosynthesis (i.e., carbon, water, and energy into biomass) is "desynthesized" through decomposition.

5.4.3.3 Burning of Forests/Woods

Burning of forests, woods, and all the biomass synthesized and accumulated by means of photosynthesis release CO_2 into the atmosphere. This occurs due to anthropogenic activities.

5.4.3.4 Fossil Fuel Combustion

Combustion of fossil fuels – coal and petroleum – in transport vehicles, industries, and households leads to enormous amounts of CO_2 entering into the atmosphere.

5.4.3.5 Hot Springs

Carbon dioxide largely due to decomposition of $CaCO_3$ gets released into the atmosphere through hot springs.

5.4.3.6 Weathering of Rocks

Carbonate-containing rocks, upon action of acids or chemicals excreted by plant roots and microorganisms and due to treatment of carbonate minerals, add to CO_2 in the atmosphere.

5.4.3.7 Volcanic Eruptions

Enormous amounts of lithospheric carbon in the form of CO_2 are released into the atmosphere in a natural way during volcanic eruptions (see Fig. 5.3).

5.4.4 Anthropogenic Effects on the Carbon Cycle

Large-scale deforestation is leading to significantly reduced rate of carbon sequestration due to which carbon dioxide is not becoming part of life (through photosynthesis) and is accumulating in the atmosphere. Further, modern agriculture dependent on intensive ploughing and irrigation is also one of the causes of increased rates of carbon emission. Extensive and intensive ploughing increases direct carbon emission from the soils directly and indirectly by making soils prone to erosion. Increased irrigation frequency increases carbon emission by enhancing decomposition of soil organic matter. The conventional agriculture gives no room for trees and focuses on monocultures of the food crop annuals. This leads to decreased efficiency of carbon sequestration of the modern agriculture.

Overexploitation of fossil fuels giving way to ever-increasing combustion of coal and petroleum products is responsible for ever-increasing carbon emissions and thus for ever-increasing CO_2 concentration in the atmosphere.

All these anthropogenic activities are adding to atmospheric CO_2 and other greenhouse gases resulting in global warming and consequent climate change. Gradually increasing atmospheric CO_2 levels owing to anthropogenic activities are gradually triggering mechanisms detrimental to life.

5.5 Oxygen Cycle

Oxygen is indispensable for life as it generates energy by oxidizing carbohydrates in the cells of the organisms. Oxygen is very reactive. It quickly combines with a variety of the atoms to form a variety of compounds – the oxidized compounds – occurring all over the Earth. Both carbon dioxide and water have oxygen as their constituents. Therefore, both water cycle and carbon cycle involve the oxygen cycle also.

5.5.1 Sources of Oxygen

Oxygen occurs in all the four components of the Earth's environment, viz., atmosphere, hydrosphere, lithosphere, and biosphere.

Oxygen in its molecular form (O_2) makes up 20.84% of the atmosphere by volume. Other than its molecular form, oxygen also exists in the atmosphere in its various other forms, e.g., O_3, CO_2, SO_2, NO, N_2O, H_2O vapor, etc.

- About one-third of the hydrosphere by volume is made up of oxygen as a component of H_2O, dissolved O_2, and H_2CO_3.
- As much as 94% by volume of the lithosphere comprises oxygen as SiO_2 (silica minerals) and various other oxides.
- Most of the biosphere – chemically – comprises organic molecules CxHxNxOx and H_2O. Oxygen is 22% (by volume) as a constituent of the biosphere.

5.5.2 Oxygen Production

Photolysis of water during the light phase of the photosynthesis in green plants and blue-green algae results in oxygen production. A large proportion of oxygen enters into the atmosphere through photosynthesis. Photosynthesis in terrestrial and aquatic ecosystems is the major source of maintaining oxygen equilibrium in the atmosphere:

$$6CO_2 + 6H_2O + \text{light} \rightarrow C_6H_{12}O_6 + O_2$$

H_2O and N_2O also undergo photolysis due to the reaction of high-energy ultraviolet radiation releasing oxygen, which is another source of free atmospheric oxygen:

$$2H_2O + \text{uv radiation} \rightarrow 4H + O_2$$

$$2N_2O + \text{uv radiation} \rightarrow 4H + O_2$$

5.5.3 Oxygen Utilization

All plants, animals, and a variety of microorganisms utilize oxygen for respiration turning carbohydrates or other respiratory biomolecules into carbon dioxide and water. Respiration is just opposite of photosynthesis.

Decomposition of organic matter requires oxygen for the microorganisms involved in the process. Decomposers in the water bodies polluted with organic matter consume oxygen in the process of decomposition. Oxygen is also consumed in the process of composting of the organic matter.

Combustion of biomass and fossil fuels (coal and petroleum) use oxygen and in the process oxides of carbon, sulfur and nitrogen, and water are produced. Many oxides are also produced due to microbial oxidation (see Fig. 5.4).

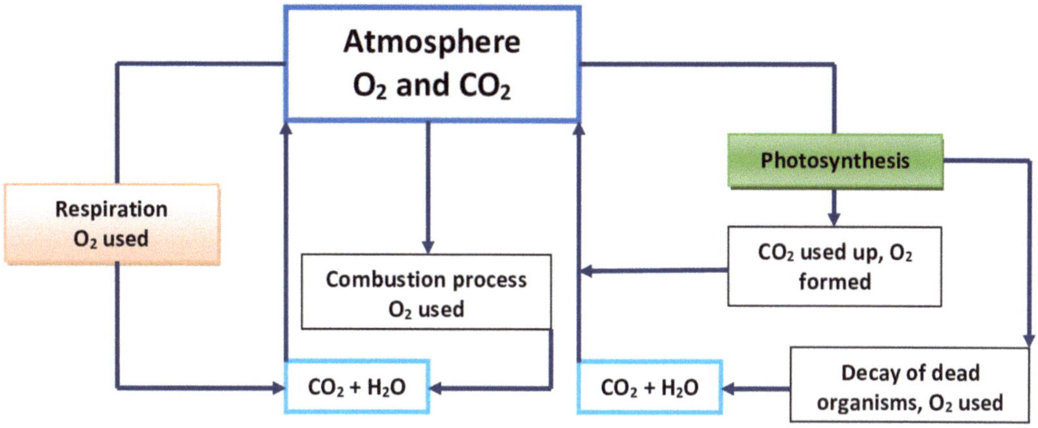

Fig. 5.4 The oxygen cycle

Atmospheric oxygen is also utilized in the formation of various oxides in the lithosphere due to surface reactions and weathering processes. Oxygen thus utilized in the oxide formation becomes part of its reservoirs in the lithosphere. Formation of the iron oxide is one of the examples:

$$4FeO + 2O_2 \rightarrow Fe_2O_3$$

5.5.4 Anthropogenic Effects on the Oxygen Cycle

Large-scale combustion of fossil fuels in the transport sector and industries is one of the root causes of oxygen consumption at the cost of switching over to global warming and consequent climate change. Ever-exceeding rates of oxygen utilization are causing havoc for life. When the oxygen cycle is affected adversely, the carbon cycle is affected simultaneously. Exit of carbon from the lithosphere and hydrosphere into the atmosphere in ever-increasing amounts is ailing our living planet and posing a question mark on life.

- Warming of water bodies is reducing their oxygen-dissolving capacity. Lack of oxygen in the water bodies is detrimental for aquatic life.
- Lack of oxygen in water bodies deteriorates the quality of drinking water, and if such water is used for drinking, it causes a number of serious water-borne diseases.
- Improper management of biodegradable wastes is also causing rampantly increased consumption of copious amounts of oxygen.
- Atmospheric pollution caused largely due to human activities reduces oxygen intake by human beings which becomes a cause for several kinds of health disorders.
- Chlorofluorocarbons (CFCs) responsible for ozone layer depletion thanks to the breakdown of ozone into oxygen molecule and nascent oxygen are produced absolutely due to human activities.

5.6 Phosphorus Cycle

Phosphorus is one of the major mineral constituents of life. As a part of the cell membrane (phospholipid molecules), phosphorus serves as a structural constituent of life. Found in appreciable amounts in bones, teeth, and shells, phosphorus provides rigidity and strength to living individuals. As a high-energy phosphate bonds in the molecules of ATP, ADP, and GTP, phosphorus is vital for the energetics of life. Thus, phosphorus plays key role in metabolism among living beings and also aids them in performing various functions. Phosphorus is indispensable in genetics as being a component of nucleotides of nucleic acids it serves encoding information in genes. Despite playing such a crucial biological role, phosphorus is not found in very high abundance in the biosphere. It is due to great biological importance that the ecologists pay high attention to phosphorus cycling (Fig. 5.5).

5.6.1 Sources and Movement of Phosphorus

Most abundant quantities of phosphorus in the biosphere occur in mineral deposits and marine sediments. Especially rich in phosphorus, the sedimentary rocks are mined for applying fertilizers in croplands. Thus, soils also contain substantial amounts of phosphorus. Phosphorus in the soils, however, is present in such chemical form that much of it remains unused by crop plants. Due to weathering phosphorus slowly releases into terrestrial and aquatic ecosystems. Released from mineral deposits into ecosystems, the phosphate ions are absorbed by the plants and, passing through food chains via trophic levels, recycles into terrestrial and aquatic ecosystems.

Large proportions of phosphorus from the terrestrial ecosystems are washed into rivers and through them eventually reach marine ecosystems. In the marine environments the dissolved phosphorus slowly keeps depositing into ocean sediments eventually transforming into phosphate-bearing sedimentary rocks. These rocks, through geological processes, can form new land (Molles 2005), which are subject to release phosphorus through weathering, thus maintaining continuity of the phosphorus cycle. In his biogeochemical analysis, Schlesinger (1991) mentioned that released from the sedimentary rocks through weathering, phosphorus has undergone at least one cycle within the global phosphorus cycle.

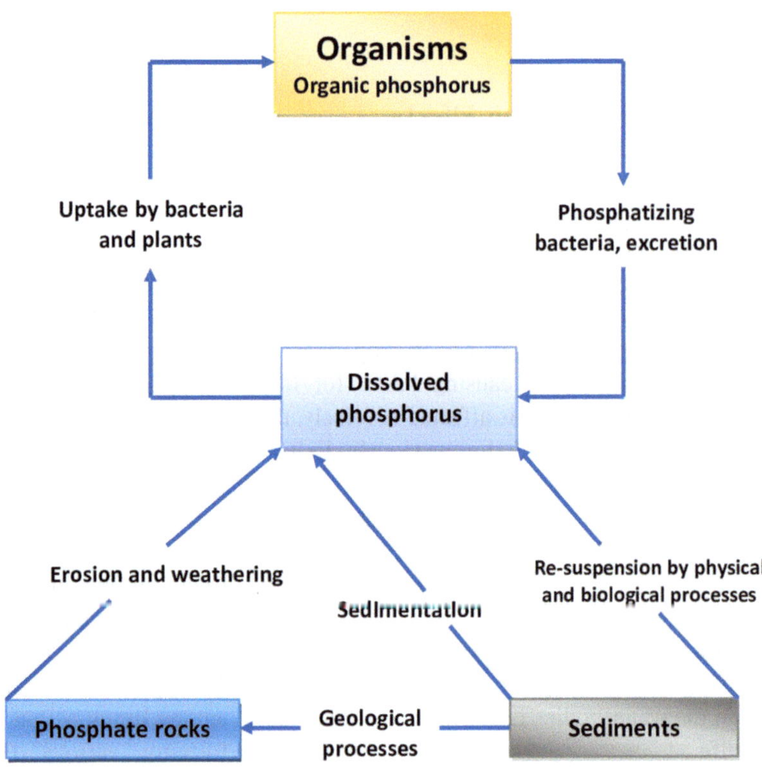

Fig. 5.5 The phosphorus cycle

5.6.2 Anthropogenic Effects on the Phosphorus Cycle

Intensive cropping under the Green Revolution cover requires repeated doses of phosphate along with other inorganic fertilizers to be applied in the soils. This has led to harping on callous mining of phosphate-rich mineral deposits in nature. Excessive use of detergents especially for washing clothes in houses has further aggravated the mining of phosphate rocks. This practice is gradually depleting the most abundant and relatively stable phosphorus reserves in nature.

Large-scale deforestation and intensive cropping followed by bumper harvests in cultivated lands are disturbing the natural phosphorus cycle.

Ever-increasing commercial harvest of fisheries from the oceans, seas, and lakes contributes to squeezing out tremendous amounts of phosphorus from aquatic ecosystems to terrestrial ecosystems, affecting the natural phosphorus cycle in several ways. In lakes that are hot spots for fishing, phosphorus output exceeds phosphorus inputs, thus disturbing overall phosphorus balance.

Pollution of water bodies, especially of lentic ecosystems, with phosphorus-rich wastewaters, animal wastes, and sewage leads to the disturbance of aquatic life and, therefore, the natural phosphorus cycle. Water runoff exacerbating soil erosion from croplands that washes phosphate fertilizers into water bodies contributes to nutrient enrichment followed by eutrophication of these waters, affecting the phosphorus cycle in an aquatic ecosystem in a big way.

5.7 Sulfur Cycle

Sulfur is a constituent of most of the proteins in which it is incorporated through two S-containing amino acids, viz., cysteine and methionine. The other amino acids that do not incorporate into protein, viz., homocysteine and taurine, also contain sulfur. Sulfur, thus, is a component of many enzymes and of some vitamins, notably vitamin B1, biotin, and pantothenic acid.

Sulfur is widely used as a source of energy by S-oxidizing chemosynthetic bacteria. These autotrophs play a key role in life processes in the biosphere. One of the rarest bio communities inhabiting hydrothermal vents on the ocean floor is based on sulfur-oxidizing microorganisms. Nitrogen-fixing bacteria are also dominantly sulfur-oxidizing. Sulfur, often in small amounts, is essential for biological functions. Disulfide bridges form a critical function in imparting many biologically important molecules' specific shape and properties.

Sulfur is also of industrial importance. For example, SO_2 is used as a bleaching agent in bleaching wood pulp for paper and fiber. Such fibers are, then, used in textiles such as linen, silk, and wool.

Sulfur, under anaerobic conditions, has potential to bind to cations of calcium to form insoluble calcium sulfate ($CaSO_4$) and to iron to form ferrous sulfide (FeS) and ferric sulfide (Fe_2S_3).

5.7.1 Sulfur Utilization

Terrestrial plants pick up sulfur in the form of sulfates (SO_4^{2-}) from the soil. Aquatic plants take up the same from the water. Some plants take up sulfur in organic form through S-containing amino acids. Animals get sulfur through the food chain. Sulfur being a component of proteins is gained by animals through the foods they derive from different trophic levels. Some animals get this mineral from water. Chemosynthesizing prokaryotes utilize sulfur and sulfur compounds to derive energy for chemosynthesis.

5.7.2 Sulfur Production

Hydrogen sulfide (H_2S) is released upon the decomposition of dead plants and animals by microbes, such as *Neurospora*, *Aspergillus* (both aerobes), and *Escherichia* (an anaerobe). In the deeper ocean zones, aquatic animals do not survive due to high concentrations of H_2S. Sulfates, under anaerobic conditions, are reduced to sulfides or elemental sulfur by some heterotrophic bacteria, like *Acetobacter*, *Desulfovibrio*, *Escherichia*, etc.

$$SO_4^{2-} + 2H^+ \rightarrow H_2S + 2O_2$$

Huge amounts of volatile dimethyl sulfide (CH_3SCH_3) are produced by some marine algae. Tiny droplets of CH_3SCH_3 serve as nuclei for water condensation into tiny droplets found in clouds. CH_3SCH_3 carries potential to significantly affect cloud cover and climate.

Flooded swamps, croplands, and bogs invite anaerobic decomposers to perpetuate. These convert organic matter into H_2S.

Sulfur bacteria, such as *Thiobacillus*, oxidize H_2S into sulfates. *Beggiatoa* bacteria, on the other hand, oxidize a part of H_2S to elemental sulfur. In this process energy is released which is used by bacteria in the process of chemosynthesis, involving reduction of CO_2 and production of glucose.

$$6CO_2 + 12H_2S \rightarrow C_6H_{12}O_6 + 6H_2O + 12S$$

The sulfur lost to the ocean depths combines with iron forming ferrous sulfide (FeS). The FeS happens to be responsible for the black color of marine sediments. The remaining sulfur passes into deep ocean sediments, which are the reservoir pools of this element.

From the oceans, seas, and lakes, the sulfur returns back to the lands through food chains, sea sprays, and geological upheavals. Dust storms and forest fires are also responsible for the exit of sulfates into the atmosphere.

Many industries using fossil fuels emit SO_2 into the atmosphere. Petroleum refineries ooze out copious amounts of SO_2 into the atmosphere. Enormous amounts of SO_2 are added to the atmosphere by transport vehicles using petrol and diesel.

SO_2 in the atmosphere is oxidized to SO_3 (sulfur trioxide) and is also converted to tiny droplets of H_2SO_4. Tiny particles of ammonium sulfates are produced with its reaction with ammonia in the atmosphere and deposit on Earth. Excessive amounts of SO_2 with moisture and rainwater convert into H_2SO_4 which comes down with rain as acid rain. Acid rain damages terrestrial vegetation, adversely affects aquatic ecosystems, animals, human beings, and buildings.

Volcanic eruptions are a natural cause of spewing sulfates into the atmosphere and soil. H_2S and SO_2 are the major sulfur compounds that enter into the atmosphere in excessive amounts upon a volcanic eruption (see Fig. 5.6).

5.7.3 Anthropogenic Effects on the Sulfur Cycle

We alter the sulfur cycle from its normal pace by accelerating production of sulfur compounds, for example:

- Coal-based power stations use huge amounts of coal to generate electricity. This contributes to enormous amounts of SO_2 entering into the atmosphere.
- Petroleum refineries add excessive amounts of SO_2 to the atmosphere.
- S-containing metallic mineral ores are converted into pure metals (Cu, Zn, Pb) during metallurgical processes that release huge amounts of SO_2 into the environment.

5.8 Summary

Nutrient cycles – which on vast geographical scale or in the context of the whole globe are referred to as biogeochemical cycles – involve the use, transformation, movement, and reuse of all nutrients that are essential for plants, animals, and microorganisms. The nutrient cycles, based on their main reservoirs, can be broadly categorized into sedimentary and gaseous cycles. Those with nutrient reservoir in the rocks are the sedimentary cycles and with reservoir in the air (and water) are the gaseous cycles. Gaseous cycles take place at a faster pace compared to sedimentary ones.

Water, through its cycles, exchanges itself in different environmental components (oceans, atmosphere, land, and living organisms) within the biosphere. Precipitation enables the water to collect in various water bodies on the planet. The hydrological cycle comprises seven major stages, viz., evaporation, condensation, sublimation,

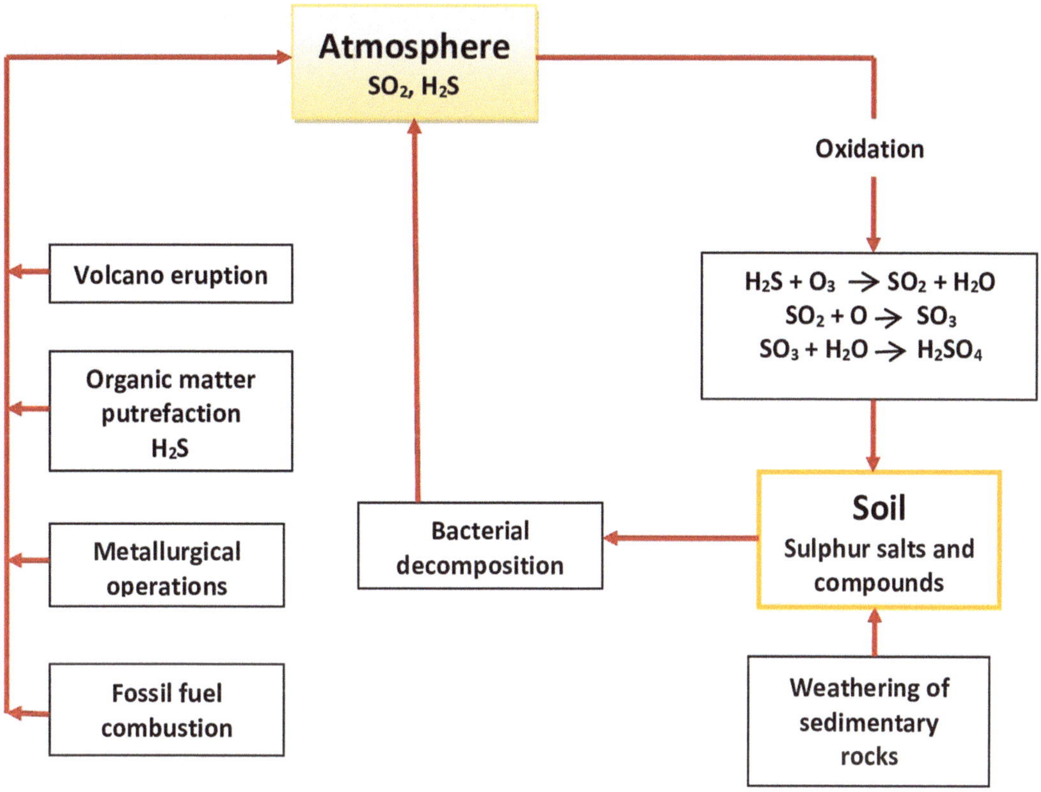

Fig. 5.6 The sulfur cycle

precipitation, transpiration, runoff, and infiltration. The residence time of a reservoir is the average age of water that resides in that reservoir. Antarctica records the longest residence time (approximately 20,000 years) and the atmosphere the shortest (only 9 days).

The nitrogen cycle, a gaseous type of the biogeochemical cycles, converts molecular nitrogen into various compounds that incorporate in structural, functional, and genetic biomolecules in living organisms. Nitrogen fixation involves conversion of molecular nitrogen into nitrites and nitrates. In nature, this conversion takes place physically as well as biologically and is also carried out industrially. Investment of substantial energy in converting molecular nitrogen into a usable form imposes limits on nitrogen fixation into organic matter. Physical fixation, or atmospheric fixation or non-biological fixation of the atmospheric nitrogen, takes place due to lightning. Of the total nitrogen fixed, only about 5%–8% is fixed through non-biological or physical means. Biological nitrogen fixation (BNF) is an energy-demanding process which the nitrogen-fixing prokaryotes invest at 16 moles of ATPs per molecule of N_2. Ammonia formed is converted into amino acids which serve as building blocks for subsequent protein synthesis. BNF involves free-living bacteria, aerobic and anaerobic bacteria, and cyanobacteria. Symbiotic nitrogen fixation provides very important pathways to nitrogen fixation. Several species of fungi and nodule-forming bacteria are the two major sources of N_2 fixation. In industrial nitrogen fixation, what is called the Haber-Bosch process, nitrogen and hydrogen (from petroleum or natural gas) are combined into ammonia at very high temperature (400 °C) and high pressure (200 atmospheres). The industrial process contributes nearly 30% of the total nitrogen fixed. Denitrification, the final step of the nitrogen cycle, recycles the nitrogen in its molecular form into its pool, the atmosphere.

Though only in trace amounts in the atmosphere, the inorganic carbon plays a key role to create and sustain life on Earth. Photosynthesis fixes carbon into life. Within organisms carbon moves from one biomolecule to another. From the autotrophs (the photosynthesizers or producers), carbon moves to herbivores and from there to carnivores, up to the top carnivores, through foods. Deposited in the soil, carbon becomes a culture for soil organisms, a medium of the detritus food chain and is also stored in humus and other carbon compounds and mixtures. In aquatic ecosystems inorganic carbon enters through dissolution. Carbon in deeper layers remains stored for a pretty long period of time. Respiration by all organisms, decomposition of organic matter, burning of forests, woods and all biomass, combustion of fossil fuels, and decomposition of $CaCO_3$ and volcanic eruptions are the major sources of CO_2 production.

Oxygen is indispensable for life as it generates energy by oxidizing carbohydrates in the cells of the organisms. Oxygen is very reactive, quickly combines with a variety of the atoms to form a variety of compounds – the oxidized compounds – occurring all over the Earth. Both water cycle and carbon cycle involve the oxygen cycle as well. Other than its molecular form, oxygen also exists in the atmosphere in its various other forms, e.g., O_3, CO_2, SO_2, NO, N_2O, H_2O vapor, etc. Photolysis of water during the light phase of the photosynthesis in green plants and blue-green algae results in oxygen production. A large proportion of oxygen enters into atmosphere through photosynthesis. H_2O and N_2O also undergo photolysis due to the reaction of high-energy ultraviolet radiation releasing oxygen. All plants, animals, and a variety of microorganisms utilize oxygen for respiration turning carbohydrates or other respiratory biomolecules into carbon dioxide and water. Decomposition of organic matter, composting of the organic matter, combustion of biomass, and fossil fuels consume oxygen.

Phosphorus is one of the major mineral constituents of life. The most abundant quantities of phosphorus in the biosphere occur in mineral deposits and marine sediments. Released from mineral deposits into ecosystems, the phosphate ions are absorbed by the plants and, passing through food chains via trophic levels, recycles into terrestrial and aquatic ecosystems. In the marine environments the dissolved phosphorus slowly keeps depositing into ocean sediments eventually transforming into phosphate-bearing sedimentary rocks.

Sulfur is a component of many proteins, including enzymes, and of some vitamins, notably vitamin B1, biotin, and pantothenic acid. It is widely used as a source of energy by S-oxidizing chemosynthetic bacteria. One of the rarest bio communities inhabiting hydrothermal vents on ocean floor is based on sulfur-oxidizing microorganisms. Nitrogen-fixing bacteria are also dominantly sulfur-oxidizing. Terrestrial plants pick up sulfur in the form of sulfates (SO_4^{2-}) from the soil. Aquatic plants take up the same from the water. Some plants take up sulfur in organic form through S-containing amino acids. Animals get sulfur through the food chain. Flooded swamps, croplands, and bogs invite anaerobic decomposers to perpetuate which convert organic matter into H_2S. The sulfur lost to the ocean depths combines with iron forming ferrous sulfide (FeS). The remaining sulfur passes into deep ocean sediments, which are the reservoir pools of this element. From the oceans, seas, and lakes, the sulfur returns back to the lands through food chains, sea sprays, and geological upheavals.

5.9 Exercises

5.9.1 Multiple-Choice Questions

1. Which of the following nutrient cycles is of gaseous type?
   (a) Carbon cycle
   (b) Sulfur cycle
   (c) Phosphorus cycle
   (d) Magnesium cycle
2. Which of the following cycles is "perfect"?
   (a) Nitrogen cycle
   (b) Oxygen cycle
   (c) Carbon cycle
   (d) All of the above
3. How many stages does the hydrological cycle comprise?
   (a) 4
   (b) 7
   (c) 9
   (d) 10

4. About how much solar energy on Earth is used in driving the hydrological cycle?
   (a) 60%
   (b) 50%
   (c) 33%
   (d) 22%
5. How long is the residence time of water in the oceans?
   (a) 1100 years
   (b) 2200 years
   (c) 3200 years
   (d) 5200 years
6. The residence time of water is longest and shortest, respectively, of
   (a) Antarctica and atmosphere
   (b) groundwater and rivers
   (c) atmosphere and soil
   (d) glaciers and lakes
7. Of the total geospheric carbon, the inorganic carbon accounts for about
   (a) 20%
   (b) 40%
   (c) 60%
   (d) 80%
8. As much as 94% of the lithosphere by volume comprises
   (a) carbon as $CaCO_3$
   (b) oxygen as SiO_2
   (c) nitrogen as NO_3
   (d) phosphorus as P_2O_5
9. About one-third of the hydrosphere by vol ume is made up of oxygen as component of
   (a) H_2O
   (b) dissolved O_2
   (c) $CaCO_3$
   (d) H_2O, dissolved O_2, and H_2CO_3
10. How many moles of ATP are hydrolyzed in the following reaction?
   $N_2 + 8H^- + 8e^- \rightarrow 2NH_3 + H_2$
   (a) 8
   (b) 12
   (c) 16
   (d) 20
11. The net energy input for biological nitrogen fixation is
   (a) 13 kcal/mole
   (b) 147 kcal/mole
   (c) 160 kcal/mole
   (d) 173 kcal/mole
12. Conversion of ammonia to nitrites and then to nitrates is what is called as
   (a) Ammonification
   (b) Denitrification
   (c) Nitrification
   (d) Nitrogen fixation
13. Name the prokaryotes involved in the conversions represented by the following reactions:
   $2NH_3 + O_2 \rightarrow 2NO_2^- + 2H^+ + 2H_2O$ [1]
   $2NO_2^- + O_2 \rightarrow 2NO_3^- +$ energy [2]
   (a) *Nitrosomonas* in reaction [1], *Nitrobacter* in reaction [2]
   (b) *Nitrobacter* in reaction [1], *Nitrosomonas* in reaction [2]
   (c) *Nitrococcus* in reaction [1], *Clostridium* in reaction [2]
   (d) *Clostridium* in reaction [1], *Nitrococcus* in reaction [2]
14. In the human body about 80% of phosphorus is found in
   (a) DNA and RNA
   (b) lipids
   (c) bones and teeth
   (d) ATP, ADP, and GTP
15. The largest source of phosphorus on the planet is the
   (a) soil
   (b) water
   (c) algae
   (d) rocks formed from the Earth's crust
16. Elemental sulfur in the soils is not directly utilized by the plants. In order to be taken up by the plants, it is converted by chemolithotrophic bacteria into
   (a) sulfates
   (b) sulfites
   (c) hydrogen sulfide
   (d) sulfuric acid
17. Which amino acids contain sulfur?
   (a) Methionine
   (b) Cysteine
   (c) Both of the above
   (d) None of the above
18. The black color of marine sediments is due to
   (a) Ferrous sulfide (FeS)
   (b) Ferric sulfide (Fe_2S_3)
   (c) Hydrogen sulfide (H_2S)
   (d) Ferrous sulfate ($FeSO_4$)

19. This volatile compound is attributable to sulfur in the hydrosphere and is produced thanks to the role of certain marine algae. It carries potential to significantly affect cloud cover and climate. What is this?
    (a) H_2S
    (b) SO_2
    (c) CH_3SCH_3
    (d) H_2SO_4
20. Anthropogenic interventions leave their impact on
    (a) carbon cycle
    (b) oxygen cycle
    (c) nitrogen cycle
    (d) all the nutrient cycles

Answers: 1-a, 2-d, 3-b, 4-c, 5-c, 6-a, 7-d, 8-b, 9-d, 10-c, 11-b, 12-c, 13-a, 14-c, 15-d, 16-a, 17-c, 18-a, 19-c, 20-d.

5.9.2 Short-Answer Questions

1. Differentiate between the gaseous and sedimentary cycles.
2. What do you mean by imperfect and perfect cycles?
3. Why are the sedimentary cycles slower than the gaseous ones?
4. How does an element flow and recycle in an ecosystem?
5. Name the stages in the hydrological cycle.
6. How much energy is invested in driving the water cycle of the Earth?
7. Name the stages in the nitrogen cycle.
8. Atmospheric nitrogen is fixed in three ways. Name them.
9. Where is phosphorous found in our body?
10. What are the different sources of sulfur?

5.9.3 Long-Answer Questions

1. How does the water cycle contribute to the ecological integrity of the biosphere? Discuss with reasoning.
2. Describe the carbon cycle of planet Earth? How does the carbon cycle play a crucial role in regulating the climate of the Earth?
3. Explain the nitrogen cycle through a flow diagram. How does the nitrogen cycle play a key role in improving soil fertility and ecosystem productivity?
4. Examine the anthropogenic effects on carbon, nitrogen, and phosphorus cycles.
5. Write short notes on the following:
    (a) Oxygen cycle
    (b) Sulfur cycle
    (c) Residence time of water in different sources

References

Falkowski P, Scholes RJ, Canadell J, Canfield D, Elser J, Gruber N, Hibbard K, Högberg P, Linder S, MacKenzie FT, Moore B, Pedersen T, Rosenthal Y, Seitzinger S, Smetacek V, Steffen W (2000) The global carbon cycle: a test of our knowledge of Earth as a system. Science 290(5490):291–296

Holmes R (2008) The age of wonder. Pantheon Books. ISBN 978-0-375-42222

Molles MC (2005) Ecology: concepts and applications. McGraw Hill, Boston, 622 pp

Pidwirny M (2006) The hydrologic cycle. In: *Fundamentals of physical geography*, 2nd edn. http://www.physicalgeography.net/fundamentals/8b.html

Sclesinger WH (1991) Biogeochemistry: An Analysis of Global Change. Academic Press, New York

Singh V, Rastogi A, Nautiyal N, Negi V (2017) Livestock and climate change: the key actors and the sufferers of global warming. Indian J Animal Sci 87(1):11–20

Smith B, Richards RL, Newton WE (2004) Catalysts for nitrogen fixation: nitrogenases, relevant chemical models and commercial processes. Kluwer Academic Publishers, Dordrecht

Section II

Natural Resources

6 The Natural Resources: Introduction

The Earth is full of resources. Resources are there in reserves, in usage, services, and natural phenomena. Since the beginning of the Industrial Age, natural resources are not just regarded as the means of survival and sustainability of life; they are looked at from their specific values, such as how can they be transformed into various value-laden products, brought into services, and used in economic growth. Although everything on Earth – living or nonliving, visible and non-visible, usable directly or indirectly – is a resource, in this section, we shall focus on the resources that are of socioeconomic importance.

Which country would be branded as a developed country today? The one possessing a large enough proportion of the world's natural resources. Natural resources are an indicator of socioeconomic development. The larger the area under natural resources in the possession of a country, the greater the opportunities for the country to be socioeconomically more progressive than the others. Most of the prevailing conflicts among nations, among different communities in a nation, are on account of natural resources. Depletion of natural resources at a faster pace in our times is one of the most serious issues facing our world.

6.1 What Is a Natural Resource?

A natural resource has been defined in various ways depending upon one's perspective, such as the following:

- Anything that can be utilized by human beings for their welfare
- Materials of value to a particular human culture
- A stock extractable from nature that can be utilized directly or by transforming into other forms/products
- A component or part of the environment extracted for a specific use or for supporting life
- Aspects of the natural environment from which goods and services can be obtained and produced
- Resources (actual and potential) supplied by nature
- Something that occurs in the natural state and has economic value
- A new or reserve supply that can be drawn upon when needed
- The material source of wealth
- The natural wealth of a country
- Industrial materials and capacities supplied by nature

V. Singh, *Textbook of Environment and Ecology*, https://doi.org/10.1007/978-981-99-8846-4_6

Science and technology have changed the meaning of natural resources. What was not considered a resource earlier has been brought into some kind of use for human comforts or socioeconomic welfare directly or upon being transformed into useful products.

Sometimes, a natural resource is analyzed from a cultural perspective. A natural resource may be of the critical need for a culture, while it might hardly be of any use by the other. For instance, uranium is a precious resource for communities dependent on electricity generated by nuclear power. However, the same is not a resource for the forest dwellers of tribal cultures who depend only on firewood as an energy resource. Similarly, ozone, a harmful gas in the troposphere, is of no human use, but the same is of vital significance for its functions in the stratosphere. The nature of the work of a society and its development level determines the kind of natural resources it would rely upon.

Nonetheless, from an ecological perspective, everything in nature, irrespective of its direct use for human beings, is a natural resource. It might not be of any human use or may carry no consumptive use value; it serves as an integral element of the Earth. Ecological integrity involves everything out there in nature. Anything might be useless or even harmful from our perspective, but the same is crucial in the evolutionary development of the living planet and, therefore, vital for human evolution. What is a resource from our perspective is not isolated from all other components of nature and is not "drifted" away from the overall integrity of the biosphere. Everything that is out there in nature, therefore, is phenomenal for the evolution of life with the human species as one of the integral parts of it.

What cannot be regarded as a resource is a pollutant created as a result of human activity. For example, the pollutants generated through transport vehicles are harmful and not a natural component of the environment, so cannot be regarded as a resource. However, the transport vehicles themselves and the metals and timber they are made up of are important resources used for people's movement and transportation of goods. Similarly, domestic gadgets and machines manufactured for human comforts serve as resources, but the pollutants, such as chlorofluorocarbons, hydrofluorocarbons, ozone gas, etc., they generate are not.

In human systems, something as such might be condemned as a pollutant, but when used in a process or transformed to yield something of socioeconomic use, it is regarded as a resource. For example, cow dung or livestock wastes are pollutants causing a nuisance in human habitats. However, when the same is used as manure or for producing biogas, they are transformed into resources. Similarly, garbage causes pollution and health hazards in cities and towns and as such cannot be considered a resource. But the same when used for electricity generation or recycling to produce new items is regarded as a resource.

6.2 Types of Natural Resources

A natural resource can be a (i) biophysical substance (e.g., all foods, water, soil, forests, livestock, etc.), (ii) an energy unit (e.g., fossil fuel energy, chemical energy, electricity, solar energy, wind energy, etc.), and (iii) a phenomenon, or a process (e.g., rain, photosynthesis, pollination, etc.). Further, the resources of the planet are classified on various bases, notably: origin, availability or abundance, chemical nature, and occurrence (Table 6.1).

6.3 Depletion of Natural Resources

Depletion of natural resources – that is, consumption rates faster than resource replenishment – is one of the most serious tragedies of our times. Reduced proportion or exhaustion of a resource imposes restrictions on human welfare programs and blocks paths to a sustainable future. With growing human needs for better, more comfortable, and luxurious lifestyles coupled with advancements in science and technology providing ample ways of exploitation, nonrenewable natural resources are being depleted at a faster

Table 6.1 Classification of natural resources

Category	Main characteristics	Examples
I. Origin		
1. Biotic resources	Basically organic in nature: (i) Renewable: living beings (ii) Nonrenewable: fossil fuels	Renewable: biodiversity (plants, animals, and microorganisms) including all food crops, food-yielding wild plants, trees, livestock, poultry, etc. Nonrenewable: coal, petroleum, natural gas, etc.
2. Abiotic resources	Basically inorganic in nature, emanating from nonliving matter	Oxygen, water, minerals, sandstone, etc.
II. Availability or abundance		
1. Inexhaustible resources	Unlikely to be exhausted despite use in any measure; unlimited amount in nature	Light, solar power, wind energy, tidal energy, geothermal energy
2. Exhaustible resources	Likely to be depleted and then exhausted due to overexploitation (i) Nonrenewable: lack ability of recycling or replacement, uneven distribution in the world, made available by mining, digging, and processing (ii) Renewable: inherent capacity of being replenished, reproduced, and/or recycled within reasonable a reasonable time length	Nonrenewable: fossil fuels, minerals, soil, etc. Renewable: forests, water, biogas, etc.
III. Chemical nature		
1. Inorganic resources	Inorganic in nature	Water, rocks, sand, minerals, etc.
2. Organic resources	Organic in nature	Forests, foods, microbes, fossil fuels, alcohol, etc.
3. Mixed resources	Combination of inorganic and organic components	Soil
IV. Occurrence		
1. International resources	Available in all countries, have no boundary	Light, air, water, soils
2. Multinational resources	Shared in more than one country	Some rivers, lakes, migratory birds
3. National resources	Confined to a particular country	River waters, lakes, fauna, flora, humans, endemic species/genotypes, genetic resources, animal breeds

pace. Even renewable-type exhaustible resources are being exploited at rates faster than their capacity for regeneration.

Rapid growth in the industry, transport, and trade sectors has led to the exploitation of nonrenewable energy resources to the extent that the whole globe is being increasingly loaded with pollution driving adverse changes in the planetary climate system. Careless use and mismanagement of the natural resources following their overexploitation snatch away prospects of socioeconomic development and set out a dismal environmental scenario. Developing countries are compelled to constantly intensify resource exploitation to meet the needs of their burgeoning populations. Competition for economic growth among developed countries is also one of the leading causes of continuously exacerbating rates of resource consumption. The whole world, in essence, is advancing towards a state of resource depletion and consequent disasters.

6.3.1 Lifetime and Depletion Time of Natural Resources

There are two aspects of resource depletion, viz., lifetime and depletion time. Lifetime refers to the length of time within which the availability of the resource for consumption by human beings is assured on a global basis. The depletion time of a resource denotes the time length within which resource availability would be reduced to the extent that it will not be available for fulfilling human requirements.

There are three types of resource depletion patterns: (i) rapid depletion time, (ii) extended depletion time, and (iii) indefinite depletion time. The periods within which the resource availability is ensured for use are determined by the management (conservation and utilization) practices, lifestyles, and living philosophy of a community.

The rapid depletion pattern is a boom and bust–type depletion pattern that involves an unrestricted pattern of resource extraction with a lot of wastage at every stage. The extended resource depletion patterns are maintained on account of proper management of the resource involved at the points of extraction, concentration, and manufacture with reduced wastage at every point, thus extending the depletion period of the resource. The indefinite resource depletion involves various conservation strategies resulting in the resource availability being extended indefinitely. The main characteristics of the three depletion patterns are illustrated in Fig. 6.1. If these patterns are plotted on a production-time basis, they would be akin to those depicted in Fig. 6.2.

6.3.2 Causes of Resource Depletion

There could be several factors contributing to natural resource depletion. The major ones are as depicted in Fig. 6.3. All these causal factors are linked with each other. One factor influences the others resulting in the intensification of resource depletion.

6.3.2.1 Uneven Geographical Distribution of Resources

The uneven geographical distribution of resources is one of the biggest causes of resource exploitation. Some of the resources are concentrated in a few countries or a particular region, while the same is lacking in other parts of the world or are only in meager amounts. For example, most of the world's gold and platinum are found in Africa more than in other regions of the world. Similarly, petroleum resources are con-

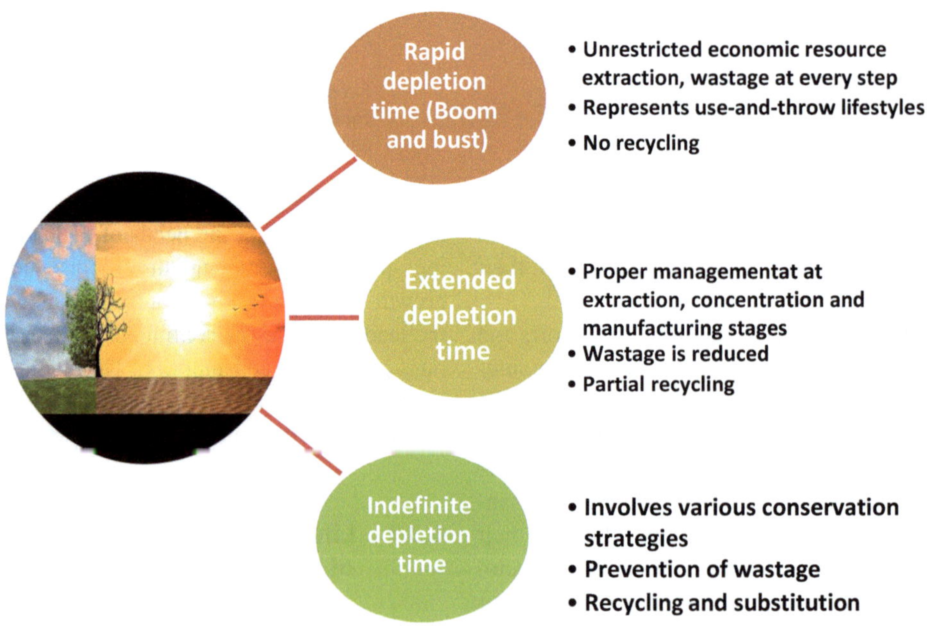

Fig. 6.1 Major characteristics of the resource depletion patterns

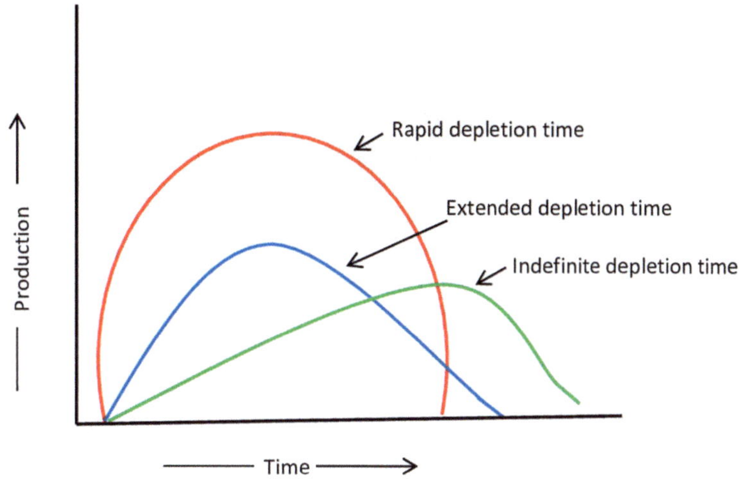

Fig. 6.2 The resource depletion patterns

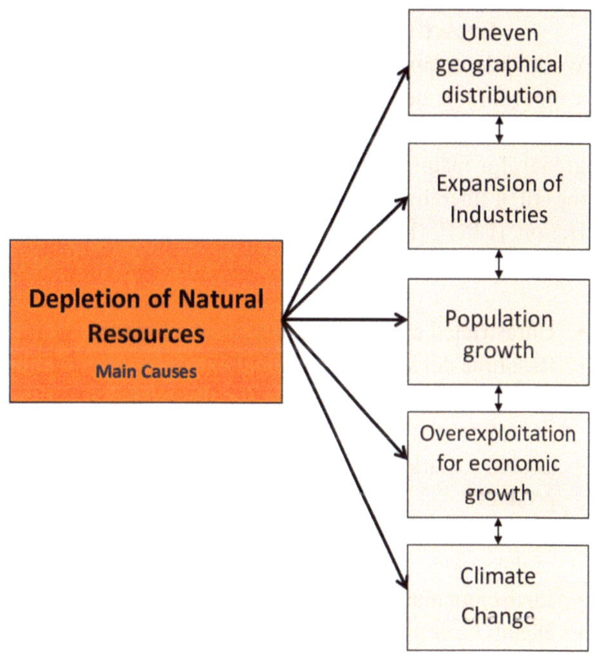

Fig. 6.3 Various factors leading to resource depletion

centrated in the Gulf countries, while many other countries only poorly represent these resources or are lacking them. It is due to this uneven geographical distribution of the resources that they have been plundered and overexploited by colonial states in the past.

6.3.2.2 Expansion of Industries

Intensification of industrial development has caused a spurt in the demand for natural resources. To sustain, industrial units need a constant electricity supply for which fossil fuels are burnt in powerhouses, nuclear fission is carried out in nuclear reactors, and high dams are constructed on rivers in fragile mountain ecosystems. To meet their unabated demands of raw materials, forests are cleared, trees are felled, minerals are dug out, and freshwaters are consumed. In addition, in the processes of product manufacturing, these industrial units spew out deadly pollutants further contributing to resource depletion.

6.3.2.3 Population Growth

Population growth is one of the biggest menaces to natural resources. Forest areas are being converted into cultivated lands to fulfill populations' essential needs. Further, forests, pasturelands, and agricultural lands are converted into concrete jungles to provide dwelling facilities for growing populations. Industries are expanded to fulfill people's needs for so many essential goods and luxurious items. Growing populations are central to a vicious cycle of resource depletion.

6.3.2.4 Overexploitation for Economic Growth

Worldwide competition for economic growth or socioeconomic development is at the cost of natural resources. Maximization of economic growth for maintaining high GDP is a mantra in the era of liberalization, privatization, and globalization (LPG). The emerging scenarios in the LPG era are as follows:

- Unrestricted exploitation of natural resources
- Resource depletion to a critical extent
- Diversification of industries for the exploitation of as many resources as possible
- Global markets packed with industrial products
- Increasing gap between the rich and the poor of the world
- Environmental pollution and climate change
- Health hazards (the Covid-19 pandemic being a glaring example)
- So many other negative consequences to be faced in the future

6.3.2.5 Climate Change

Climate change, largely an outcome of anthropogenic activities, exerts physiological stresses on biotic resources and creates adverse environmental conditions suppressing ecological regeneration and growth of the resources (Singh et al. 2017; Singh 2020). Climate change also adversely affects the supply of abiotic resources. For example, increased evaporation rates due to elevated environmental temperatures cause arid conditions and a shortage of water. Increasing ice melt rates cause sea level increases and inundation of coastal areas.

6.4 Conservation of Natural Resources

Protection, conservation, and augmentation of natural resources are not only necessary but imperative for humanity in our times. As the natural resources are not in a healthy state, their management should be given priority. A natural resource is not isolated from others in nature. They are all tied to ecological integrity. If one resource is depleted, the others have to be adversely affected. For example, the destruction of a forest ecosystem exerts an adverse impact on water sources, the water cycle, wildlife, plant biodiversity, soil fertility, etc. The extraction of mineral resources would lead to the destruction of natural forests, grasslands, water sources, etc.

The first principle of natural resource management should be based on the ecological integrity of the biosphere. Ecological integrity is rooted in photosynthesis (Singh 2019, 2020). Fortification of photosynthesis is vital for generating foods, enhancing soil fertility, biodiversity conservation, water resource protection, and maintaining carbon balance and climate patterns (Singh 2020). However, the fortification of photosynthesis itself depends on green plants – that is, natural forests. Operationalization and maintenance of ecological integrity are the basis of resource conservation and augmentation.

There must be a balance between extraction rates and resource regeneration rates and the latter should be higher than the former. In the case of nonrenewable resources, however, consumption rates should be as low as possible. The principle of the 3Rs – reduce, reuse, and recycle – should be applied in respect of nonrenewable resources. Replacement of nonrenewable and polluting energy resources with inexhaustible and renewable ones is promising and one of the essential conditions for a healthy environment.

Natural resources should fulfill the needs, not greed, of the present generations and they must be managed (conserved and utilized) in such a way that they could fulfill the needs of future generations. National governing bodies, international organizations, and United Nations' concerned agencies must evolve concrete and workable strategies for the protection, conservation, and enhancement of natural resources for a secure and sustainable future. Conservation strategies for individual natural resources are discussed in the following chapters.

6.5 Summary

A natural resource is anything that can be utilized by human beings for their welfare. It is a natural wealth that can exist in the form of material, energy, and a process. Science and technology have changed the meaning of natural resources. Sometimes, a natural resource is analyzed from a cultural perspective. The nature of the work of a society and its development level often determines the kind of natural resources they would rely upon. From an ecological perspective, everything in nature, irrespective of its direct use for human beings, is a natural resource as it serves as an integral element of the Earth and is essential for the very ecological integrity of nature. What cannot be regarded as a resource is a pollutant created as a result of human activity, but the same when used in a process or transformed to yield something of socioeconomic use is regarded as a resource. A natural resource can be a biophysical substance (e.g., all foods, water, soil, forests, livestock, etc.), an energy unit (e.g., fossil fuel energy, chemical energy, electricity, solar energy, wind energy, etc.), and a phenomenon, or a process (e.g., rain, photosynthesis, pollination, etc.). Further, the resources of the planet are classified on various bases, notably: origin, availability or abundance, chemical nature, and occurrence.

Depletion of natural resources – that is, consumption rates faster than resource replenishment – is one of the most serious tragedies of our times. Rapid growth in the industry, transport, and trade sectors has led to the exploitation of nonrenewable energy resources to the extent that the whole globe is being increasingly loaded with pollution driving adverse changes in the planetary climate system. Competition for economic growth among developed countries is also one of the leading causes of continuously exacerbating rates of resource consumption and depletion. There are two aspects of resource depletion, viz., lifetime and depletion time. The former refers to the length of time within which availability of the resource for consumption by human beings is assured on a global basis, while the latter denotes the time length within which resource availability would be reduced to the extent that it will not be available for fulfilling human requirements. There are three types of resource depletion patterns: rapid depletion time, extended depletion time, and indefinite depletion time. The major reasons for natural resource depletion include uneven geographical distribution of resources, expansion of industries, population growth, overexploitation for economic growth, and climate change.

Protection, conservation, and augmentation of natural resources are not only necessary but imperative for humanity in our times. The first principle of natural resource management should be based on the ecological integrity of the biosphere rooted in photosynthesis. Fortification of photosynthesis is vital for generating foods, enhancing soil fertility, biodiversity conservation, water resource protection, maintaining carbon balance and climate patterns, etc. There must be a balance between extraction rates and resource regeneration rates and the latter should be higher than the former. The principle of the 3Rs – reduce, reuse, and recycle – should be applied in respect of nonrenewable resources. Replacement of nonrenewable and polluting energy resources with inexhaustible and renewable ones is promising and one of the essential conditions for a healthy environment. Natural resources should fulfill the needs, not greed, of the present generations and they must be managed (conserved and utilized) in such a way that they could fulfill the needs of future generations.

6.6 Exercises

6.6.1 Multiple-Choice Questions

1. Which of the following is a nonrenewable resource?
   (a) Hydroelectric energy
   (b) Gold
   (c) Forests

(d) Livestock
2. Which of the following groups represents inexhaustible energy?
   (a) Solar energy, wind power, tidal energy
   (b) Geothermal energy, wind power, solar energy
   (c) Tidal energy, geothermal energy, wind power
   (d) All of the above
3. Which of the following is a correct match?
   (a) Solar energy – inexhaustible energy
   (b) Coal – non-renewable energy
   (c) Biogas – renewable energy
   (d) All of the above
4. Identify the mismatch.
   (a) Forests – a renewable resource
   (b) Fossil fuels – a renewable resource
   (c) Fish – a renewable resource
   (d) Water – a renewable resource
5. Identify the renewable but an exhaustible resource.
   (a) Forest
   (b) Fossil fuels
   (c) Solar power
   (d) Geothermal energy
6. Which of the following would be categorized as an abiotic resource?
   (a) Minerals
   (b) Livestock
   (c) Fisheries
   (d) Wildlife
7. Which of the following groups represents inorganic resources?
   (a) Forest, fish, livestock
   (b) Biodiversity, wildlife, fossil fuels
   (c) Water, rocks, sand, minerals
   (d) Microbes, alcohol, petroleum
8. Which of the following resources has come into being through photosynthesis?
   (a) Forests
   (b) Alcohol
   (c) Fossil fuels
   (d) All of the above
9. Which natural resources are international types?
   (a) Those found in all the countries of the world
   (b) Which have no national boundaries
   (c) Both of the above
   (d) None of the above
10. Identify an international resource.
   (a) Sunlight
   (b) Lakes
   (c) Uranium
   (d) Seafood
11. Identify a multinational resource.
   (a) Some rivers
   (b) Migratory birds
   (c) Some lakes
   (d) All of the above
12. The length of time within which the availability of the resource for consumption by human beings is assured on a global basis is referred to as
   (a) The lifetime of a resource
   (b) Depletion time of a resource
   (c) Uncertainty of a resource
   (d) The dynamism of a resource
13. The time length within which resource availability would be reduced to the extent that it will not be available for fulfilling human requirements is referred to as
   (a) The lifetime of a resource
   (b) Depletion time of a resource
   (c) Uncertainty of a resource
   (d) The dynamism of a resource
14. Which of the following does not represent a resource depletion pattern?
   (a) Rapid depletion time
   (b) Extended depletion time
   (c) Definite depletion time
   (d) Indefinite depletion time
15. This depletion time involves unrestricted resource extraction and wastage at every stage. What type of resource depletion pattern it is?
   (a) Indefinite depletion time
   (b) Extended depletion time
   (c) Rapid depletion time
   (d) None of the above
16. Proper management at extraction, concentration, and manufacturing stages, reduced wastage, and partial recycling characteristics

represent which of the resource depletion pattern?
(a) Rapid depletion time
(b) Indefinite depletion time
(c) Extended depletion time
(d) None of the above

17. It involves various conservation strategies, prevention of wastage, recycling, and substitution. What type of resource depletion time it is?
(a) Rapid depletion time
(b) Extended depletion time
(c) Indefinite depletion time
(d) All of the above
18. Most of the world's gold is found in
(a) Australia
(b) Africa
(c) North America
(d) South America
19. Most of the world's petroleum resources are concentrated in
(a) Gulf countries
(b) Antarctica
(c) Australia
(d) North America
20. What problem relating to natural resources has been largely responsible for the overexploitation of certain precious natural resources of the world by invaders or colonialists?
(a) Expansion of industries
(b) Uneven geographical distribution of resources
(c) Population growth
(d) Overexploitation for economic growth

Answers: 1-b, 2-d, 3-d, 4-b, 5-a, 6-a, 7-c, 8-d, 9-c, 10-a, 11-d, 12-a, 13-b, 14-c, 15-c, 16-c, 17-c, 18-b, 19-a, 20-b.

6.6.2 Short-Answer Questions

1. What do you mean by natural resources?
2. Mention examples of the resources that occur as a substance or material, as an energy form, and as a phenomenon.
3. What is an inexhaustible resource?
4. What resources emerge through photosynthesis?
5. What is fossil fuel? Write examples of fossil fuels.
6. Which resources are categorized as renewable and exhaustible?
7. Differentiate between international, multinational, and national resources.
8. What do you mean by inorganic and organic resources?
9. Write examples of the resources concentrated in a particular region or a group of countries.
10. How does population growth impact natural resources?

6.6.3 Long-Answer Questions

1. What is the importance of natural resources?
2. Write an account of the depletion of natural resources with a focus on resource depletion patterns.
3. Describe the problems associated with natural resources and the main causes of resource depletion.
4. Why is natural resource conservation an imperative of our times?
5. Write short notes on the following:
(a) Exhaustible and inexhaustible energy resources
(b) Lifetime and depletion time of resources
(c) Climate change influences on natural resources

References

Singh V (2019) Fertilizing the universe: a new chapter of unfolding evolution. Cambridge Scholars Publishing, London. 263 pp.
Singh V (2020) Environmental plant physiology: botanical strategies for a climate smart planet. Taylor and Francis (CRC Press), Boca Raton/New York/London. 214 pp.
Singh V, Rastogi A, Nautiyal N, Negi V (2017) Livestock and climate change: the key actors and sufferers of global warming. Ind J Animal Sci 87(1):11–20

7 Water Resources

Water is the elixir of life. Ours is a water planet, and water is a precondition for the Earth to be a living planet. Water is an inorganic compound pervading most of the Earth defining the living planet. Water resources – the oceans, seas, lakes, reservoirs, wetlands, ponds, rivers, streams, springs, and aquifers – are not exclusively water. They inhabit a variety of organisms. Water resources, thus, constitute wonderful ecosystems. Organisms belonging to all kingdoms ranging from the size of a bacterium to a blue whale are found in water (or aquatic) ecosystems. A life dependent on photosynthesis as well as on chemosynthesis prevails in water habitats. Various life forms perpetuating in aquatic ecosystems also serve as precious resources of the Earth.

Water is a dynamic natural resource. It always stays in a cycle, the water cycle or the hydrological cycle. In the hydrosphere, it is in the liquid phase. In the poles and high mountains, water exists in solid (snow or ice) form, and in the atmosphere, it is in vapor (gaseous) form. And yet, there is another crucial resource of water, the groundwater that invariably exists in liquid form.

It is the freshwater on which humanity directly depends for critical needs and economic development. However, the proportion of the freshwater is relatively small and most of it is frozen as polar and glacial ice and is not directly available for human use. Nevertheless, the frozen water is of key importance as, upon melting slowly, it unceasingly flows through rivers and streams paving pathways for multiple socioeconomic activities.

Interestingly, water is the only inorganic compound that exists in all three states of matter: solid, liquid, and gas (vapor). These states of water are vital for the role this compound plays on the living planet.

A pretty huge amount of foods vital for human consumption and economic systems is harvested from aquatic ecosystems. Apart from this, water systems are intensively exploited for food production in terrestrial ecosystems (irrigation of crops), navigation purposes, livestock, cleanliness, sanitation, energy generation, and so many other economic activities ranging from cooling the surrounding air to warming up residences.

7.1 Waters for Life

Water is an integral component of life (Fig. 7.1). Most of the Earth is water. Most of life is water. Most of every living organism is water. The first and foremost phenomenon of life, i.e., photosynthesis, occurs with water as an inevitable input. No water, no photosynthesis. Water, in essence, is the key environmental component that strikes and maintains the ecological integrity of the living planet (Singh 2020).

Water is a source of diatomic oxygen (O_2) production through photolysis, the light-dependent reaction, the light phase, or the photochemical phase of photosynthesis. Photolysis takes place in the thylakoids of blue-green algae and the chloroplasts of green algae and plants. Serving as

V. Singh, *Textbook of Environment and Ecology*, https://doi.org/10.1007/978-981-99-8846-4_7

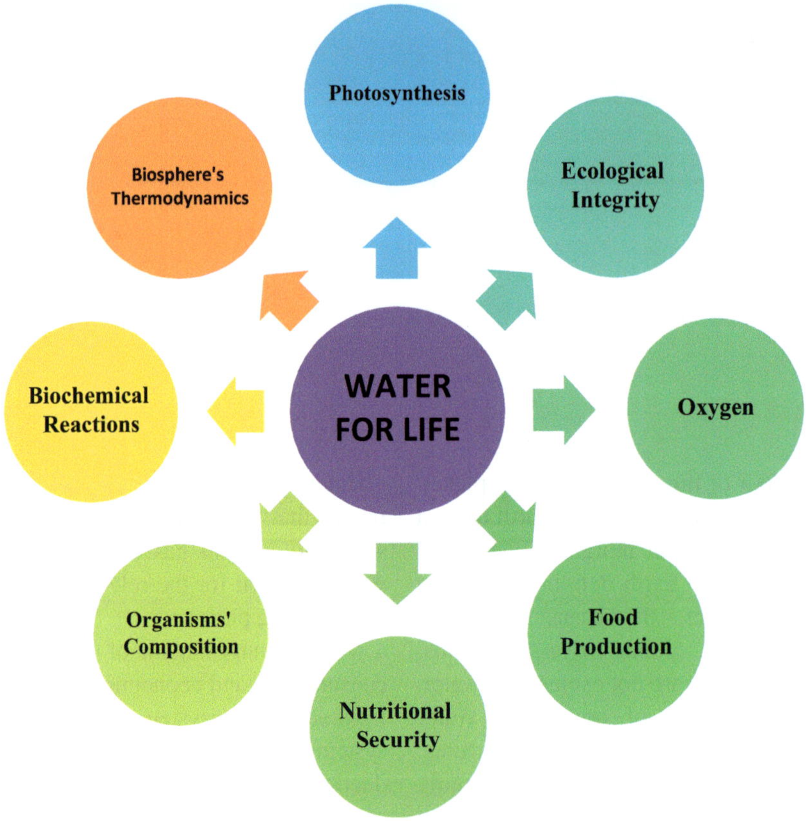

Fig. 7.1 Water for life: essential attributes of water for sustaining life

a substrate for photolysis with subsequent production of O_2, water plays a pivotal role in returning O_2 to Earth's atmosphere needed for the respiration by living organisms.

As mentioned earlier, water is the major component in the constitution of organisms' structures. A large chunk of the proportion of water in the biomass of living organisms reveals the importance of this compound for life. Considerable decrease of water in the composition of living organisms happens to be dangerous for their lives. A compromise with the consumption of water in required quantities costs life.

Water as an integral component in the structure and composition of living organisms serves as the basis for biochemical reactions. All the enzymes involved in the biochemical reactions function only in an aquatic medium. The reactions leading to activating energy systems vital for the maintenance of the entire metabolism and sustenance of life are water-dependent.

Water is indispensable for food production. Most of the water usage is annexed for agricultural production. A stunningly large proportion of the freshwater – approximately 70% – is used only in agricultural production. Not only cropping, but all other economy-related activities, such as animal husbandry, poultry, fishery, piggery, sericulture, etc., are water-intensive. Food security depends solely on the availability of and accessibility to water. Water security is a prerequisite of food security.

Digestion, absorption, transport, and assimilation of nutrients take place due to water. All the enzymes necessary for digestion and assimilation of nutrients function only in the presence of water. Absorption of nutrients occurs only due to water. Transport of nutrients from the digestive system to different body parts is carried away through the blood which is largely composed of water (Singh 2020). Thus, water is essential for the nutrients to play their role in metabolism.

Biosphere's existence and sustenance are thanks to the unique thermodynamic features of water. Water plays a key role in the regulation of thermodynamics. The unique properties of water hold the temperature of the biosphere in a range appropriate for the maintenance of organisms' physiology.

7.2 Waters of the Water Planet

Numerous water bodies prevailing on Earth have their characteristics varying in size, volume, depth, dynamics (lotic or lentic), water quality, types of organisms within, etc. Most of the water on the water planet is saline water extending over some 70% of the planet's surface and comprising 97.5% of the total planet's water. Freshwater constitutes only 2.5% of the total water most of which (about 69%) is concentrated in the glaciers and mountain caps. The distribution of the Earth's water resources is shown in Fig. 7.2.

Most of the saline water (96.5%) is concentrated in five oceans of the planet, namely, the Pacific, Atlantic, Indian, Southern, and Arctic Oceans. Little less than 50% of the total area and 50% of the water volume of the oceans is covered by the Pacific Ocean alone. Being the largest water body, the Pacific Ocean covers a nearly one-third area of the planet's surface. As much as 1,334,840,900 km^3 volume of water is encompassed in the five oceans (Table 7.1). Mariana Trench is the deepest point, 11,033 m, in the Pacific Ocean. The second largest ocean, the Atlantic Ocean, covers about the one-fifth surface area of the Earth with Puerto Rico Trench as the deepest point (8376 m).

The Indian Ocean, the third largest ocean on the planet, is surrounded by the Arabian Peninsula and Southeast Asia towards the north, Africa in the west, and Australia in the east. The major seas of the Indian Ocean include the Gulf of Kutch, Arabian Sea, Persian Gulf, Gulf of Oman, Gulf of Aden, Gulf of Khambat, Gulf of Mannar, Laccadive Sea, Bay of Bengal, and Andaman Sea. The fourth largest ocean in the world, the Southern Ocean, comprises the waters around Antarctica. The smallest (as well as the shallowest) ocean, the Arctic Ocean, is located around the Arctic Circle.

There are seven major seas of the Earth: Arabian, Caspian, Caribbean, Gulf of Mexico, Mediterranean, Persian Gulf, and South China Sea. Among them, the Arabian Sea in the Indian Ocean is the largest one covering an area of 3,862,000 km^2, followed by the South China Sea with an area of 3,500,000 km^2. The Caribbean Sea is the deepest one with a maximum depth of 7686 m. With a coverage area of 251,000 km^2 and a maximum depth of only 90 m, the Persian Gulf is the smallest and shallowest sea. Larger and deeper than the Caspian Sea and the Persian Gulf, the Sea of Japan is often not counted among the "Seven Seas" of the world (Table 7.2).

The seas of the world are extensively used for economic activities and national strategies in different continents of the globe. Providing livelihoods to millions of people throughout the world, almost all the seas are also the root cause of political strains among many countries sharing these seas.

There are 10 major lakes: the Caspian Sea, Lake Superior, Lake Victoria, Lake Huron, Lake Michigan, Lake Tanganyika, Lake Baikal, Great Bear Lake, Lake Malawi, and Great Slave Lake.

Many water bodies contain excessive salt concentrations. Salt, especially the common salt NaCl, is one of the essential products of certain water bodies, mainly lakes or lagoons. Salinity and temperature determine the movement of large masses of water and their characteristics, including marine flora and fauna (Sharma 2013).

Salinity percentage in water bodies ranges from as high as 44% in the Don Juan Pond, a salt lake in Antarctica, to 27% in the Great Salt Lake in the USA. And yet, some of the salt lakes, like the Dead Sea in Israel, are the water sources to cure some skin diseases. The Dead Sea is the lowest place on Earth (about 427 m below the mean sea level). Millions of people from around the world take a bath in the Dead Sea and also do mud treatment on their skin. The mud of the Dead Sea is also used as a cosmetic (Singh 2018).

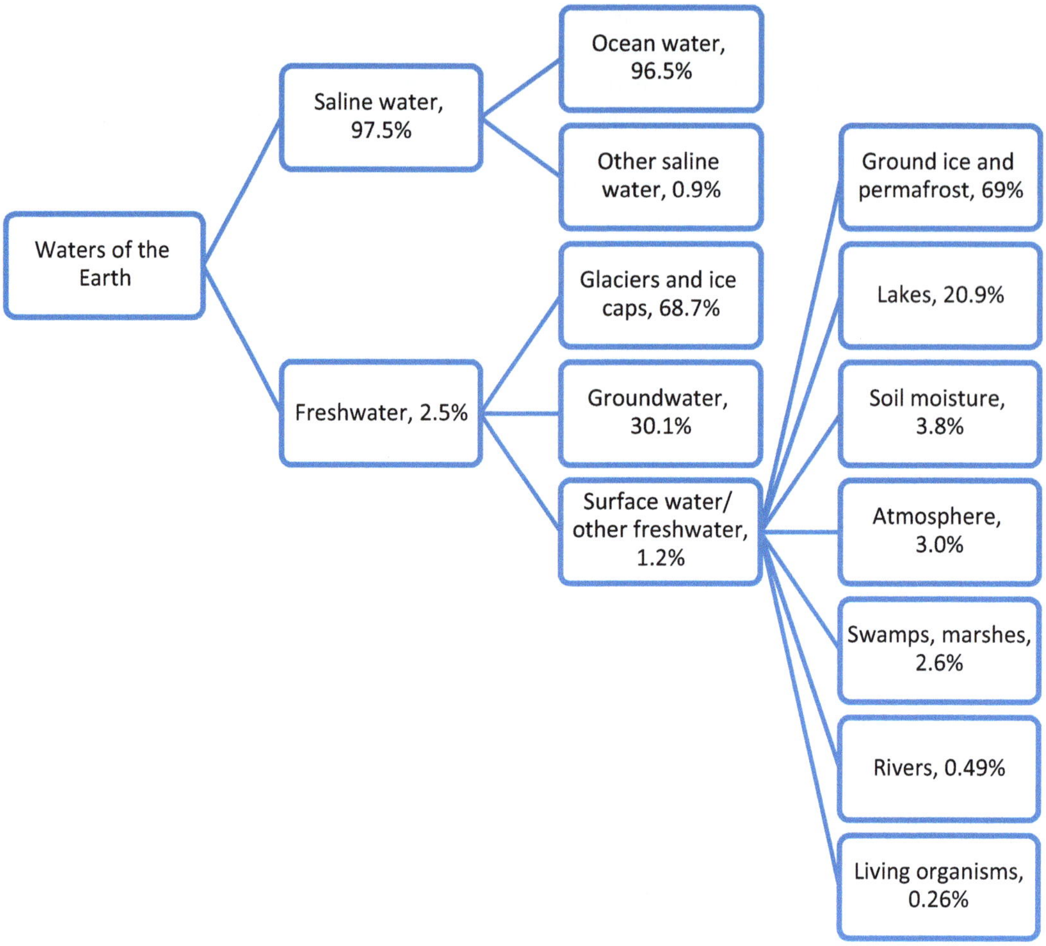

Fig. 7.2 Distribution of the Earth's water reserves

Table 7.1 The Earth's oceans

Rank	Ocean	Area (km^2) (%)	Volume (km^3) (%)	Average depth (m)	Coastline (km)
1	Pacific Ocean	168,723,000 (46.6%)	669,880,000 (50.1%)	3970	135,663
2	Atlantic Ocean	85,133,000 (23.5%)	310,410,900 (23.3%)	3646	111,866
3	Indian Ocean	70,560,000 (19.5%)	264,000,000 (19.8%)	3741	66,526
4	Southern Ocean	21,960,000 (6.1%)	71,800,000 (5.4%)	3270	17,968
5	Arctic Ocean	15,558,000 (4.3%)	18,750,000 (1.4%)	1205	45,389

Source: Compiled from Singh et al. (2012)

Table 7.2 The seas of the world

Rank	The sea	Area (km^2)	Maximum depth (m)	Main islands
1	Arabian Sea	3,862,000	4652	Kavaratti (India), Masirah (Oman), and Socotra (Yemen) islands
2	South China Sea	3,500,000	5016	Spratly, Paracel, Pratas, Riau, and Palawan islands
3	Caribbean Sea	2,754,000	7686	Greater Antilles and Lesser Antilles
4	Mediterranean Sea	2,510,000	5267	Cyprus, Malta, Cecily, Ibiza, Balearic islands of Spain, and Crete (Greece)
5	Gulf of Mexico	1,550,000	4384	Arrecife Alacranes (group of coral islands)
6	Sea of Japan	977,980	3742	Sakhalin, Rishiri, Oshima, Tsushima, Rebun, Okushiri, Kojima, Russky, Sado, and Ulleung
7	Caspian Sea	371,000	1025	Bulla, Pirathi, and Nargin islands
8	Persian Gulf	251,000	90	Bahrain, Dalma (UAE), Tarout (Saudi Arabia), Kish (Iran), Bubyan (Kuwait), and Greater and Lesser Tunb

Source: Compiled from Singh et al. (2012)

7.3 Third Pole of the Earth

Himalayan glaciers cover about three million ha or 17% of the global mountain area – the largest bodies of ice outside the polar caps. The Himalayas, thus, are also popularly called the Third Pole of the Earth (Singh et al. 2011). The total area of the Himalayan glaciers is 35,110 km^2. The total ice reserves of these glaciers are 3735 km^3, which is equivalent to 3250 km^3 of freshwater. Himalayan mountains serve as a source feeding the nine giant river systems of Asia: the Indus, Ganga, Brahmaputra, Irrawaddy, Salween, Mekong, Yangtze, Yellow, and Tarim (IPCC 2007; Singh et al. 2011).

The Himalayas serve as a lifeline for more than 500 million inhabitants of the region, or about 10% total regional human population (Singh et al. 2011, 2012). The glacier-fed perennial rivers originating in the Himalayas are sources of drinking water, crop irrigation, and livelihoods for billions of people in the foreland areas. Having its origin in the Central Himalayan mountains in the Uttarakhand state of India, the holy Ganga River makes the largest water system in the Indian subcontinent. It is thanks to the benevolence of the Ganga waters that the Indo-Gangetic plains are perhaps the most densely populated region in the world.

7.4 The Running Water Bodies

The rivers, streams, canals, rivulets, falls, geysers, and springs are the running water bodies or lotic ecosystems comprising freshwaters. Running water has the power to purify itself and is considered to be the most liked drinking water. The running surface water is easy to be exploited for domestic water supplies and irrigation. Most of the world's cities harboring huge populations are located on the banks of large rivers. The lotic ecosystems harbor a variety of organisms, especially the fishes that survive and reproduce only in freshwaters. A sizeable proportion of the world's human population depends on the lotic ecosystems for livelihoods.

A river, a large stream of water flowing from high land to low land, is the most crucial source of potable water as well as of numerous other socioeconomic benefits, including the generation of energy, the hydroelectricity. A river is fed by a glacier, rainwater, a spring, or a lake and after flowing up to a certain distance eventually merges into an ocean.

The Nile River in Africa (6690 km), the Amazon River in South America (6300 km), the Yangtze River in Asia (5530 km), the Mississippi-Missouri River in North America (4090 km + 3770 km), the Murray-Darling River in Australia (3720 km), and the Volga River in Europe (3700 km) are some of the world's largest rivers.

7.5 Aquifers and Groundwater

A water body of porous rocks or sediment saturated with groundwater constitutes an aquifer. With the precipitation seeping through the soil, the groundwater enters an aquifer. This water moves through the aquifer and resurfaces through springs and wells. Aquifers can be of two types: confined (having a layer of impenetrable rock or clay above them) and unconfined (lying below a permeable layer of soil).

Groundwater (used to describe precipitation that filters the soil beyond the surface and collected in empty spaces underground) is of the prominent resources of fresh drinkable water in many regions of the world and constitutes little less than 10% of the total freshwater resources. Of the total water supplies, the share of the groundwater is about 30–50 times more than that of the surface water supplies. When it rains, most of the water flows down to oceans through rivers and streams. The same thing happens with the melting of snow and ice. However, a proportion of the rainwater and melted snow water is absorbed below the surface of the Earth through soil pores and cracks. This is the groundwater. Additionally, a proportion of the groundwater may come from streams rising from molten rock materials deep inside the Earth.

The groundwater is contained in aquifers, a highly permeable layer of sediment or rock-containing water. Layers of sand and gravel make good aquifers. On the other hand, clay and crystalline rocks, such as granite, make no good aquifers owing to their poor permeability characteristics. There are two types of aquifers, viz., (i) unconfined aquifers and (ii) confined aquifers. The former is covered by permeable Earth materials and is rechargeable by rainwater and/or melted snow, while the latter is confined within two impermeable layers of rocks and rechargeable only in the areas in which the aquifers meet the land surface. In many cases relating to the confined aquifers, it is often realized that the recharging area is several kilometers away from the place where a well to harness groundwater is located.

7.5.1 Advantages of Dependence on Groundwater

Dependence on groundwater is comparatively more advantageous than on surface water. The following are the advantages worth to be counted:

- In many regions of the world, groundwater is easily accessible and can be readily made available quite near the site of use. You can dig a well near your field for crop irrigation. You can install a hand pump in your house to directly avail water for drinking and other domestic purposes.
- Groundwater is likely to be pure and contamination-free and its quality is quite uniform throughout. Thus, it is good for drinking purposes, health, and sanitation.
- Groundwater resources are safe from seepage as well as from evapotranspiration losses.
- There is hardly anything like a "lean period" associated with groundwater use. In other words, unlike surface water resources, groundwater does not become scarce during a certain period of the year. Areas dependent on groundwater normally do not suffer from drought spells or acute water shortages in a certain season. Of course, this state of the groundwater level is conditioned by a balance between the rates of depletion and recharge of groundwater resources.

7.5.2 Effects of Groundwater Overuse

Being the largest source of freshwater supplies worldwide, the groundwater is bound to be depleted at rates faster than it is replenished. Groundwater depletion is primarily caused by pumping out of the water on a sustained basis for use of various purposes. The major consequences of the overexploitation of groundwater as being witnessed in several regions of the world are as follows:

- Excessive depletion of groundwater for irrigation, industry, drinking, and other domestic purposes gradually results in the lowering of the water table, which, in turn, causes all the problems arising out of water scarcity. The worst is likely to occur when wells go dry leading to a decline in agricultural production and crop failure. The worst is likely to be in arid and semiarid regions of the world.
- The lowering of the water table imposes the need for extra energy costs in pumping out water. Water supply dependent on the sources with low water table ceases to be cost-effective and sustainable.
- The exploitation of groundwater over recharge rates of the groundwater resources leads to the compaction of sediments in the aquifers resulting in ground subsidence. As a consequence, the sinking of the overlying land surface takes place causing damage to buildings, obstruction in water supplies, reversal of the flow of canals and sewers, etc.
- Groundwater and surface water systems are connected. With excessive depletion of groundwater sources, diminished water supplies from surface water sources – the freshwater lakes, rivers, and streams – are often recorded.
- Excessive exploitation of groundwater may also become a cause of water pollution through

increased seeping of heavy metals, such as arsenic, and many kinds of other deleterious pollutants. In coastal areas excessive pumping out of groundwater causes salt water to move down and mix with the freshwater, thus contributing to badly affecting water quality.

7.6 Water for Economy

All the economic sectors of the world are integrated with water. Water scarcity, water pollution, deterioration in water quality, and overall mismanagement of water resources are the factors that affect the economy of a country or a region. Let us, first of all, look into the various socioeconomic attributes of water (Fig. 7.3). These economic attributes of water are different from those vital for the very sustenance of life we have already discussed.

The freshwater requirement in adequate quantities is a must for these sectors to function and perform economically. But where does this freshwater in enormous quantities to support these economic activities come from? The credit goes mainly to the water cycle. Water, as we discussed at the beginning of this chapter, prevails in the natural cycle. There is a fixed lifetime of water in its natural resources as well as in the atmosphere. It is this hydrological cycle that helps us receive water supplies the natural way on a regular basis. As much as some 430,000 km^3 of water evaporates from the oceans annually. A large chunk of this evaporated water, about 390,000 km^3, precipitates over the oceans, making up the water losses from the largest water bodies. The rest of the water mass precipitates over land portions of the Earth as rain and snow.

7.7 The Natural Water Supplies

The hydrological cycle operating in the biosphere is the key to water supplies to all ecosystems and their organisms. This cycle is also vital for sustainable water supply systems managed by human beings. There are two natural aspects of the regular supply of water through the hydrological cycle:

(i) The saline water in the oceans and seas unfit for drinking and other domestic, agricultural, and industrial uses is received on land in the form of pure drinkable water through surface and groundwater resources.
(ii) The precipitation received as snow serves as a long-term deposit of pure and healthy water.

Is the water availability for supporting and sustaining the global human population of about 7.3 billion enough? The natural annual water supplies of 40,000 km^3 imply that as much as 5480 m^3 of water is available per person per year (Rajagopalan 2017). This figure exceeds the threshold value of 1700 m^3 per person per year. Despite this seemingly satisfactory situation concerning natural water availability per capita per annum, most of the world experiences water scarcity continuously or during certain periods of the year. It implies that water availability need not be equivalent to accessibility to water. Again, water is not just to be meant for consumption by human societies. Water has to fulfill the requirement of all living organisms, for the entire biodiversity of the Earth to blossom and prevail.

Fig. 7.3 Socioeconomic attributes of water resources

With an unabated spurt in human populations and consequent increase in the demand for food production using enormous amounts of water, inventions of ever more water-dependent economic avenues, and water-wasting lifestyles, our future will be caught in the brutal grips of water crises. As water scarcity is intensifying and the vicious drought-flood cycle continues to prevail year after year, the shadow of water-borne crises seems to be enlarging in size, leading to complete paralysis of the socioeconomic systems in many regions of the world.

7.8 Water Use Pattern

How much water do we need? Villagers in India use quite less amount of water for their daily needs compared to their urban counterparts. In most rural areas, domestic water supplies are a private matter. People manage domestic water on their own depending upon the local natural water resources, for example, water from wells, hand pumps, springs, rivers, etc. Urban areas depend on public water supply systems almost invariably.

Humanity the world over uses little over half of the accessible water supplies. Towards the middle of the third decade of the current century, this share is likely to increase to about 70%. Such a high degree of water consumption by human species alone would throw many terrestrial and aquatic organisms onto pathways to extinction. Our economy is largely based on water consumption. Generally, the higher the rates of water consumption, the more impressive the economic progress. What would be the state of the economy if water consumption goes on increasing and one day this natural resource becomes scarce to a critical extent? And many examples of this state of critical water scarcity are already visible in numerous areas of the world.

Agriculture is a water-guzzling sector consuming a whopping share of freshwater. As India is largely an agrarian economy, the percentage of water used in the agriculture sector is staggeringly high. Modern agriculture involving high-yielding varieties of cereals and vegetable production are especially water-guzzling. Freshwater consumption by industry in industrialized countries is also a matter of concern.

7.9 Desalination of Seawater

Seawater, because of its mineral component, is unfit for drinking, other domestic purposes, and irrigation. Desalination is a process by which we can remove minerals and salts dissolved in seawater and make the saline water worth drinking and for other domestic purposes. Saline water of the seas is rainfall-independent water available abundantly that can be supplied for human use after desalination particularly in human-inhabited areas lacking freshwater resources, such as rivers, freshwater lakes, and groundwater. However, as the desalination process consumes vast amounts of energy and is pretty expensive, large-scale water supplies dependent on this process are not feasible everywhere. The current emphasis, therefore, is on making the desalination process more cost-effective and feasible.

Desalination is more prevalent in North Africa, Middle East, subtropical islands, and the Canary Islands where it is the only way to obtain drinkable water. About 16,000 desalination plants operating across 177 countries generate approximately as much as 95 million m^3 of freshwater per day (Jones et al. 2019).

7.10 The Global "Water Wars"

It is often echoed in various schools of thought that the Third World War will be fought over water. We need not endorse this prediction, but, surely, the situation over water sharing between two or more countries and between various pockets or areas in a country will become grimmer and more chaotic in the years to come. Even wars over waters in the future cannot be ruled out.

The water cycle, streams, and rivers know no political boundaries. Currently, more than two hundred running water bodies are shared by two or more countries. A warlike situation often prevails over water sharing in the Middle East. River basins of the Jordan, Nile, Tigris, and Euphrates rivers are the sources of major discord among Middle East countries. Fierce battles over the sharing of Jordan River water between Syria, Israel, and Jordan are there in historical records.

Israel is pretty conscious about the management of its resources and hardly uses Jordan River water for domestic supplies and irrigation. Its neighbors, Jordan and Syria, want to overexploit the river. Israel's Sea of Galilee is a unique example of water management. Israel is always conscious of the amount of water drawn from the Sea of Galilee. People of Israel talk about the level of water in the Sea of Galilee as the people talk about the "Sensex" of share markets (Singh 2018).

Turkey has a notorious plan to construct a chain of dams over the Tigris-Euphrates Rivers to generate hydroelectricity and sell water to Middle East countries. This plan could exacerbate tensions among many Middle East countries.

The Ganga-Brahmaputra Basin becomes a source of conflicts between India, Bangladesh, and Nepal. Nepal has an eye on the huge potential of the basin to generate hydroelectricity. Worried about the floods the Ganga-Brahmaputra Rivers bring, Bangladesh emphasizes the management of river waters to minimize floods in the neighboring countries.

7.11 Damming the Rivers: Advantages and Disadvantages

One of the techniques to manage water resources is to construct artificial water reservoirs. This is done by damming a river and thus obstructing water flows and collecting water in a reservoir behind the dam. Although this is done for the primary purpose of hydroelectricity generation, the artificial reservoir so created serves many purposes.

The construction of high dams on the rivers in mountain areas and thus prevention of the flow of water has been one of the major development agendas in India. Several high dams and, as a consequence, an equal number of water bodies (artificial lakes) have come into existence. Dams like Bhakra-Nangal, Heerakud, Damodar, Nagarjuna Sagar, and Sardar Sarovar Dam are often projected as proud symbols of India's socioeconomic development. India also ventured into building up one of the world's highest dams, the Tehri Dam, towards the beginning of the twenty-first century, and that too is seismologically one of the most interesting regions of the Himalayas, despite the warning of experts. Tehri Dam, with a height of 260.5 m, obstructs the flow of the Bhagirathi and Bhilangana rivers in the Central Himalayas, inundating an area of about 50 km^2 and drowning the historic city of Tehri in the Garhwal region of India's Uttarakhand state.

While several socioeconomic benefits are accrued from high dams, in most cases they have turned into an ecological disaster. Earlier, the Aswan Dam across the Nile River was a center of controversy. Completed in 1970, this huge dam was castigated as an ecological disaster. Many scientists, environmentalists, geologists, and seismologists had warned of the disastrous consequences of the Tehri dam before it came into being.

The bright side and the dark side of high dams have been compared in Fig. 7.4. Negative consequences of high dams for use of water resources include the submergence of agricultural land, industrial areas, and inhabited areas – cities, towns, and villages. Destruction of native biodiversity including wild animals is obvious when a large area of land is inundated by a reservoir. Displacement of masses from their native cultural settlements to distant places is often managed to compromise with local people's rights to their cultural ethos.

The potential risks of high dams are sometimes unimaginable. These dams in the fragile Himalayan mountains if damaged intentionally by a nefarious group or terrorists or by an enemy country, the disaster caused to life and property will be incalculable. A disaster can also take place due to a high-intensity earthquake the Himalayan areas are quite prone to. Again, an extremely huge amount of water in an artificial lake behind a high dam might exert pressure on the land causing what is called reservoir-induced

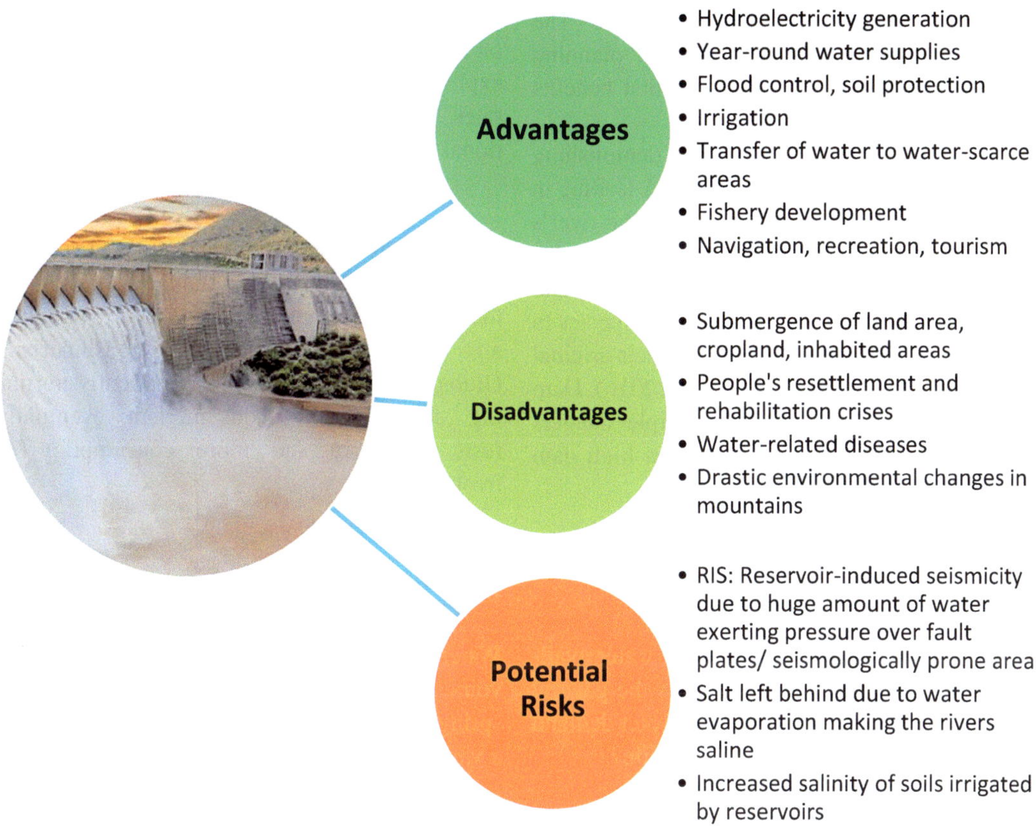

Fig. 7.4 The bright and dark sides and potential risks of high dams

seismicity (RIS) leading to even a high-intensity earthquake (Singh et al. 1995). A reservoir in a temperate climatic region can alter environmental qualities and cause various types of skin and breathing problems.

Workable alternatives to high dams are available in our contemporary world. Instead of a few large dams, a chain of small dams can be constructed at appropriate sites. These dams are safe from every angle. Further, electricity can also be generated following the run-of-the-river technique. Development of alternative sources of electricity using solar power, wind power, tidal power, geothermal power, etc. and hydropower generation using small- and medium-sized dams and run-of-the-river engineering may help avoid our dependence on the construction of mega dams. In fragile mountain areas, such as in the Himalayas, our development focus must be on afforestation and regeneration and conservation of natural forests. A dense cover of forests in the mountains, in addition to contributing to climate regulation, will be vital for maintaining rainfed rivers evergreen, coping with water scarcity during the off-monsoon season. The development of minor irrigation systems in the mountains would be instrumental in making water available for irrigation purposes.

7.11.1 Dam Removal

It is being increasingly realized that damming a river invites large-scale environmental and ecological disruptions. Environmental safety and ecological security overweigh economic benefits. Therefore, dam removal is now regarded as an essential process for ecological restoration in the course of rivers and streams. Dam removal or demolition is a planning process that can take many years and is also quite expensive. The underlying principle of dam removal planning policy is that safety and environmental benefits are priceless and most important.

Currently, many countries are demolishing dams as a country policy, and the USA leading in the process. When a dam is removed the river's natural hydrology is restored, habitat for native river species above the dam is reestablished, and natural flow patterns vital for the native species to flourish downstream are resumed to their original configuration (Cho 2011; McCombs 2016). Dam demotion also rules out the multiple dangers impending due to the existence of a high dam often in fragile mountain ecosystems.

7.12 Causes of Water Scarcity

Water demand is rapidly increasing; water availability is rapidly decreasing. That is the general water scenario of the world. Why is water demand increasing in every sphere of human life?

Spurt in Human Population The larger the number of people, the more the demand for water for drinking, bathing, washing, cooking, kitchen, and other domestic chores.

The Green Revolution Agriculture Agricultural practices involving high-yielding varieties (HYVs) of food crops demand excessive amounts of water for irrigation. Vegetable cultivation is especially a water-guzzling practice. Food production is directly proportional to water consumption. As the population increases, so does the demand for food. Consequently, agriculture becomes more intensive, leading to higher water consumption.

Rapid Industrialization Ever-increasing industrialization to raise the gross domestic product to attain economic targets and satiate human greed is at the cost of the planet's water resources. The meat industry, textile industry, paper industry, and construction industry are especially voracious for water consumption.

Rapid Urbanization Growing like mushrooms on the face of the Earth, the cities and towns are exploiting water resources at unprecedented rates. Water sources will go on depleting proportional to the rate of urbanization.

Consumerism-Oriented Lifestyles The lifestyle characteristic of high standards of living of the contemporary civilization is evolving which are based on insatiable consumerism. Globalization processes are impatiently nurturing consumerism. Consumerism eventually rests on more and more consumption of freshwater.

7.13 Summary

Water resources – the oceans, seas, lakes, reservoirs, wetlands, ponds, rivers, streams, and springs – are not exclusively water. They inhabit a variety of organisms. A life dependent on photosynthesis as well as on chemosynthesis prevails in water habitats. Various life forms perpetuating in aquatic ecosystems also serve as precious resources. Water is a dynamic natural resource. The various attributes of water are photosynthesis, ecological integrity, biosphere's thermodynamics, biochemical reactions, organisms' structure and composition, oxygen, food production, and nutritional security.

Most of the water on the water planet is saline water extending over some 70% of the planet's surface. And most of the saline water is concentrated in five oceans of the planet, namely, the Pacific, Atlantic, Indian, Southern, and Arctic Oceans. Little less than 50% of the total area and 50% of the water volume of the oceans is covered by the Pacific Ocean alone. Being the largest water body, the Pacific Ocean covers a nearly one-third area of the planet's surface. There are seven major seas of the Earth: Arabian, Caspian, Caribbean, Gulf of Mexico, Mediterranean, Persian Gulf, and South China Sea. There are 10 major lakes: the Caspian Sea, Lake Superior, Lake Victoria, Lake Huron, Lake Michigan, Lake Tanganyika, Lake Baikal, Great Bear Lake, Lake Malawi, and Great Slave Lake. Salinity percentage ranges from as high as 44% in the Don Juan Pond, a salt lake in Antarctica, to 27% in the Great Salt Lake in the USA.

Himalayan glaciers cover about three million ha or 17% of the global mountain area – the largest bodies of ice outside the polar caps. The Himalayas, thus, are also popularly called the Third Pole of the Earth. The total area of the Himalayan glaciers is 35,110 km^2. The Himalayas serve as a lifeline for more than 500 million inhabitants of the region, or about 10% total regional human population. The glacier-fed perennial rivers originating in the Himalayas are sources of drinking water, crop irrigation, and livelihoods for billions of people in the foreland areas. Having its origin in the Central Himalayan mountains in the Uttarakhand state of India, the holy Ganga River makes the largest water system in the Indian subcontinent. It is thanks to the benevolence of the Ganga waters that the Indo-Gangetic plains are perhaps the most densely populated region in the world.

The rivers, streams, canals, rivulets, falls, geysers, and springs are the running water bodies or lotic ecosystems comprising freshwaters. A river, a large stream of water flowing from high land to low land, is the most crucial source of potable water as well as of numerous other socioeconomic benefits, including the generation of energy, the hydroelectricity. A river is fed by a glacier, rainwater, a spring, or a lake and after flowing up to a certain distance eventually merges into an ocean. India is one of the exceptional examples of the rivers and river systems in the world.

Groundwater, one of the prominent resources of fresh drinkable water in many regions of the world, constitutes little less than 10% of the total freshwater resources. Of the total water supplies, the share of the groundwater is about 30–50 times more than that of the surface water supplies. Being the largest source of freshwater supplies worldwide, the groundwater is bound to be depleted at rates faster than it is replenished.

All the economic sectors of the world are dependent on water. Water scarcity, water pollution, deterioration in water quality, and overall mismanagement of water resources are the factors that affect the economy of a country or a region.

The hydrological cycle operating in the biosphere is the key to water supplies to all ecosystems and their organisms. With an unabated spurt in human populations and consequent increase in the demand for food production using enormous amounts of water, inventions of ever more water-dependent economic avenues and water-wasting lifestyles, our future world will be caught in the brutal grips of water crises. As water scarcity is intensifying and the vicious drought-flood cycle continues to prevail year after year, the shadow of water-borne crises seems to be enlarging in size, leading to complete paralysis of the socioeconomic systems in many regions of the world.

Humanity the world over uses little over half of the accessible water supplies. Towards the middle of the third decade of the current century, this share is likely to increase to about 70%. Agriculture, industry, and the domestic sector are the three major sources dependent on freshwater for survival and sustenance. Among these, agriculture alone consumes as high as 70% of the total global freshwater, on average. Consumption of water by the agriculture sector in India is nearly 80%, exceeding the average of industrialized countries (30%). Industry worldwide consumes 22% of the water on average, with India's share being just 15%, much less than that of the industrialized countries (59%). Freshwater consumption by the domestic sector, on average, is 8%. This sector in India uses only 5%, much less than the average of the industrialized countries (11%). Water demand is rapidly increasing; water availability is rapidly decreasing. That is the general water scenario of the world.

7.14 Exercises

7.14.1 Multiple-Choice Questions

1. Diatomic oxygen is produced by cyanobacteria and green plants as a result of
   (a) photolysis of water
   (b) photolysis of CO_2
   (c) oxidation of CO_2
   (d) photoreduction of water

2. Where do the lowest places on the surface of the Earth and in the ocean lie, respectively?
   (a) The Mariana Trench and the Dead Sea
   (b) The Puerto Rico Trench and the Mariana Trench
   (c) The Dead Sea and the Mariana Trench
   (d) The Mariana Trench and the Puerto Rico Trench
3. This water body covers about 47% of the total water surface and 50% of the total volume of the planet's total water. This is the
   (a) Arctic Ocean
   (b) Pacific Ocean
   (c) Indian Ocean
   (d) Atlantic Ocean
4. The Laccadive Sea is part of the
   (a) Arctic Ocean
   (b) Pacific Ocean
   (c) Atlantic Ocean
   (d) Indian Ocean
5. Which of these water bodies has maximum depth?
   (a) Caribbean Sea
   (b) South China Sea
   (c) Bay of Bengal
   (d) Gulf of Mexico
6. Where is the Sea of Galilee?
   (a) Jordan
   (b) Syria
   (c) Israel
   (d) Egypt
7. The Don Juan Pond is a/an
   (a) artificial pond as large as a lake
   (b) salt lake with the highest salinity in Antarctica
   (c) pond created behind Aswan Dam in Egypt
   (d) wetland on an island in the Bay of Bengal
8. In China, the Salween River is known as the
   (a) Tsangpo River
   (b) Yellow River
   (c) Yangtze River
   (d) Nu River
9. Which of the following Deccan rivers flows westward?
   (a) Narmada
   (b) Krishna
   (c) Cauvery
   (d) Mahanadi
10. The city of Leh in the Indian Union Territory of Ladakh is located near the River
    (a) Sindhu
    (b) Chenab
    (c) Beas
    (d) Ravi
11. What is true about groundwater?
    (a) Groundwater can never be polluted.
    (b) Surface water availability exceeds that of groundwater.
    (c) Heavy metals cannot contaminate the groundwater despite its over-exploitation.
    (d) Groundwater resources are safe from evaporation losses.
12. Globally, which sector uses the largest proportion of available freshwater?
    (a) Domestic
    (b) Industry
    (c) Agriculture
    (d) Energy
13. Which of the following products produced for consumption uses the maximum amount of water per unit mass?
    (a) Rice
    (b) Sugar
    (c) Beef
    (d) Potato
14. What is not a correct statement about water?
    (a) Water on planet Earth is in fixed quantity, so per capita, water availability can never go down.
    (b) Most of the world's water resources are increasingly becoming polluted.
    (c) Overexploitation of groundwater can result in land subsidence.
    (d) The salinity and temperature of a water body determine its movement.
15. Which water body represents a lotic ecosystem?
    (a) A lake
    (b) A river
    (c) A reservoir
    (d) A pond

16. Which of the following is the most critical factor responsible for widespread water crises?
    (a) Drought
    (b) Dams
    (c) Landslides
    (d) Population explosion
17. Which of these rivers flows in three countries?
    (a) The Indus
    (b) The Ganga
    (c) The Krishna
    (d) The Chambal
18. The Aswan Dam is constructed on the
    (a) Jordan River
    (b) Euphrates
    (c) Nile
    (d) Huang Ho
19. The Tehri Dam obstructs the flow of two rivers creating a vast reservoir of water. These rivers are
    (a) Bhagirathi and Mandakini
    (b) Bhagirathi and Alaknanda
    (c) Bhagirathi and Bhilangana
    (d) Bhagirathi and Pinder
20. Which of the following is not a cause of water crises?
    (a) Forestation
    (b) Population growth
    (c) Agriculture
    (d) Industry

Answers: 1-a, 2-c, 3-b, 4-d, 5-a, 6-c, 7-b, 8-d, 9-a, 10-a, 11-d, 12-c, 13-c, 14-a, 15-b, 16-d, 17-a, 18-c, 19-c, 20-a

7.14.2 Short-Answer Questions

1. How does water play a role in food security?
2. There can be no nutritional security without water. How come?
3. How does water contribute to producing diatomic oxygen?
4. Which types of organisms participate in the photolysis of water?
5. Mention the names of the planet's oceans in decreasing order of the total water volume they comprise.
6. Mention the names of any four water bodies comprising very high salinity percentages.
7. How is water availed annually in a natural way?
8. How can overexploitation of groundwater pollute this precious resource?
9. What are the solid reserves of water on Earth?
10. Name major water systems of India.

7.14.3 Long-Answer Questions

11. The Earth is a water planet. The water planet is a living planet. Explain this statement with deep reasoning.
12. The Himalayan mountains are often called the "Third Pole" and "Water Towers for Mankind." What are your arguments in this regard?
13. What are the pros and cons of high dams? Explain with examples.
14. Describe various reasons for water scarcity in the world. What are your suggestions to reverse the water crises intensifying year after year?
15. Write short notes on the following:
    (a) Groundwater resources
    (b) Economic role of water resources
    (c) Water wars and water conflicts

References

Cho R (2011) Removing dams and restoring rivers. News from the Earth Institute, August 29, 2011

IPCC (2007) Climate change 2007. The physical science basis, contribution of working group I to the fourth assessment report of the Intergovernmental Panel on Climate Change. Cambridge University Press, Cambridge, 996 pp

Jones E, Qadir M, van Vliet MTH, Smakhtin V, Kang S-m (2019) The state of desalination and brine production: a global outlook. Sci Total Environ 657:1343–1356

McCombs ES (2016) Partnering to remove the Shuford Dam. American Rivers, August 12, 2016

Rajagopalan R (2017) Environmental studies: from crisis to cure. Oxford University Press, New Delhi, 334 pp
Sharma PD (2013) Environmental biology and toxicology. Rastogi Publications, Meerut, 570 pp
Singh V (2018) The speaking stones. Notion Press, Chennai
Singh V (2020) Environmental plant physiology: botanical strategies for a climate smart planet. CRC Press (Taylor and Francis), Boca Raton, 330 pp
Singh V, Sharma RJ, Kumar A (eds) (1995) Ecological carnage in the Himalaya. IBD, Dehradun, 178 pp
Singh V, Nautiyal N, Apparusu SK, Rawat MSS (eds) (2011) Climate change in the Himalayas: preserving the third pole for cooling the earth. Indus Publishing Company, New Delhi, 302 pp
Singh V, Kumar B, Nautiyal N, Rastogi A, Shankhwar AK (eds) (2012) Climate change and hydrosphere: the water planet in crises. Biotech Books, New Delhi, 258 pp

Land and Soil Resources

8

Land and soil are the two integrated resources so vital for the socioeconomic progress of a country, society, or individuals. What is most critical about the Earth is that it has a limited land area. Further, all the land of the planet is not utilizable and would hardly be referred to as a resource. Soil is an indistinguishable component – the uppermost layer – of the land. Soil is an integral part of the land, but the uppermost layer of all the land need not be soil. In numerous places, the land is composed of rocks and other non-soil materials. In many areas, regions, and mountain-tops, the land's uppermost layer is permanently covered with snow and ice. Soil is a "product" of weathering of the rocks. Thus, at some point of time in the distant future there would be soil where there are rocks today.

Land and soil are the natural resources on which the future welfare of our world depends. Management of these two resources is of critical importance to manage most other natural resources and elicit socioeconomic advantages while ensuring ecological security and environmental safety.

8.1 Land as a Resource of All Resources

Land resources include all those features and processes of the land, which, in some way, is used to fulfill certain human needs (Vink 1975). Earth is the most wonderful and the lone living planet of the solar family. Endowed with water resources on 70% of its surface, the planet is popularly called Blue Beauty. According to modern estimates, the surface area of the Earth is approximately 5.1×10^8 km^2. The land area of the Earth is approximately 149 million km^2. In our Solar System, the Earth is comparatively larger than Mercury, Venus, and Mars, but smaller than Jupiter, Saturn, Uranus, and Neptune.

The land of the Earth is the superlative resource, a treasure of natural resources on Earth. It accommodates all other terrestrial resources: forests, grasslands, wetlands, biodiversity, wildlife, agricultural land, livestock, minerals, soil, and human resources. Land accommodates even surface water resources and also serves as a source for the exploitation of groundwater. Land, in essence, is the most important natural resource for mankind and is a resource of all natural resources. The more land area available in a country, the greater the potential for its socioeconomic development.

The land is the major constituent of the lithosphere making up about 29% of the Earth's surface. The land is covered with natural forests, rangelands, grasslands, deserts, snows and ice, wetlands, agriculture, and human settlements – the villages, the cities, and the towns. Most human activities are concentrated on the land of

V. Singh, *Textbook of Environment and Ecology*, https://doi.org/10.1007/978-981-99-8846-4_8

the Earth. Most of the livelihood systems are structured on land.

The components of the natural land unit – that is, physical, biological, environmental, infrastructural, social, and economic – include numerous, varied, and complex resources that humanity overwhelmingly depends upon. Human activities are so intensive that most of the ecosystems on Earth are now transformed. We can call them anthropogenic ecosystems. Human dimensions are so staggering that very little is left on the land that could be witnessed in a natural state. The land is such a resource that cannot be stretched. This limitation leaves limited scope for the other resources to stretch despite the best-known management practices brought into practice.

8.2 Land Use Pattern

Land use pattern is not uniform throughout the globe. It varies from one country to another. As land is put to different uses, it depends on the cultural ethos, planning, programs, and socioeconomic targets of a country and how a governing system puts its land area into different uses. The average land use of the global land surface is depicted in Fig. 8.1. Most of the land surface throughout much of human history has been covered by natural forests. Human societies have changed the Earth's natural scenario dramatically by transforming forest lands into cultivated areas over the last few centuries. About half of the world's habitable land, which measures about 71%, is currently used for agriculture. About 10% of the land surface is covered by glaciers and 19% comprises barren land, including deserts, dry salt flats, exposed rocks, dunes, beaches, etc. About 37% of the habitable land is covered by forests and 11% by shrubland and grasslands; nearly 1% is occupied by freshwaters, such as lakes, rivers, streams, etc.; and 1% constitutes human settlements (rural, urban, and industrial areas) and other constructed areas, such as roads, railways, airports, and other human infrastructures (Ritchie and Roser 2019).

India, an agrarian economy, has as much as 44% of land under agriculture, 15% under permanent pastures, 12% under dense forests, 12% as culturable wasteland, 7% as barren and degraded land, and 6% urban land (Fig. 8.2). Further, the land use is not a fixed pattern. It not only varies spatially but undergoes changes from year to year and from time to time in the same year.

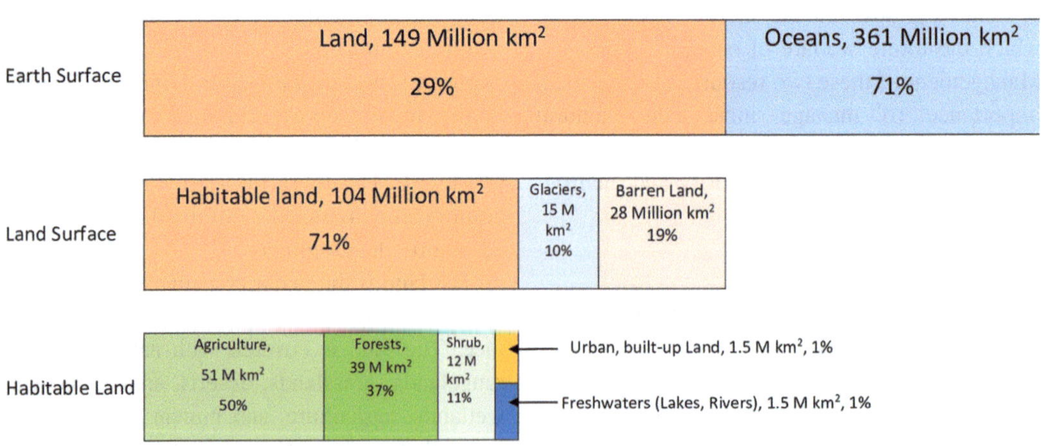

Fig. 8.1 Land use at global level based on Ritchie and Roser (2019)

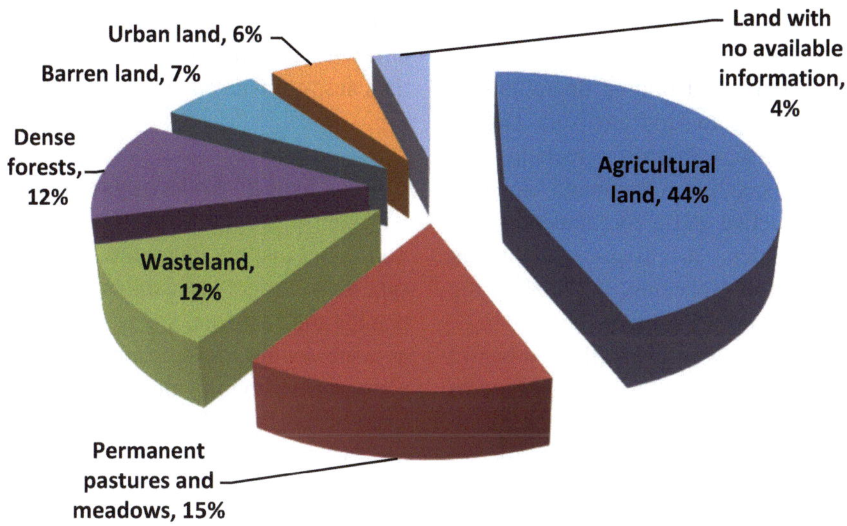

Fig. 8.2 Land use pattern in India (values are approximate and subject to temporal variation)

8.3 Land Degradation

Land degradation, according to Conacher and Conacher (1995), is a process in which the value of the biophysical environment is affected by a combination of human-induced processes on the land. Johnson et al. (1997) define land degradation as any change or disturbance to the land perceived as deleterious or undesirable. Land degradation inevitably leads to soil erosion or loss of soil fertility or a substantial decline in its productive capacity.

Land degradation is generally related to human activities, not to natural factors. In our times, most of the natural factors leading to land degradation are often influenced and/or exacerbated due to erroneous human activities. Land degradation appears to be a permanent feature in our times and also one of the most serious problems leading to grave and multiple consequences – ecological, environmental, and socioeconomic. According to an estimate, a large chunk of agricultural land – approximately 40% – is degraded (Sample 2007). The Special IPCC Report (IPCC 2019) further makes us cautious about land degradation:

- About a quarter of the Earth's ice-free land area is subject to human-induced degradation.
- Soil erosion from agricultural fields is estimated to be currently 10–20 (no tillage) to more than 100 times (conventional tillage) higher than the soil formation rate.

8.3.1 Causes of Land Degradation

Land degradation is a global problem. This problem is largely attributable to erroneous anthropogenic activities. Clearance of vegetation or deforestation is one of the dominant causes. Many land degradation factors trigger disruptive processes following deforestation as we have discussed in the chapter on forest resources. Land use change – that is, conversion of forest lands into pastures, agricultural land, industrial areas, construction sites, etc. – invites land degradation in several ways. Intensive terracing of mountain slopes, as evident in the Himalayas, results in extreme rates of soil erosion and destabilization of fragile mountain ecosystems. Too much tillage of agricultural lands is one of the prime reasons for soil erosion by winds and water and the rapid loss of soil carbon. Intensive farming involving monocultures robs the soils of their nutrients. Inappropriate irrigation leads to rapid soil erosion and soil carbon loss and gradually results in soil salinity. Overgrazing damages land by trampling vegetation, soil compaction, and soil erosion.

Quarrying stones and mining operations are extremely damaging to land and soil. Derelict land or mine spoils are difficult to reclaim. Road construction, especially on mountain slopes, is extremely destructive and a root cause of catastrophic landslides in fragile mountains. Vegetation clearance, blasting, bulldozing, digging, and leveling operations using heavy machinery during road construction, urbanization, and area development for industries, tourist activities, etc. leave deep wounds on the surface of the land. The area of the land covered for construction purposes turns forever into unproductive concrete jungles culminating in total ecological ruin.

8.3.2 Climate Change and Land Degradation

Climate change is emerging as one of the causes of various environmental disruptions, including land degradation. According to the IPCC Special Report on Climate Change and Land (IPCC 2019):

- Area of drylands in drought has increased by 1% per year between 1961 and 2013.
- In 2015, as many as 500 million people lived within areas that experienced desertification between the 1908s and 2000s.
- South and East Asia, the Middle East, and Sahara in Africa are the regions where the highest number of people were affected by desertification.
- Desertification has also been experienced in other dryland regions.
- The climate change effects are more pronounced in the areas already degraded and desertified.

The rise in sea level, which is attributable to climate change, inundates low-lying island areas and river deltas and becomes a potential hazard for land degradation. Errant weather cycles, increased frequency of storms, torrential rains, a vicious cycle of droughts and floods, etc. are the other climate change-borne factors that are potential causes of land and soil degradation.

8.4 Soil as an Ecosystem

Soil is not just the uppermost layer of the lithosphere. Soil is not just a mixture of so many compounds and elements. Soil is also not just to serve as a substratum for the plants to grow. Soil is an ecosystem harboring numerous organisms in a reciprocal relationship with each other and with their physical environment. The soil ecosystem is the largest on the land of the Earth as well as the most complex one. It is also the most active and yet the most vulnerable ecosystem. The soil community with microorganisms as the dominant living beings comprises autotrophs (both photosynthetic and chemosynthetic) as well as heterotrophs (herbivores, carnivores, and omnivores). All kinds of organisms belonging to all kingdoms and ranging from bacteria to mammals thrive in the soil ecosystem. Food chains (mainly detritus but also grazing ones) and complex food webs and nutrient cycles operate in a soil ecosystem ensuring nutrient and energy transfer among organisms. Soil biodiversity is richer and far more fascinating than that of other ecosystems on the Earth. The "invisible world" of the soil is more interesting and intriguing than the visible world (Singh 2020).

The biotic component of the soil ecosystem baffles the human mind in several ways (after Singh 2020):

- There is maximum life (no. of organisms) per unit area and unit volume of the soil.
- Biomass of soil microorganisms alone is more than that of all human beings, elephants, and whales of the Earth put together.
- A pinch of fertile soil has in it numerous microorganisms that are more than all human beings on the planet.
- Most of the organisms belonging to three out of five kingdoms of living organisms, viz., Monera, Protista, and fungi, inhabit the soil ecosystem.

- A teaspoon of soil from a native grassland ecosystem contains:
- Bacteria numbering 600–800 million from as many as 10,000 species, 10,000 protozoa, 20–30 nematodes from 100 species, and several kilometer-long fungi.

8.5 Soil as the Foundation of Terrestrial Life

Soil is the basis of all terrestrial life. All organisms living within the soil ecosystem and above-soil surface depend on soil for their survival and sustenance. The above-soil life is rooted in the soil ecosystem (Singh 2020). Since all the components of the Earth's environment – viz., lithosphere, hydrosphere, atmosphere, and biosphere – are interrelated, soil also plays a significant role in aquatic life. The bulk of the mineral nutrients that plants and, through the food chain, all animals need come from the soil through nutrient cycles. Microscopic life forms that could also be found in the atmosphere are often laden with soil particles. To appreciate how the soil serves as the foundation of terrestrial life, let us marvel at the soil profile – the vertical section of the soil from the surface to the rock material that depicts all of its horizons. Soil structure undergoes gradual changes as we move deeper. In the field of soil science, experts typically classify soil into four distinct horizons, viz., O, A, B, and C, following changes with depth (Fig. 8.3).

The O, i.e., organic, horizon makes the topmost layer of the profile. The O-horizon is composed of fallen leaves, twigs, bark, and various plant materials. It extends a bit deeper and contains partially decomposed organic matter, a result of the activities of soil organisms like bacteria, fungi, and animals, including mites, nematodes, and burrowing mammals.

The A-horizon consists of materials like clay, silt, sand, and organic matter percolating from the O-horizon. This horizon is vibrant with microbial activities. Burrowing animals, especially earthworms, contribute to mixing the organic matter of the O-horizon in the A-horizon. This horizon is rich enough in mineral nutrients essential for plants. Iron, aluminum, clays, silicates, and humus gradually leach from this layer into the B-horizon.

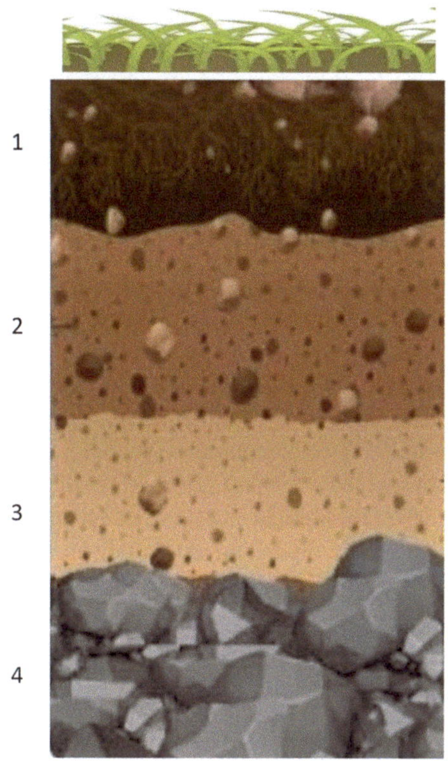

Fig. 8.3 Soil profile: (1) the O-horizon (organic), (2) the A-horizon or topsoil, (3) the B-horizon or subsoil, and (4) the C-horizon or weathered parent material and bedrock

The B-horizon contains humus, clays, and other mineral nutrients transported from the A-horizon. Distinct color and banding pattern to this horizon are imparted due to the deposition of these materials. In addition, the roots of numerous plants also lie on this horizon.

The lowermost portion of the soil profile, the C-horizon, comprises weathered plant material. As the weathering in this horizon is incomplete, this may comprise several rock fragments. Below the C-horizon lies unweathered parent material, often the bedrock.

The soil structure and composition – and, therefore, its functioning – are all in a state of flux due to several operating factors, especially, as Jenny (1980) summarized, soil organisms, topography, parent material, and time. As a complex and dynamic and, of course, as the largest terrestrial ecosystem, the soil accommodates most of the terrestrial life within and provides the basis for all above-soil life. All the world's civilizations are rooted in soil. Human history is replete with the dire fact that a civilization that does not respect soil ultimately vanishes from the face of the Earth.

8.6 Soil Types of the World

The diversity of soil (or pedodiversity) is the manifestation of the diverse natural features of the Earth (Singh 2020). Geographical positioning, climatic region, geologic features, environmental factors, seasons of the year, hydrologic factors, and vegetation type are the major factors influencing soil types and soil ecology. Apart from these natural factors, human activities also influence soil characteristics and soil ecology to a great extent.

Soils of the world have been divided into six major categories: sandy, silty, clay, loamy, peaty, and chalky. These soil types are made up of three major soil types, viz., sandy, silty, and clay. Some interesting features of these soil types are described in Table 8.1.

Soil is the richest habitat in terms of the measure of its biodiversity. Soils host a quarter of our planet's biodiversity. A scenario of the diversity of life within the soil habitats of the world presented in a nutshell by FAO (2015) that baffles the human mind is as follows:

- More than 1000 species of invertebrates may be found in a single square meter of forest soil.
- World's several terrestrial insect species dwell in soil for at least some stage of their life cycle.
- One gram of soil might contain millions of individuals and thousands of bacteria species.
- Healthy soil may contain thousands of species of bacteria and actinomycetes, hundreds of species of fungi, tens of nematode species, 50–100 species of insects, 20–30 species of mites, several species of earthworms, and several species of vertebrate animals.
- Soil contains the organisms with the largest area; e.g., a single colony of the honey fungus (*Armillaria ostoyae*) may cover approximately 9.0 km^2.

Table 8.1 Major types of the world's soils

Soil type	Main characteristics
Sandy soil	Largest particles among all the soil types Gritty and dry feel upon touching the soil Huge spaces between the soil particles Poor water-holding capacity Quick draining of water through the soil Difficult for the plant roots to search water in the soil Poor plant growth in the soils
Clay soil	Very good water-retentive quality Quite smooth when dry, but sticky when wet Tiny particles not to allow much air to pass through High capability to hold plenty of nutrients within due to slow drainage property Good for plant growth
Silty soil	Much smaller particle size than those of sandy and clay soils Somewhat smooth touch Dirt left on skin if in contact with the silty soil Good in its water-retentive quality Poorly aerated
Peaty soil	History of peaty soil-type formation traced to some 9000 years ago when glaciers began to melt rapidly Black and dark brown in color Soft and somewhat spongy to touch Very rich in organic matter High water-retentive quality May get saturated with water Good for plant growth Support plant growth even during drought spell
Chalky soils or basic soils	Alkaline in nature Soft in touch Soft rocks made up of chalk breakdown easily contributing to the formation of chalky or saline soils Quick drainage of water Water-retention capacity very poor Quick drying up of the soil Fertile for certain plants Certain soil nutrients, e.g., iron, rendered unavailable for the plants due to high alkalinity

(continued)

Table 8.1 (continued)

Soil type	Main characteristics
Loam soil	"Perfect" in itself with the balanced proportion of all the three major soil types, i.e., sand 40%, silt 40%, and clay 20%, along with humus Dark in color Soft, brittle, and powdery in touch Very high water-holding capacity Preserves moisture even during a dry spell, often enough to sustain plant growth Often characterized with appropriate proportion of organic matter, essential plant nutrients, and moisture Considered ideal soil type for crop cultivation

Source: Based on Singh (2020)

Land-use practices are the major determinants of soil biodiversity. The soil organisms maintain the key services such as nutrient cycling, gaseous exchange, carbon sequestration, enhanced nutrient supplies to plants, and, consequently, maintenance of plant growth and production performance (Singh 2020).

8.7 Soil Erosion

Soil erosion is the displacement, loss, or removal of the upper fertile layer of the soil and is a form of soil degradation. The dynamic activity of erosive agents is responsible for soil erosion inevitably leading to a decrease or total loss of soil fertility in the affected area or sites. The erosive agents of soil erosion are winds, water, glaciers (ice), snow, animals, and humans. Based on the main agent of erosion, soil erosion is often categorized as wind (or aeolian) erosion, water erosion, glacial erosion, snow erosion, zoogenic erosion, and anthropogenic erosion. All other causes of soil erosion other than anthropogenic are natural ones but are significantly influenced by anthropogenic activities.

Soil erosion rate is not a fixed phenomenon; it may be slow enough that goes unnoticed and leaves its effect after a long time, or it may go on at an alarming rate revealing an immediate catastrophic effect. In addition to decreased soil fertility and reduced production potential (for example, agricultural lands), soil erosion may also cause air and water pollution, siltation of dams, damage to drainage networks, sinkholes, etc.

8.7.1 Physical Processes of Soil Erosion

Physical processes of soil erosion involve wind force, mass movement, rainfall and surface runoff, rivers and streams, and floods. Strong winds exert force causing soil erosion. These are especially the main features of arid and semiarid regions. Wind erosion is accelerated by human activities, such as indiscriminate deforestation, land cultivation, urbanization, etc. Mass movement involves the downward and/or outward movement of rocks and soil mainly on account of gravitation force applied on steep slopes in mountain and hill areas. Rainfall and consequent surface runoff exert physical forces leading to four kinds of soil erosion, viz., sheet erosion, gully erosion, splash erosion, and rill erosion. Rivers and streams when flowing through narrow valleys cause soil erosion. The sediment moves downward with the stream as well as it erodes sloppy lands extending channels into hills and mountains. Extremely high rates of water flow due to torrential rains, glacier rains, dam bursts, etc. create floods running with huge force causing soil erosion at an alarming rate.

8.7.2 Factors Affecting Soil Erosion

8.7.2.1 Topography

The topography of the land determines the speed and consequent force of the surface runoff water and the resultant erosivity of the runoff. Mountains, hills, and upland areas are more prone to soil erosion. The steeper the slope, the faster the speed and force of the surface runoff and the greater the erosivity of the runoff. Soil erosion further exacerbates if the sloppy terrain is not covered by vegetation. Steeper terrain is also more sensitive to mudslides and catastrophic landslides due to gravitational force.

8.7.2.2 Vegetation

An area devoid of vegetation is more prone to soil erosion. Vegetative cover on the soil is phenomenal for preventing or reducing soil erosion as:

(i) It provides shelter to the soil from direct exposure to winds and thus protects the soil from wind erosion.
(ii) It maintains soil moisture not allowing desiccation of the land thereby reducing erosivity.
(iii) It decreases surface runoff by increasing soil permeability to rainwater.
(iv) The roots of the plants bind the soil together forming a solid mass not susceptible to water and wind erosion.
(v) It builds up a microclimate conducive to soil conservation.

8.7.2.3 Soil Structure and Composition

Soils with a greater proportion of clay, rather than sand and silt particles, are less susceptible to wind and water erosion. It is because clay plays a crucial role in binding soil particles together, thus helping increase soil resistance to water and wind erosion. Soils with a higher proportion of organic matter are more resistant to erosion. It is because organic materials coagulate soil colloids, making stronger and more stable soil structures (Blanco and Lal 2010). Wet and saturated soils tend to absorb less amount of rainwater and thus contribute to increased runoff and higher rates of soil erosion.

8.7.2.4 Climate Change

The amount, intensity, and frequency of precipitation often determine the rates of soil erosion especially in the hills and mountains and in the areas devoid of vegetation or in the regions with sparse vegetation. With factors like land use change not considered, Pruski and Nearing (2002) expect approximately a 1.7% change in soil erosion for each 1% change in precipitation owing to climate change. When precipitation occurs at a time when agricultural lands have no crop cover, rates of soil erosion would be higher. Climate change is contributing to the desiccation of the lands, including rangelands and forested regions making the soils more prone to wind and water erosion. Increased rates of glacier melt due to global warming are giving way to increased water erosion. Swelling of rivers and increase in sea levels are adding to the factors exacerbating soil erosion.

8.7.2.5 Anthropogenic Activities

Anthropogenic activities, in fact, fuel all the natural operating factors and make the state of soil erosion grimmer. Dominant anthropogenic activities leading to soil erosion in enormous amounts and on a large scale are deforestation, agriculture, and road urbanization (Fig. 8.4). Worst hit by the intensifying human activities leading to soil erosion and consequent loss of soil fertility and production potential, our agricultural land is posing one of the most serious challenges to humanity.

8.8 Checking the Soil Erosion

Prevention of soil erosion is not only necessary but also imperative for us to usher in a sustainable and food-secure future. There could be many measures of checking the erosion that would vary according to the area, circumstances, and specific purpose. However, the most common and worth applying are as follows.

8.8.1 Controlling Livestock Grazing

As we have already seen overgrazing by livestock is one of the major reasons for soil erosion, controlling grazing wherever it is posing a problem is necessary. The number of livestock grazing a piece of land should be within the limits of the carrying capacity of the grazing land. Further, there should be rotational-cum-deferred type of grazing so that the regenerative capacity of the grazing land is not compromised. To maintain and enhance the productive potential and carrying capacity of grazing land, its sound ecological management would be promising for soil conservation.

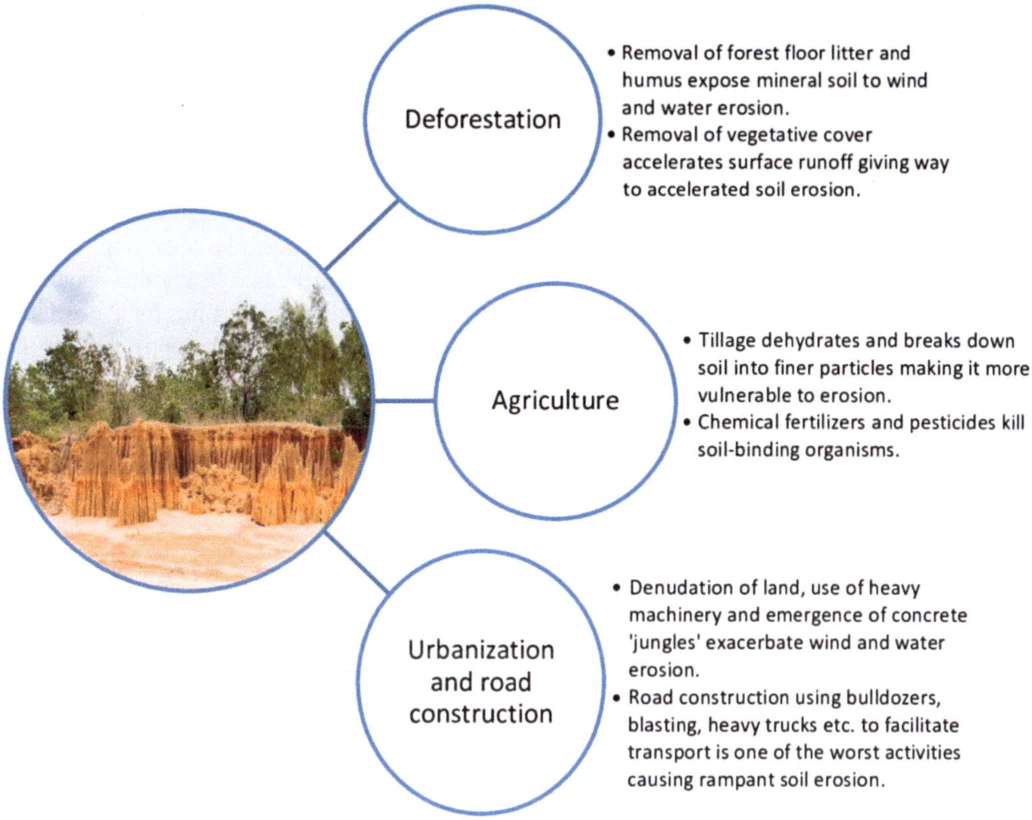

Fig. 8.4 Human-induced actions aggravating soil erosion

8.8.2 Plantation of Trees

As we know, natural forests are the best bet for soil, water, and biodiversity conservation, plantation of bare lands and forest areas with just thin stands of trees at a massive scale and transformation of such land areas into forests will be a sound ecological, sustainable, and most promising solution for soil erosion. Deforested lands, low-density forests, rangelands, grasslands, and grazing lands can also be made ecologically vibrant if their exploitation is ceased and are left for "convalescence" for their natural ecological regeneration.

Dense vegetation cover on land minimizes the force exerted by raindrops, thus preventing soil erosion. Plants' roots hold the soil particles together thus phenomenally contributing to soil conservation. Further, forest floor litter and humus that builds up due to organic matter shed by plants in the soil further reduce soil erosion. Looking at the phenomenal functions of vegetation cover following large-scale plantation and natural ecological generation of deforested lands and low-density forests, we can conclude that from the point of soil conservation and/or preventing soil erosion:

(i) Multistory forests are better than the single-story ones.
(ii) More dense forests are better than low-density ones.
(iii) Biodiversity-rich forests are better than the ones with a monoculture of trees.
(iv) The grazing lands or rangelands interspersed with trees and shrubs are better than those exclusively with grasses and herbaceous plants.
(v) Agroforestry systems are better than the cultivated lands growing only annual crops.

8.8.3 Growing Cover Crops

Growing cover crops such as cowpea, groundnut, and beans provides better protective cover for soil conservation. Cover crops cover most of the soil that many crop annuals cannot do. The role of minimizing soil erosion in the cultivated lands is in addition to many other ecological and socio-economic benefits ensued by cover crops.

8.8.4 Mulching

Mulching of plants helps prevent soil erosion by:

(i) Reducing evaporation and not letting drying the soil and thus preventing wind erosion
(ii) Maintaining soil moisture and reducing irrigation needs, thus minimizing water erosion
(iii) Enriching the soil with organic matter (in the case of organic mulch) and helping bind soil particles together, thus helping reduce wind and water erosion
(iv) Improving soil structure of clay soils and increasing moisture holding capacity of sandy soils
(v) Reducing soil compaction and decreasing surface runoff
(vi) Keeping the soils warmer in winter, thus breaking down the freezing-thawing cycle

8.8.5 Terracing of Slopes

Terrace farming is extensively practiced in the hill and mountain areas of Asian countries, especially in rice-producing countries, such as India, Nepal, China, Bhutan, Philippines, Indonesia, Vietnam, etc. Terracing of the slopes is an art of engineering in which "steps" known as terraces are carved out on hill and mountain slopes that are used for crop cultivation. Terracing of the steep slopes slows down surface runoff, thus protecting the soils getting washed away by the rains.

8.8.6 Contour Plowing

Contour plowing practiced in terrace farming involves the plowing of land across a slope following its elevation contour lines. This is a soil conservation technique in terrace farming. The contour plowing creates a water break that phenomenally helps reduce the formation of gullies and rills when there is a heavy downpour. The contour lines allow more time for the water to settle into the soils of the sloppy terrain, thus playing a crucial role in preventing or minimizing soil erosion.

8.9 Soil-Water-Biodiversity Depletion

Soil, water, and biodiversity are intertwined together. One is pushed into a perilous state by the destruction of the other. When one is ameliorated, the other ameliorates simultaneously. Each of the three provides a base for the other. The relationship between the three is cyclic, not linear.

Nutrient-rich soil which attains its fertility due to the presence of water in it in appropriate amounts nourishes, conserves, and sustains an enormous measure of biodiversity both within and above-soil surface. The biodiversity, in turn, nourishes soil with organic matter the plants photosynthesize. This matter feeds the populations of numerous detritivores. Nitrogen-fixing microorganisms prospering within soil help feed the plants the building base of the proteins. Soil- as well as above-soil biodiversity depends on water conservation which goes hand in hand with that of the soil.

When one out of the soil-water-biodiversity "trinity" debases, the other two simultaneously get debased. Erosion of the soil, which is the topic under discussion here, therefore, phenomenally contributes to the erosion of the other two. Should we want to conserve one, the other two have to be simultaneously conserved.

8.10 Desertification

Desertification has been defined variously. According to the Princeton University Dictionary, desertification is the process of fertile land transforming into a desert typically as a result of deforestation, drought, or improper/inappropriate agriculture. As desertification involves ecological, environmental, and anthropogenic dimensions, mere physical expansion of deserts on land is not what desertification precisely is. The widely recognized definition of desertification is presented by the United Nations Convention to Combat Desertification (UNCCD): "Land degradation in arid, semi-arid and dry sub-humid regions resulting from various factors, including climatic variations and human activities."

A significant proportion of the Earth's land part is desert. Throughout the geological history of the living planet, all the deserts, as the one shown in Fig. 8.5, have emerged as a result of natural factors. In recent decades, however, desertification has been largely accelerating due to human activities. Therefore, desertification is a matter of deep concern. If it keeps extending unabatedly on the land surface, desertification would put a question mark not only on our materialistic progress but also on our food security and survival.

Desertification, in a sense, is the creeping of death over land. The Hindi word for a desert is *Marusthal*, which means a dead land. Death may not be permanent. In agriculture, the land dies almost every year or after every crop season. When the bulk of the nutrients is squeezed out following every crop season, the soil of a field attains "temporary" death. By feeding with nutrients in adequate amounts, the soil again gets a lease of life with the onset of a new crop season. Thus, the game of desertification of agricultural fields is played by human beings. Our modern agriculture depends not on nature's nutrient cycles, but on chemical fertilizers available in the market. Agriculture which humanity depends upon is in the cruel clutches of desertification almost everywhere. In the long run, despite huge amounts of external inputs applied in agriculture, death would creep on cultivated land irreversibly.

Desertification is not a spontaneous process. It is a dynamic process and is visible only after a pretty long period. Desertification is the opposite of ecological succession, but dynamic succession. Succession dynamically advances towards a state of ecological climax, desertification towards disclimax (Fig. 8.6). Desertification is, however, faster than succession. It takes hundreds of years for a bare land to successionally attain ecological climax. However, the state of disclimax may be reached in a few years if human activities are intensively ecocidal.

Some facts and figures about desertification, deserts, and desert inhabitants in the drylands of

Fig. 8.5 Desert signifies the encroachment of lifelessness on the landscape

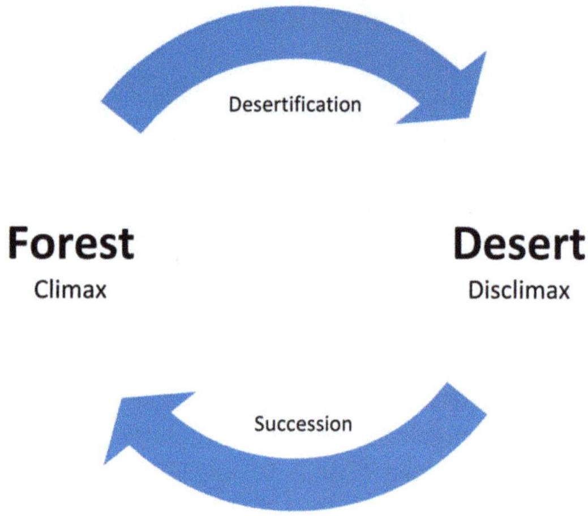

Fig. 8.6 Succession versus desertification: the processes destined to ecological climax and arid land, respectively

the world are as follows (Johnson et al. 2006, Bauer 2007, World Bank 2009, UNCCD 2017):

(i) The most affected areas by desertification are drylands that alone occupy about 40% land area of the Earth.
(ii) The drylands are the habitats supporting more than 2 billion people across the world.
(iii) Out of the world's drylands, an area ranging between 6 and 12 million km^2 is affected by desertification.
(iv) About 10–20% of the drylands are already degraded.
(v) Among all the dryland inhabitants, 1–6% live in desertified areas
(vi) Continuing desertification threatens more than a billion people living in the drylands across the world.

The main hot deserts of the world are (i) the Sahara Desert in North Africa, extended over 9.100,000 km^2, the largest in the world; (ii) the Great Victoria Desert in south and western Australia, extended over 647,000 km^2; (iii) the Arabian Desert in Arabian Peninsula spread over 2,600,000 km^2; (iv) the Syrian Desert in Middle east spread over 49,000 km^2; and (v) the Kalahari Desert in southwestern Africa, covering an area of 570,000 km^2.

Death Valley, the hottest place on Earth, is located in the Mojave Desert bordering the Great Basin Desert between California and Nevada in North America. Average temperatures in Death Valley reach 46.7 °C. The hottest-ever atmospheric temperature recorded in the Hot Valley is 56.7 °C (134 °F) on July 10, 1913, at Greenland Ranch, now known as Furnace Creek, which was the highest ever on Earth. The highest ground surface temperature ever recorded in Death Valley was 93.9 °C (201 °F) on July 15, 1972, at Furnace Creek, which was the highest ever on Earth.

The Namib Desert in Namibia and Southern Angola is distinctively known for one of the most peculiar, amazing, fascinating, unique, and wonderful plant species, *Welwitschia mirabilis*, with the longest life span in the Plant Kingdom: between 400 and 1500 years. This weird plant was "discovered" and named after Austrian botanist Dr. Friedrich Welwitsch in Namib Desert in 1859.

Outside the polar region, the Atacama Desert is the driest on the planet. The average annual rainfall is just 15 mm and some of the places in the desert receive only 1–3 mm of rain in a year. Even Death Valley in the Namib Desert is 50 times less arid than the Atacama Desert of Chile. The Atacama Desert soil is also compared with that of Mars.

Contrary to the Atacama Desert, the Sonoran Desert is the wettest one. Spread in Northwestern Mexico and Southwestern US, the Sonoran Desert experiences two wet seasons: from December to March and July to September due to regional summer monsoon. It is why among all the hot deserts on Earth, the Sonoran Desert is the wettest one.

Marathon des Sables, or MdS – that is, Marathon of the Sands – is one of the prominent events organized in the Sahara Desert. A six-day 251-km-long ultramarathon, the MdS is celebrated every year in southern Morocco since 1986 and is regarded as the toughest foot race witnessed on the planet. The credit for beginning this annual event goes to French concert promoter Patrick Bauer who himself covered on foot 350 km of the Sahara Desert in 12 days in 1984.

8.10.1 Cold Deserts

So far our discussion has been focused on hot deserts that are located in tropical or subtropical regions/areas of the world. There is another conspicuous category of deserts known as cold deserts. These are extremely cold because of the very low temperature prevailing throughout the year. The temperature in these deserts generally ranges between −20 and 40 Celsius, which might come down to −50 °C and in some places even less than that. The lowest temperature at ground level ever recorded on Earth was −89.2 °C at the Vostok Station in Antarctica, on July 21, 1983, according to the World Meteorological Organization (WMO).

Winters are long (9 months or more) and summers are very short (3 months or less). Water available in directly usable liquid form is extremely scarce. Precipitation occurs in the form of snow. Apart from extreme types of cold deserts like Antarctica, the Arctic, and the high peaks of some mountain ranges, there are other comparatively less cold deserts. These are home to some algae, lichens, ephemeral flowering plants, and other herbaceous plants.

Plants in these habitats are specially adapted to harsh climatic conditions and complete their life cycle in a short period of a few weeks when the environmental conditions are favorable for their specific physiological needs (Singh 2020). Some migratory birds and mammals are also found in the cold deserts.

The geographical expanse of the cold deserts on Earth is much larger than those of the hot deserts. For example, Antarctica embracing polar ice and tundra on the South Pole of the Earth is the largest desert on Earth, followed by the Arctic desert on the North Pole interspersed with tundra-type vegetation. The geographical area of both deserts is approximately 28,000,000 km^2. The driest area on Earth – the Valley Area – where there has been no rain for about 2 million years lies in Antarctica.

The major cold deserts of the world are (i) Antarctica (polar ice and tundra) spread over 14,000,000 km^2, the world's largest cold desert; (ii) Arctic (polar ice and tundra) extended over 13,985,000 km^2; (iii) Gobi desert in Mongolia and China covering an area of 1,300,000 km^2; (iv) Great Basin in Western United States spreading over 411,000 km^2; and (v) Atacama Desert on the coast of Peru and Chile covering an area of 140,000 km^2, one of the driest areas on Earth.

8.10.2 Effects of Desertification

Desertification results in reduced land productivity. There can be a 10–15% decline in crop productivity when desertification is only moderate. Under severe desertification, crop decline may be more than 50%. Thus, desertification is a danger to food security. People living in desert areas are faced with the risk of utter poverty and food insecurity.

In uncultivated areas, like in forests and rangelands, desertification reduces vegetation density, alters vegetation composition, declines biodiversity, lowers the water table, causes soil salinity, breaks down food chains, hampers nutrient cycles, and reduces primary productivity. As a result of such changes in the ecosystems reeling under desertification, the weather cycle is badly influenced and climate changes are exacerbated.

An increase of 25% has been recorded in global dust emissions between the late nineteenth century and our present times (Stanelle et al. 2014). Increases in the amounts of dust and loose sand, frequency of dust storms, air pollution, open water pollution, respiratory diseases, etc. have gone hand in hand with the intensification of desertification.

When dust particles enter the atmosphere due to dust storms, they scatter incoming solar radiation as a result of which there could be deforming and shortening of the life span of clouds, leading to a substantial decrease in precipitation. The ecological, climatic, and socioeconomic changes brought about by desertification are interrelated (Fig. 8.7).

8.10.3 Causes of Desertification

The causes of desertification, soil erosion, and land degradation are common and interlinked. The natural mechanisms that result in soil erosion and land degradation also lead to desertification of the land. As mentioned earlier in this chapter, all the deserts in the geological history of the Earth had emerged thanks to natural factors. In our contemporary world, however, desertification has anthropogenic roots. This is the only deep concern we need to be focused on and chalk out all workable strategies to combat, curb, and reverse desertification.

Removal of vegetation from the surface of the land, or deforestation, is the dominant reason for triggering desertification. Overgrazing by livestock is another reason for vegetation removal and land degradation. Deforestation, without a shadow of a doubt, has become an essential condition for socioeconomic progress in our times. Agriculture, road construction, urbanization, industrialization, and almost all socioeconomic developments cost forests directly or indirectly. All these activities without which our contemporary world cannot do cannot go without devouring the forests of the Earth.

Agriculture, without which our survival would become questionable, is another major reason to

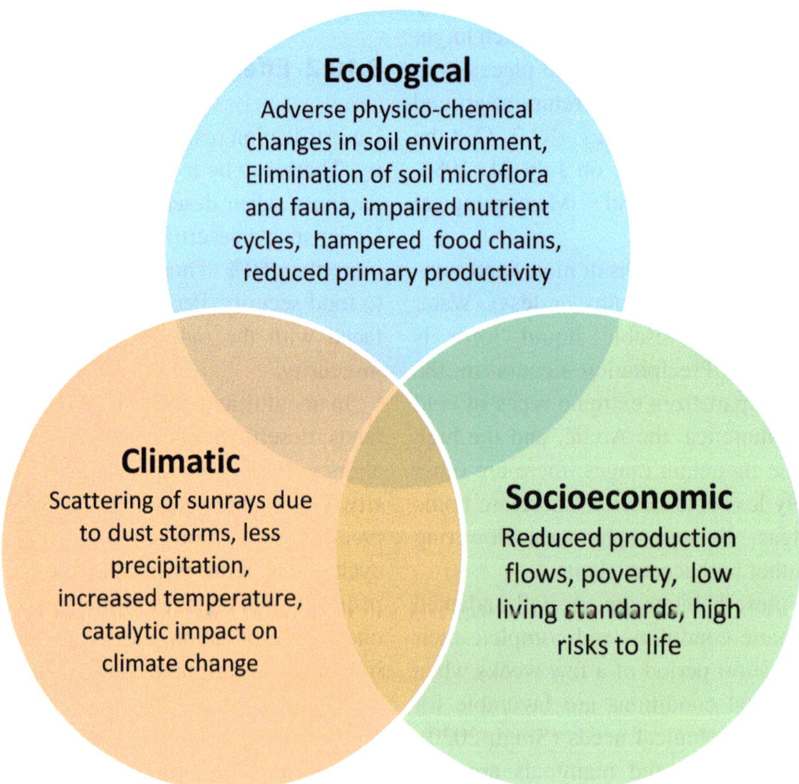

Fig. 8.7 Ecological, climatic, and socioeconomic transformations resulting from desertification

trigger desertification. Delinked from forests and without farming systems, modern agriculture depends on a set of inputs that invariably exacerbate desertification. These are chemical fertilizers and a variety of pesticides that wipe out soil microflora and fauna and change soil structure and composition. Intensive tillage of the soils, heavy machinery and equipment, trend of monocultures, overexploitation of soils, pollution of soils, etc. all culminate into the processes that turn the land into a desert, slowly but steadily.

8.10.4 Measures to Combat Desertification

Desertification takes only the upper fertile layer of the land in its fold. Deep below, the land is pregnant with nutrients. If those nutrients are not available on the land surface, i.e., in the soil, in required amounts, survival or growth of the plants and regeneration of vegetation would be substantially hindered. There could be nutrients in plentiful amounts, but due to certain factors, like a high proportion of salts in the soil, i.e., soil salinity, water potential would be reduced to the extent that soil water and nutrients dissolved in it would not be available for the plants and vegetation would not grow on such soils. Thus, unless and until some extremely adverse factors, like extreme temperature, are not operating in a region or a geographical area, reversal of desertification and/or mitigation of desert is not an impossible task. Cold deserts of Antarctica and the Arctic must prevail and be preserved as they are, as they are of vital significance for the planet Earth and are of critical value for the life within the biosphere. We need to apply our strategies of mitigation in hot deserts of the tropical/subtropical climate zones or of reversal of desertification in these regions of the world. These are the real challenges for us.

Soil fertility, in essence, is a function of soil microbial activities and is phenomenally rooted in photosynthesis. Organic content in the biosphere is all on account of photosynthesis (and chemosynthesis). Plants manufacture organic matter through photosynthesis. The organic matter reaches soil through plant roots, leaves, and other parts (also through microbial biomass and partly through dead and decomposing soil animals). Soil organisms feed on this organic matter (detritus) and play their role in nitrogen fixation (nitrogen-fixing bacteria), biodegradation, and mineralization.

Thus, if we boost photosynthesis, on temporal and spatial scales – i.e., its efficiency as well as the overall amount of synthesized matter – problems of desertification would be solved over time. It must be made clear that photosynthesis is merely a key phenomenon and a principle to reverse and mitigate desertification. A quantum boost of photosynthesis will be possible only through green plants. Therefore, we need to manage the land-based systems that accommodate, nourish, and sustain green plants: trees, shrubs, and herbs. To operationalize and manage to boost of photosynthesis:

(i) Protect, regenerate, and conserve existing forest areas
(ii) Take up afforestation program at a massive scale
(iii) Plant trees and cover as much land area with vegetation as possible
(iv) Develop farming systems, manage soil fertility, enhance agrobiodiversity, and operationalize principles of agroecology and sustainability in agriculture
(v) Reclaim deserts and implement technologies that could transform deserts into fertile lands
(vi) Manage water resources in such a way that the conserved water is brought in use into the desertification combating processes

The process of desertification reversal and mitigation of desertification and reclamation of deserts is slow but possible. This is not just a task of some organizations in a few countries. Desertification is a global issue. All the countries of the world need to come together to resolve desertification-related issues and solve one of the burning global problems on a priority basis. The only internationally binding framework aimed at addressing desertification problems is the United Nations Convention to Combat Desertification

(UNCCD). Emanating from Rio Conference Agenda 21, the convention was adopted in Paris, France, on June 17, 1996, and brought into effect in December 1996. However, not much progress towards the realization of its objects and goals is visible. The global community has to build up the willpower necessary to implement workable strategies for combating desertification.

8.11 Summary

Land and soil are the natural resources on which the future welfare of our world depends. Land resources include all those features and processes of the land which may be used to fulfill certain human needs. The surface area of the Earth is approximately 5.1×10^8 km^2. The land area of the Earth is approximately 149 million km^2. The land accommodates all other terrestrial resources: forests, grasslands, wetlands, biodiversity, wildlife, agricultural land, livestock, minerals, soil, human resources, and even surface waters. The components of the natural land unit – that is, physical, biological, environmental, infrastructural, social, and economic – include numerous, varied, and complex resources that humanity overwhelmingly depends upon. Land use pattern is not uniform throughout the globe. As land is put to different uses, it depends on the cultural ethos, planning, programs, and socioeconomic targets of a country and how a governing system puts its land area into different uses. About 30% of the land is under forest cover, 26% pasture land, 11% cropland, and 33% under snow, ice, rocks, deserts, and inhospitable areas like tundra. Land use is not a fixed pattern. It not only varies spatially but undergoes changes from year to year and from time to time in the same year.

Land degradation is a process in which the value of the biophysical environment is affected by a combination of human-induced processes on the land. Land degradation is generally related to human activities, not to natural factors. About a quarter of the Earth's ice-free land area is subject to human-induced degradation. Land degradation is a global problem. This problem is largely attributable to erroneous anthropogenic activities, such as deforestation, intensive tillage, raising monocultures, land use change, mining, quarrying, etc.

Climate change is emerging as one of the causes of various environmental disruptions, including land degradation. Soil is an ecosystem harboring numerous organisms in reciprocal relationships with each other and with their physical environment. The soil ecosystem is the largest on the land of the Earth as well as the most complex one. It is also the most active and yet the most vulnerable ecosystem. Soil scientists generally categorize soil into four discrete horizons, viz., O, A, B, and C, following changes with depth, all varying in composition. The soil structure and composition – and, therefore, its functioning – are all in a state of flux due to several operating factors. The diversity of soil (or pedodiversity) is the manifestation of the diverse natural features of the Earth. Geographical positioning, climatic region, geologic features, environmental factors, seasons of the year, hydrologic factors, and vegetation type are the major factors influencing soil types and soil ecology. Human activities also influence soil characteristics and soil ecology to a great extent. Soils of the world have been divided into six major categories: sandy, silty, clay, loamy, peaty, and chalky. These soil types are made up of three major soil types, viz., sandy, silty, and clay. Soil is the richest habitat in terms of the measure of its biodiversity. Land use practices are the major determinants of soil biodiversity.

Soil erosion is the displacement, loss, or removal of the upper fertile layer of the soil and is a form of soil degradation. All other causes of soil erosion other than anthropogenic are natural ones but are significantly influenced by anthropogenic activities. The factors affecting soil erosion include topography, vegetation, soil structure and composition, and climate change. Prevention of soil erosion is not only necessary but also imperative for us to usher in a sustainable and food-secure future. There could be many measures of checking the erosion that would vary according to the area, circumstances, and specific purpose. Controlling grazing by livestock, plantation of trees (afforestation and reforestation), growing cover crops, mulching of croplands, terracing of sloping lands, contour plowing, etc. are the main measures to prevent soil erosion. There exist soil-water-biodiversity interrelationships in nature. Conservation of the one influences the conservation of the others.

Desertification is the process of fertile land transforming into a desert typically as a result of deforestation, drought, or improper/inappropriate agriculture. As desertification involves ecological, environmental, and anthropogenic dimensions, mere physical expansion of deserts on land is not what desertification precisely is. Desertification is not a spontaneous process. It is a dynamic process and is visible only after a pretty long period. Desertification results in reduced land productivity. The causes of desertification, soil erosion, and land degradation are common and interlinked. The natural mechanisms that result in soil erosion and land degradation also lead to desertification of the land. Remedial measures for preventing land degradation, soil erosion, and desertification are also common.

8.12 Exercises

8.12.1 Multiple-Choice Questions

1. Which of the following planets of the solar family is smaller than the Earth?
   (a) Mercury
   (b) Venus
   (c) Mars
   (d) All of the above
2. Which of the following planets of the solar family is larger than the Earth?
   (a) Jupiter
   (b) Saturn
   (c) Neptune
   (d) All of the above
3. According to modern estimates, the surface area of the Earth is approximately
   (a) 5.1×10^8 km^2
   (b) 5.1×10^6 km^2
   (c) 5.1×10^4 km^2
   (d) 5.1×10^2 km^2
4. The land area of the Earth is approximately
   (a) 149 million km^2
   (b) 139 million km^2
   (c) 129 million km^2
   (d) 119 million km^2
5. The land use pattern of India has most of the land area devoted to
   (a) urban area
   (b) agriculture
   (c) pasture land
   (d) dense forests
6. According to an estimate, approximately what proportion of the Earth's land is degraded?
   (a) 10%
   (b) 20%
   (c) 40%
   (d) 80%
7. Which of the following agricultural practices is promising for minimum soil erosion?
   (a) Monocultures
   (b) Conventional tillage
   (c) No-till
   (d) Contour tillage
8. Which of the following statements is correct?
   (a) Soil is an ecosystem.
   (b) The biomass of the soil microorganisms exceeds that of all human beings, elephants, and whales of the world put together.
   (c) A teaspoon of soil from a native grassland ecosystem contains several kilometer-long fungi.
   (d) All of the above.
9. The uppermost layer in the soil profile is
   (a) A-Horizon
   (b) B-Horizon
   (c) C-Horizon
   (d) O-Horizon
10. What horizon in the soil profile is also known as "organic?"
    (a) A-Horizon
    (b) B-Horizon
    (c) C-Horizon
    (d) O-Horizon

11. What type of soils has the largest spaces between its particles and the lowest water-holding capacity?
    (a) Sandy
    (b) Clay
    (c) Silty
    (d) Loam
12. This soil is said to be "perfect" in itself with the balanced proportion of all the three major soil types, i.e., sand 40%, silt 40%, and clay 20%, along with humus. This is
    (a) Loam soil
    (b) Silty soil
    (c) Clay soil
    (d) Sandy soil
13. Which of the following does cause soil erosion?
    (a) Wind
    (b) Water
    (c) Glaciers
    (d) All of the above
14. Plowing of land across a slope following its elevation contour lines is known as
    (a) slope plowing
    (b) contour plowing
    (c) regenerative plowing
    (d) dynamic farming
15. Pedodiversity includes
    (a) bacteria
    (b) nematodes
    (c) burrowing mammals
    (d) All of the above
16. What American desert is the Hot Valley located in?
    (a) Mojave Desert
    (b) Great Basin Desert
    (c) Colorado Desert
    (d) Sonoran Desert
17. The driest nonpolar desert in the world is
    (a) Gobi Desert
    (b) Atacama Desert
    (c) Libyan Desert
    (d) Sahara Desert
18. Where is the Marathon des Sables held?
    (a) Kalahari Desert
    (b) Thar Desert
    (c) Sahara Desert
    (d) Arabian Desert
19. The only internationally binding framework aimed at addressing desertification problems is the
    (a) FAO
    (b) IPCC
    (c) UNCCD
    (d) UNFCCC
20. UNCCD headquarters are located in
    (a) Paris, France
    (b) Bonn, Germany
    (c) Nairobi, Kenya
    (d) Rio de Janeiro, Brazil

Answers: 1-d, 2-d, 3-a, 4-a, 5-b, 6-c, 7-c, 8-d, 9-d, 10-d, 11-a, 12-a, 13-d, 14-b, 15-d, 16-a, 17-b, 18-c, 19-c, 20-b

8.12.2 Short-Answer Questions

1. Name the resources dependent on land.
2. What do you mean by land use?
3. Which sector uses most of the land in India?
4. What do you mean by land degradation?
5. Why should soil be regarded as an ecosystem?
6. Name the various horizons of the soil profile.
7. Name the five types of soil found in the world.
8. How are soil, water, and biodiversity conservation interrelated?
9. Name five hot deserts of the world.
10. The expanse of cold deserts on Earth is larger than those of hot deserts. How?

8.12.3 Long-Answer Questions

1. What is land degradation? Describe various causes of land degradation.
2. Soil is the foundation of all terrestrial life. Respond in tune with your perspective.
3. What are the physical factors and main causes of soil erosion? What are the measures to check soil erosion?

4. What is desertification? What are the various causes of desertification? What are your suggestions about combating desertification?
5. Write short notes on the following:
   (a) Soil as an ecosystem
   (b) Climate change and land degradation
   (c) Cold deserts

References

Bauer S (2007) Desertification. In: Thai KV, Rahm D, Coggburn JD (eds) Handbook of globalization and the environment. Taylor and Francis, Boca Raton, pp 77–94

Blanco H, Lal R (2010) Water erosion. In: Principles of soil conservation and management. Springer, p 29

Conacher A, Conacher J (1995) Rural land degradation in Australia. Oxford University Press, South Melbourne, p 2

FAO (2015) Soils and biodiversity: soils host a quarter of our Planet's biodiversity – 2015 international year of soils. FAO, Rome

IPCC (2019) Summary for policymakers. In: Shukla PR, Skea J, Buendia EC, Masson-Delmotte V, Pörtner H-O, Roberts DC, Zhai P, Slade R, Connors S, van Diemen R, Ferrat M, Haughey E, Luz S, Neogi S, Pathak M, Petzold J, Pereira JP, Vyas P, Huntley E, Kissick K, Belkacemi M, Malley J (eds) Climate change and land: An IPCC special report on climate change, desertification, land degradation, sustainable land management, food security, and greenhouse gas fluxes in terrestrial ecosystems. IPCC

Jenny H (1980) The soil resource. Springer-Verlag, New York

Johnson DL, Ambrose SH, Bassett TJ, Bowen ML, Crummey DE, Isaacson JS, Johnson DN, Lamb P, Saul M, Winter-Nelson AE (1997) Meanings of environmental terms. J Environ Qual 26:581–589

Johnson PM, Maryland K, Paquin M (2006) The United Nations convention to combat desertification in global sustainable development governance. In: Johnson PM, Maryland K, Paquin M (eds) Governing global desertification: linking environmental degradation, poverty and participation. Ashgate Publishing Ltd, Hampshire, pp 1–10

Pruski FF, Nearing MA (2002) Runoff and soil loss responses to changes in precipitation: a computer simulation study. J Soil Water Conserv 57(1):7–16

Ritchie H, Roser M (2019). Land use published online as OurWorldInData.org. Retrieved from: https://ourworldindata.org/land-use

Sample I (2007). Global food crisis looms as climate change and population growth strip fertile land. The Guardian, August 31, 2007

Singh V (2020) Environmental plant physiology: botanical strategies for a climate smart planet. Taylor and Francis (CRC Press), Boca Raton, 216 pp

Stanelle T, Bey I, Raddatz T, Reick C, Tegen I (2014) Anthropogenically induced changes in twentieth century mineral dust burden and the associated impact on radiative forcing. J Geophys Res Atmos 119(23):13526–13546

UNCCD (2017) UNCCD: impact and role of drylands. UNCCD, Bonn

Vink APA (1975) Land resources. In: Land use in advancing agriculture: advanced series in agricultural sciences, vol 1. Springer, Berlin

World Bank (2009) Gender in agriculture sourcebook. World Bank Publications, Washington, DC, p 545

9 Forest Resources

Forests are the cradle of terrestrial life on Earth. They constitute a heterogeneous and complex community dominated by trees. There was a time when most of the Earth's land was covered with natural forests and we evolved knowledge, wisdom, and all cultural ethos and social virtues amid *Aranya* culture – the forest culture. Extraordinarily rich treasures of wisdom, our Vedas, Upanishads, and epics of Ramayana and Mahabharata have been composed amid forests. Forests are not just the sources of biophysical materials of economic importance. They are the ecosystems where the roots of life are nurtured, where the canopies of our hope expand, and where our sustainability and happiness are hidden. Forests, in essence, serve to nurture the processes of the evolution of life.

Environmental conditions and climate crises are getting worse with the gradual disappearance of forests in our times. There are many factors responsible for the shrinkage of forests. Almost all socioeconomic activities of human beings cost forests directly or indirectly. The human factor is, thus, the dominant – rather exclusive – factor giving way to the disappearance of the forests.

9.1 Forest Ecosystem Functions

Forest ecosystems perform the functions that are vital for the sustenance of life on Earth. All the terrestrial species – lower as well as higher forms of life – have evolved in the environment created by forests. The forests created a kind of environment that showers benevolence on all the terrestrial species. While materialistic economists focus on the products obtainable from the forests that have high market value, the functions vital for life and the life processes they perform are often neglected. Forest functions can be divided into three categories, viz., (i) productive functions, (ii) protective functions, and (iii) regulative functions. Of course, the three functions are inseparable from each other. In the ecological analysis of forest functions, we must look at and appreciate them with a perspective of ecological integrity (Fig. 9.1).

9.1.1 Productive Functions

Productive functions include the products of socioeconomic importance that the forests provide. The most common products extractable from a forest are timber, fuel wood, fodder, foods (fruits, edible buds and flowers, underground stems, edible leaves, pods, seeds, mushrooms, honey, etc.), and many other compounds, such as essential oils, alkaloids, latex, Ayurvedic medicines, etc.

The productivity of a natural forest is many times higher than that of cultivated annual crops. A variety of the foods obtainable from natural forests are more nutritive than the cultivated ones. Forests provide a livelihood for a very large

V. Singh, *Textbook of Environment and Ecology*, https://doi.org/10.1007/978-981-99-8846-4_9

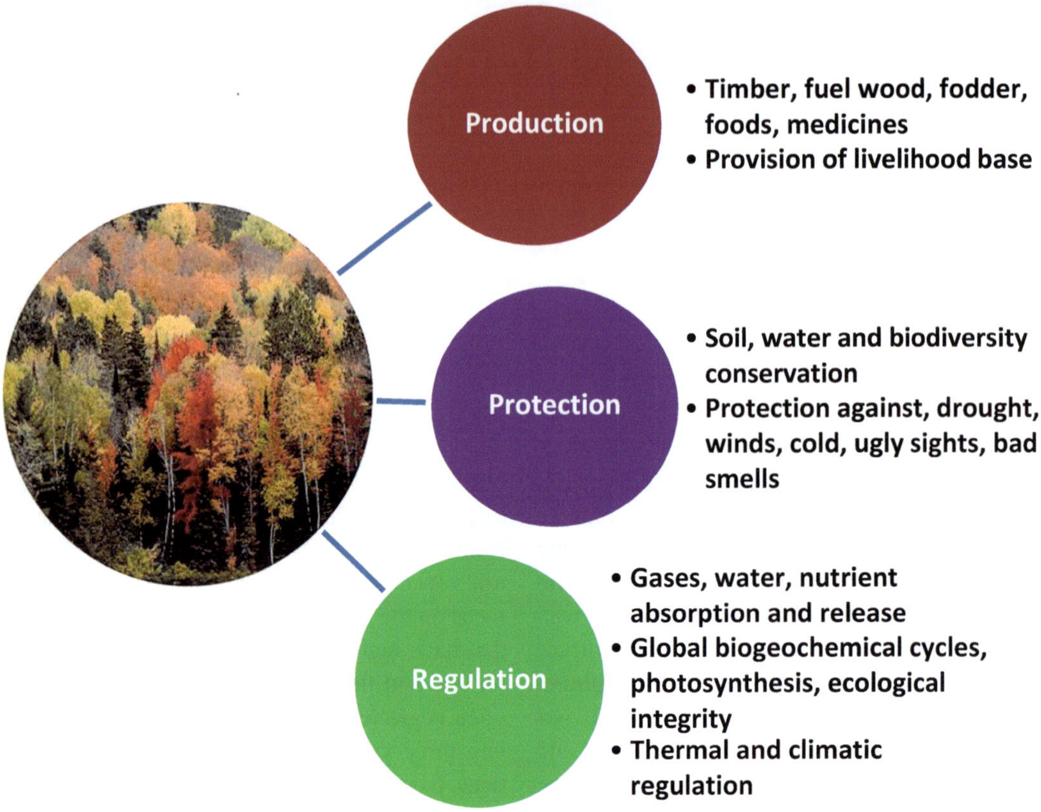

Fig. 9.1 Functions of a forest ecosystem

number of people, especially the tribal communities and other forest dwellers.

9.1.2 Protective Functions

Conservation of soil, water, and biodiversity are the forest functions that impart a high degree of ecosystem stability holding a forest in its ecological climax. A forest protects soil erosion, enriches the soil with plant nutrients, and maintains soil fertility. It slows down water runoff, controls water flow into rivers and streams, and filters water into the ground, recharging the groundwater resources. A natural forest provides shelter to a variety of plants, animals, and microorganisms and thus serves as a rich repository of nature's biodiversity.

The protective functions of a forest have a positive bearing on agriculture and other land-based livelihood systems. Further, a forest plays a crucial role in the prevention of drought and protection against wind, cold, noise, intensive radiation, ugly sights, and bad smells.

9.1.3 Regulative Functions

A forest maintains regulation of many environmental factors. Absorption and release of gases (especially of CO2 and O2), water, mineral elements, and radiation energy are the various regulative functions a forest performs. Forests of the Earth play a crucial role in global biogeochemical cycles as well as in the water cycle. Through photosynthesis, the plants absorb atmospheric carbon and soil water and convert solar energy into biochemical energy. This is the most critical phenomenon through which the forests serve to fulfill the energy and nutrient needs of the entire animal kingdom.

The forests of the planet play a critical role in regulating atmospheric temperature, creating a microclimate, and contributing to building up a macroclimate vital for the biosphere to proliferate with innumerable varieties of organisms.

9.2 Aesthetic and Spiritual Values of Forests

The cultural values of a forest are often rated lower than economic and ecological ones. However, among the social values of the forests, aesthetic values are rated of relatively high importance (Lim et al. 2015). A society, as Everard et al. (2016) unveil a fact, operates within an "ethical envelope" framing its norms and expectations. One of the basic aspects of life is woven around beauty. Beauty is the very gist of the evolution of life on Earth. As a living system, for example, a forest, advances towards the climax, its beauty goes on intensifying. According to his new theory of evolution, Singh (2019) draws interconnectedness between ecological climax, beauty, and sustainability. An ecologically more stable ecosystem is more beautiful as well as more sustainable.

Beauty is an unending desire of the human soul. Beauty nurtures all our desires and nourishes the processes of our physical, intellectual, psychological, emotional, ethical, and aesthetic development. Without beauty, we shrink. All the ills of this world emerge from the societies and belief systems that give no place for plants, trees, and forests. Beauty is an inexhaustible source of creativity. All our literature, including our epics, best forms of poetry, dramas, and philosophy, have been created amid the beauty of *Aranya* (forests).

Spirituality, like beauty, is an inherent desire of our soul. This virtue of human persons ensures personal, social, moral, and ethical integrity as well as peace of mind, body, and soul. Spiritual values contribute to extending peace across the world. Forests are immaculate sources of spirituality, tranquility, and peace. Where there are forests, there is beauty. Forests inherently spell out beauty and spirituality. A tree, a forest, a landscape blossoming with greenery and very much thought of visiting a place laden with forests, like a national park, a sanctuary, or a biosphere reserve, somewhere touch us at our emotional, aesthetic, and spiritual levels.

The aesthetic and spiritual values of a forest are outside the concept of what is referred to as "ecosystem services," as Cooper et al. (2016) would claim. "Aesthetic and spiritual understandings of the value of nature lead people to develop moral responsibilities towards nature." Looking at an expanse of forest across harvests a sense of grandeur and awe. How aesthetically pleasing is all the unique and amazing biodiversity of plants and animals that a natural forest embraces! Feel the gestalt effect of a forest with colorful biodiversity of organisms through which flows a river with unceasing music of running water.

9.3 Forest Distribution

Spread over about one-third of the land area of the Earth, the forests comprise most of the biomass, accounting for about 90%. Nearly 50% of the global forests constitute tropical forests. According to FAO's definition of forests and woodlands, i.e., land under natural or planted stands of trees, whether productive or not, including land from which forests have been cleared but will be restored in the future, about 30% of the land area on Earth is under forests and woodlands. Nearly 26% of the land is under pastures and 11% includes croplands. The other type of land, like tundra, bare rock, deserts, snow, ice, etc., constitutes the largest proportion of the land, 30% (Fig. 9.2).

Deforestation at an alarming rate is one of the gravest tragedies of our times. Since 1990, the planet has lost forest cover from 1.3 million km^2, i.e., from an area larger than that of South Africa (Rooney 2019). However, many glaring examples are revealing that many countries are protecting their forest wealth to an appreciably great extent. The CEOWORLD Magazine brings to the fore a list of such countries. According to this magazine (Miller 2019), Suriname of South America is the first country

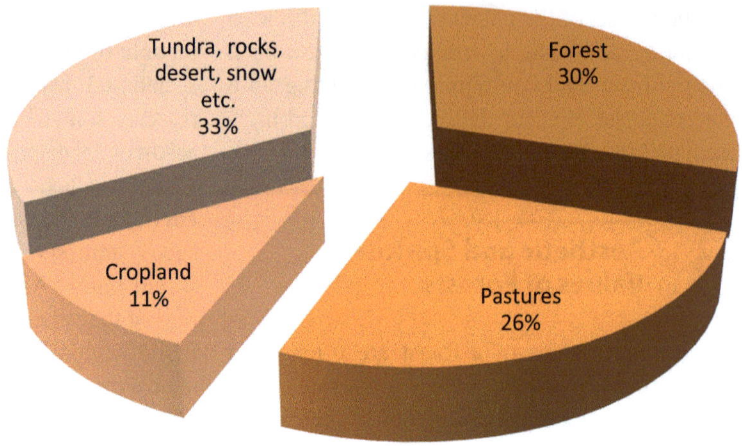

Fig. 9.2 An overview of global land use

with forest cover on as much as 98.3% of its land area. Federated States of Micronesia (91.9%), Gabon (90.0%), Seychelles (88.4%), and Palau (87.6%) are in second, third, fourth, and fifth place, respectively. Next in order are American Samoa, Guyana, and Lao PDR having 87.5%, 83.9%, and 82.1% of their land area covered with forest, respectively. Solomon Islands with 77.9% and Papua New Guinea with 74.1 land area under forest stand at next positions. These countries with an exceptionally large percentage of forests on their land are exemplars for the rest of the world.

9.4 Deforestation

All kinds of human activities about socioeconomic gains eventually inflict pressure on our forest resources. As a result, deforestation has emerged as the most serious disease our environment is infested with. Pressure on temperate forests is not as much as on tropical ones. However, whereas the world's temperate forests have lost only 1% of the cover, tropical forests have lost more than 40% of the cover. About 10 million ha of land is robbed of its forests every year. Stunningly rich in their biodiversity, the Amazon forests are disappearing at unprecedented rates. Unabated shrinkage of the natural forests is posing a threat to our survival and sustainability. The consequences of deforestation are formidable.

9.4.1 Causes of Deforestation

There are multiple factors leading to deforestation. The causes might vary from region to region. However, insatiable human greed inherent in all socioeconomic activities is the prime reason for deforestation.

9.4.1.1 Population Explosion

Ever-increasing human population is continuously engulfing forests for settlements, agriculture, industries, roads, railways, and for so many other reasons. Population explosion is particularly characteristic of poor, developing, and Islamic countries. These are also the countries where forests are hardly given priority.

9.4.1.2 Shifting Cultivation

Shifting cultivation or slash-and-burn farming or *Jhoom* is a traditional practice of agroforestry in the northeastern part of India. People burn the forests at a place and cultivate food crops there for a few years until the soil fertility is exhausted. They repeat the same practice in another forest area. This practice might be fulfilling local people's livelihood demands but at very big environmental costs.

9.4.1.3 Industrial Demand for Wood

Wood extractable from the forests is needed for furniture, boxes, plywood, house construction, paper, matchboxes, packaging material, etc. This requirement is on the increase with the expansion of industries consuming huge quantities of wood. The demand has to be met by cutting down trees and through large-scale deforestation.

9.4.1.4 Road Construction

Roads are of absolute importance in our contemporary world. Economic systems would collapse and life would come to a standstill if there are no roads. A very large network of roads is one of the major characteristics of our contemporary world that seldom comes into discussion. However, road construction processes consume extensive areas of forests, fertile land, and cultivated land. In fragile mountains, road construction is exorbitantly expensive that requires the sacrifice of forests that destabilize mountain slopes and create ecological havoc in terms of catastrophic landslides, rampant soil erosion, loss of biodiversity, drying up of natural water sources, and the like.

9.4.1.5 Mining Operations

Mining operations are extremely devastating. Mining of limestone, coal, iron ores, mica, bauxite, copper pyrites, etc. renders large areas devoid of vegetation. Desertification creeps on the land where mining operations are carried out indiscriminately. Blasting and bulldozing at mining fields scare wildlife of distant areas. Regeneration of forests becomes difficult, rather impossible, in many areas abandoned after mining activities.

9.4.1.6 Mega Dams and Hydroelectric Projects

Artificial reservoirs as a result of damming a river for implementing a hydroelectric project inundate forests in valley areas. Forests are often cleared before the land area is inundated. Once the reservoir comes into being, the drowned herbs, shrubs, trees, and forests cannot regenerate. High dams and hydroelectric projects have destroyed forests from millions of hectares so far. The most recent example is the 260.5-m-high Tehri Dam, which has created an artificial reservoir of more than 50 km^2 destroying an equal area of vegetation in the Tehri Valley of the Garhwal region of the Himalayas.

9.4.1.7 Forest Fires

Fires are the worst enemy of forests. Every year millions of hectares of forests are destroyed by fires. The forests are set on fire sometimes deliberately and sometimes accidentally. Fire destroys everything there is in a forest: trees, shrubs, ground vegetation, saplings, seeds, wild animals, and even the humus and microbial wealth of the soil. Destruction of forests by fires in the Himalayas, Australia, and the Amazon is fresh in our memories. The incidence of fires is more frequent in monoculture rather than in heterogeneous forests.

9.4.1.8 Environmental Factors

Pests, including a variety of insects, fungi, and other organisms, and adverse weather conditions, such as frost, extreme temperatures, storms, etc., also damage the forests.

9.4.2 Effects of Deforestation

Deforestation does not lead only to the disappearance of a forest but has triggering effects (Fig. 9.3), leading ultimately to an unproductive desert, adding unfavorable environmental conditions to all parts of the biosphere – the atmosphere, the hydrosphere, and the lithosphere. Deforestation blocks the vital phenomenon of life, that is, photosynthesis. Blockade or diminished rates of photosynthesis result in the gradual accumulation of CO_2 in the atmosphere, which in turn exacerbates global warming gradually precipitating a climate crisis.

It takes a forest about 2000 years from the beginning of primary succession to attain ecological climax. However, it takes only a few years for a climax forest to reach a stage of disclimax following deforestation. A stage of disclimax, that is, ecological disaster, spelled out by deforestation, translates slowly but steadily into economic disaster.

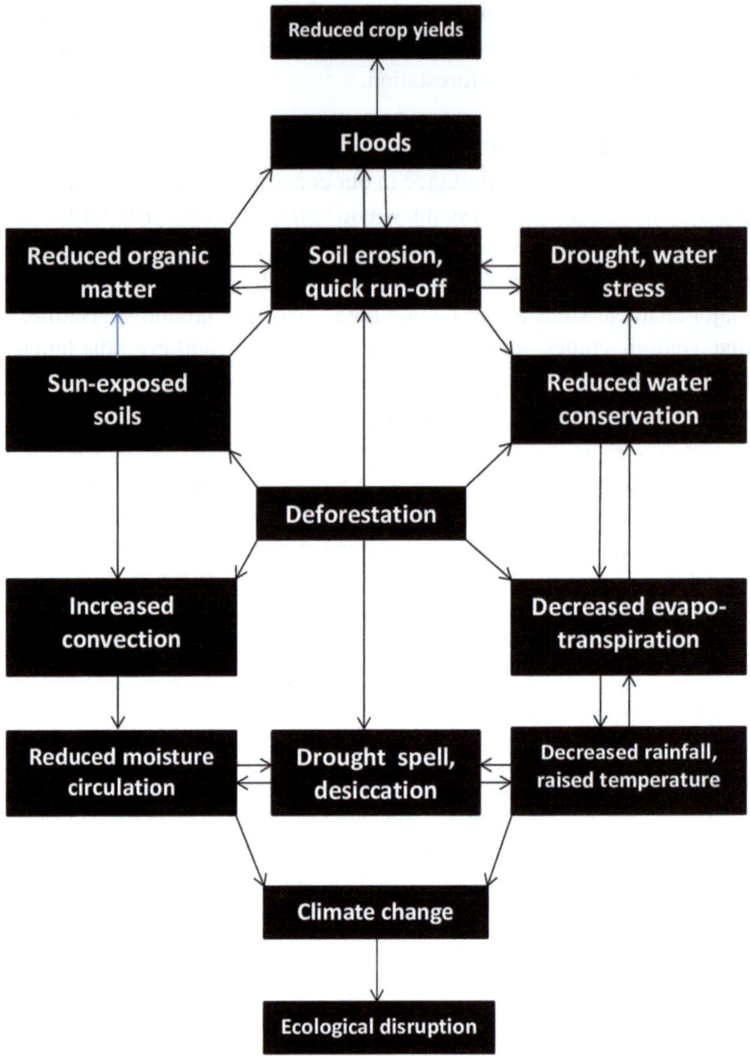

Fig. 9.3 Impacts of deforestation

9.5 Conservation of Forests

Conservation of the forest wealth of the world is not only necessary but an imperative of our times. We are living in times when ecological disaster is gradually culminating into a point of no return. In such times, protection, regeneration, augmentation, and conservation of the forests are of absolute importance.

9.5.1 Protection of the Existing Forests

The priority of forest conservation is to give due protection to the existing forests. If the existing forests are harvested or thinned for socioeconomic purposes, it will take hundreds of years for them to attain the climax stage following secondary succession. Felling and logging of trees and all other kinds of exploitation activities operating within forests must come to halt. Infiltration of poachers and vested interests in the existing forests should not be allowed at any cost. The primary purpose of providing due protection to the existing forests is to help them attain a stage of ecological climax in due course of time.

Most of the forests the world over have been reduced to thin stands of trees or monocultures of a few, often commercial, species. Canopy cover is also not enough to protect soil and play a crucial role in soil and water conservation and creating a microclimate of its own. Most of the forests have also ceased to be habitats for wild animals and diverse life forms. This all has occurred and goes on occurring due to extreme human intervention giving way to intensive forest exploitation and continuous deterioration of forest ecosystems.

The degenerated forests need to be brought into the natural process of ecological regeneration. Halted or considerably reduced human intervention is the precondition to accelerate ecological regeneration. To further accelerate this natural phenomenon, plantation of the saplings of mixed native varieties would be promising.

All the forests earmarked as protected areas – the national parks, the wildlife sanctuaries, and the biosphere reserves – must be given due protection to strictly maintain them in the state of conservation. Tourist activities in the protected areas should be kept limited and well-monitored.

One of the most interesting categories of forests is community-protected forests and all kinds of sacred groves. These forests and community activities relating to their management (conservation and utilization) must be duly recognized and encouraged. Public organizations must extend required support – financial, technical, and institutional – to the community-managed forests.

9.5.2 Social Forestry

Social forestry is a new concept implemented in India about four decades ago. This program of plantation involves forest resources, labor, and capital integrated in a way that could bring about qualitative changes in life. In many rural areas, social forestry became instrumental in bringing about an economic revolution. Social forestry is a multidimensional socioeconomic activity involving diverse land uses and different groups under diverse socioeconomic conditions.

Social forestry involves local people's participation in forestry activities right from the decision-making to utilization of the forest products. There are three dimensions of social forestry: farm forestry, rural forestry, and urban forestry. The socioeconomic objectives of social forestry are met by these three types of forestry. The 5F forest products obtainable from social forestry are foods, fodder, fuel, fiber, and fertilizer.

9.5.2.1 Farm Forestry

Farm forestry encourages farmers to raise trees on their field bunds, wasteland, and marginal land to derive extra income from farms, such as through small timber, fuel wood, and fodder for livestock. In rural areas, people were accustomed to using livestock dung as fuel. One of the objectives of farm forestry was to provide fuel wood as an alternative to dung. The dung should be converted into manure and used for raising soil fertility. Marginal land that is unfit for cultivation could also be used for plantation and turned into a source of income for farmers.

9.5.2.2 Rural Forestry

Rural forestry is a wider concept that calls for plantation on community lands, *panchayat* land, degraded forest area, roadside, canal banks, wastelands unfit for cultivation and on the periphery of ponds, and in open spaces lying barren. The rural areas are also encouraged to establish cottage industries based on forest products. Rural forestry seeks the large-scale participation of rural people to make social forestry a successful program.

9.5.2.3 Urban Forestry

Cities and towns are inhabited places mostly portraying dismal environmental scenarios. The plantation of trees for the amelioration of the urban environment should be taken on a priority basis. Plantation imparts aesthetic values to the urban environment, especially when the plants are ornamental types. Picnic spots, public parks, roadsides, city centers, clubs, etc. are the places where fruit plants, ornamental plants, flowering herbs, creepers, etc. are planted. Creating green belts around and amid urban areas, industrial and commercial areas, near hospitals and educational institutes, etc. enhances aesthetic values in addition to improving the urban environment.

9.5.3 Agroforestry

Agroforestry is a fusion of agriculture and forestry. It is a land use involving annual crops, woody species (trees, shrubs), and livestock integrated. Cropping systems involving only annual crops are more fragile than agroforestry systems. Here integration of seasonal crops with woody species of many kinds helps counter fragility and enhances the resilience and sustainability of a farming system or an agroecosystem.

Seasonal food grain crops are often shallow-rooted and they harp only on the topsoil rich in nutrients. A season of cropping exhausts nutrients from the topsoil. Therefore, fertilizers have to be applied for the crop plants of the next crop season. Integration of the deep-rooted woody perennials with shallow-rooted food grain crops contributes to maintaining the nutrient status of the topsoil. Since the deep-rooted woody perennials can depend on the moisture regime of deeper soil layers, unlike food grain crops that depend on the moisture of topsoil, irrigation requirements of an agroforestry system are comparatively less. Mulch material contributed by trees and shrubs conserves soil moisture and, thus, also reduces irrigation needs.

An agroforestry system supports a higher degree of biodiversity and is an ecologically sound, more resilient, and more sustainable system. The productivity of the system is higher than that of a cropping system involving only herbaceous plants, such as food grain crops. A variety of foods produced from trees, shrubs, and seasonal crops also becomes available. The fodder, fiber, fuel, and fertilizer needs of the farmers are also fulfilled by an agroforestry system. Higher economic returns are accrued for the farmers dependent on agroforestry.

Agroforestry is a traditional farming system. Indian farmers have been practicing and managing this system for millennia. The trend of monocultures involving only a few food grain crops has intensified after the inception of the Green Revolution when agriculture became dependent on the market. Agroforestry still prevails in many regions, such as in the rainfed areas of the Himalayan mountains.

9.6 Summary

A forest is a biotic community dominated by trees. Forest ecosystems perform the functions vital for the sustenance of life on Earth. Forest functions can be divided into three categories, viz., productive functions (products of socioeconomic importance), protective functions (conservation of soil, water, and biodiversity, etc.), and regulative functions (absorption and release of gases, water, mineral elements and radiation energy, water cycle, biogeochemical cycles, climate regulation, etc.). Forests also spell out aesthetic and spiritual values. All our literature, including our epics, best forms of poetry, dramas, and philosophy, have been created amid the beauty of *Aranya* (forests).

Spread over about one-third of the planet's land area, the forests comprise most of the biomass, accounting for about 90%. Nearly 50% of the global forests constitute tropical forests. A biome has a distinctive climate and life forms adapted to the climate. The spatial arrangement of terrestrial biomes closely correlates with climate variations, particularly the prevailing temperatures and precipitation patterns.

Different climates promote different communities. Variation in climatic patterns, thus, is expressed in varieties of communities. The plant and animal species vary in their richness from location to location within a biome which is thanks to varying environmental and climatic factors. The terrestrial biomes prevailing on the Earth are tropical rainforests, tropical dry forests, tropical savannas, deserts, Mediterranean woodland and shrubland, temperate grasslands, temperate forests, boreal forests, tundra, and mountains.

Deforestation has emerged as the most serious problem relating to our environment. Stunningly rich in their biodiversity, the Amazon forests are disappearing at unprecedented rates. The main causes of deforestation are population explosion, shifting cultivation, industrial demand for wood, road construction, mining operations, mega-dams and hydroelectric projects, forest fires, adverse environmental factors, etc. Deforestation exerts triggering effects leading ultimately to an unproductive desert, adding unfavorable environmental conditions to all parts of the biosphere – the atmosphere, the hydrosphere, and the lithosphere. Deforestation blocks the vital phenomenon of life, that is, photosynthesis.

Protection, regeneration, augmentation, and conservation of the forests are of absolute importance. First of all, protective cover should be given to all the existing forests. The degenerated forests need to be brought into the natural process of ecological regeneration. Social forestry, farm forestry, rural forestry, urban forestry, etc. are crucial programs for the regeneration of forests. Agroforestry is of great value to minimize damage due to agricultural practices involving only annual cropping systems.

9.7 Exercises

9.7.1 Multiple-Choice Questions

1. Which of the following is not a function of the forests?
   (a) Desertification
   (b) Precipitation
   (c) Climate moderation
   (d) Nutrient cycles
2. Farmers in the Himalayan mountains collect fodder from the forests to feed their livestock. This function of a forest is a
   (a) productive function
   (b) protective function
   (c) regulative function
   (d) none of the above
3. Which of the forest categories is found in India?
   (a) Temperate forests
   (b) Tropical forests
   (c) Tropical rainforests
   (d) All of the above
4. Plants, animals, and the climate of the Earth vary according to
   (a) altitude
   (b) latitude
   (c) latitude and altitude
   (d) Insolation
5. Where are tropical rainforests found?
   (a) Great Britain
   (b) South and Central America
   (c) Western Himalayas
   (d) Tibet
6. What category of forests is the richest one in biodiversity?
   (a) Temperate forests
   (b) Tropical rainforests
   (c) Alpine meadows
   (d) Coniferous forests
7. Which ecosystem is often equated with tropical rainforest in terms of the measure of its biodiversity?
   (a) Temperate forest
   (b) Coral reef
   (c) Savannah
   (d) Bamboo forest
8. Where in India are the temperate forests found?
   (a) In the Himalayas
   (b) In Madhya Pradesh and Chhattisgarh
   (c) In Bihar and Jharkhand
   (d) In the Thar desert
9. Which of the following forests would represent the most and the least biodiversity, respectively?
   (a) Conifer forests and tropical forests
   (b) Tropical rainforests and tundra
   (c) Tundra and tropical rain forests
   (d) Temperate and tropical forests
10. Which of the following Indian States/Union Territories comprise alpine pastures?
    (a) Jammu and Kashmir
    (b) Himachal Pradesh
    (c) Uttarakhand
    (d) All of the above
11. What is not a main feature of a tropical rainforest?
    (a) Abundant rainfall
    (b) Rich plant and animal diversity
    (c) Soils extremely rich in humus
    (d) Acidic soils
12. Which of the following is a community-managed forest?
    (a) Sacred grove
    (b) Panchayat forest
    (c) Community forest
    (d) All of the above
13. Shifting agriculture is prevalent in
    (a) Northeastern states of India
    (b) Uttarakhand

(c) South Indian states
(d) Madhya Pradesh and Chhattisgarh

14. Deforestation leads to decreased
(a) deserts
(b) global warming
(c) rainfall
(d) soil erosion

15. Which concept of forestry program of plantation in rural areas of India involves forest resources, labor, and capital integrated in a way that could bring about qualitative changes in life?
(a) Social forestry
(b) Economic forestry
(c) Reserve forestry
(d) Ecological forestry

16. Forestry integrated with agriculture is known as
(a) agricultural forestry
(b) agroforestry
(c) rural forestry
(d) urban forestry

17. Match the items in column A with those in column B

Column A	Column B
(a) Suriname	(i) Northeast India
(b) Tropical forests	(ii) Rich in biodiversity
(c) Slash-and-burn farming	(iii) 98% forests area
(d) Amazon	(iv) Southeast Asia
(e) Tropical rain forests	(v) About 50% of the globe

(a) a-i, b-ii, c-iii, d-iv, e-v
(b) a-ii, b-v, c-iv, d-i, e-iii
(c) a-v, b-iv, c-ii, d-iii, e-i
(d) a-iii, b-v, c-i, d-ii, e-iv

18. What would you notice if you climb a mountain that is high enough?
(a) Biological and climatic changes
(b) Plenty of the water
(c) More variety of fishes
(d) A gradually greater degree of biodiversity

19. Which of the following countries, according to FAO, has the largest cover of forests?
(a) Nepal
(b) Suriname
(c) Sri Lanka
(d) Bhutan

20. Which of the following is an attribute of the forests?
(a) Ethical values
(b) Aesthetic values
(c) Spiritual values
(d) All of the above

Answers: 1-a, 2-a, 3-d, 4-c, 5-b, 6-b, 7-b, 8-a, 9-b, 10-d, 11-c, 12-d, 13-a, 14-c, 15-a, 16-b, 17-d, 18-a, 19-b, 20-d.

9.7.2 Short-Answer Questions

1. Define a forest.
2. Give examples of the productive functions of the forests.
3. What do you mean by deforestation?
4. What is shifting cultivation?
5. What percentage of the tropical and temperate forests vanished from the globe?
6. What are the main attributes of farm forestry?
7. What do you mean by forest conservation?
8. What are the main features of the tropical savanna?
9. What are sacred groves?
10. What is agroforestry?

9.7.3 Long-Answer Questions

1. Write an account of the functions the forest ecosystems perform.
2. Draw a flow diagram and explain the impacts of deforestation.
3. About 50% of the tropical forests have vanished from the surface of the Earth. What in your view are the major reasons for this loss?
4. Discuss various ways and means of forest conservation.
5. Write short notes on the following:
(a) Causes of deforestation
(b) Aesthetic values of a forest
(c) Urban forestry

References

Cooper N, Brady E, Steen H, Bryce R (2016) Aesthetic and spiritual values of ecosystems: recognising the ontological and axiological plurality of cultural ecosystem 'services'. Ecosystem Services 21:218–229. https://doi.org/10.1016/j.ecoser.2016.07.014

Everard M, Reed MS, Kenter JO (2016) The ripple effect: institutionalising pro-environmental values to shift societal norms and behaviours. Ecosystem Services 21:230–240. https://doi.org/10.1016/j.ecoser.2016.08.001

Lim SS, Innes JL, Sheppard SRJ (2015) Awareness of aesthetic and other forest values: the role of forest knowledge and education. Society and Natural Resources 28(12):1308–1322. https://doi.org/10.1080/08941920.2015.1041659

Miller R (2019) Revealed: most forested countries in the world. CEOWORLD Magazine. September 10, 2019

Rooney K (2019) These are the world's most tree-covered countries. World Economic Forum, September 09, 2019. weforum.org/agenda/2C, Accessed October 13, 2020

Singh V (2019) Fertilizing the universe: a new chapter of unfolding evolution. Cambridge University Publishing, London. 286 pp

Agriculture and Food Resources 10

Foods are the most essential and critical need of all heterotrophs on Earth. It is the food all animals derive all necessary nutrients and energy from. Nutrients and energy are the two components of food. While the nutrients take care of body structure, the energy makes the organisms accomplish their metabolism and functions. The energy needed for the organisms is not independent; it is the energy in the chemical bonds between constituent units of the nutrients. Thus, energy vital for the body of the organisms is derived from three macronutrients in foods, namely, carbohydrates, fats, and proteins. Micronutrients, namely, vitamins and minerals, are especially for the protection of organisms.

10.1 Foods in the Wilderness

Was the primitive man a hunter and gatherer? Quite a large body of literature is replete with such arguments. But it seems less likely that in the dawn of their evolution on Earth about one million years ago, the human race killed and ate animals in the wilderness. The truth is that at no point in time in human history has the human race been a carnivorous one. Morphologically, anatomically, and physiologically, a human body was not evolutionarily designed for carnivory. If there were a race of hunters and gatherers, that should not have been a human race.

Primitive man derived their foods from the bounty of natural forests. There is a logic behind it. The land of the Earth in ancient days was laden with natural forests harboring varieties of food-providing trees, bushes, and herbaceous plants. Forests provided several types of edible fruits, nuts, pods, flowers, buds, leaves, stems, underground stems, roots, mushrooms, honey, etc. in abundance. In the wilderness what was the only danger for human beings was wild animals, especially carnivores. At that point of biological evolution, the human species was part and parcel of an integrated food chain operating in forest ecosystems. Human beings needed to protect themselves from the wild carnivores for which they needed to hunt them – for their safety, not for food. If there were plentiful plant-based foods all around to fulfill the needs of the beings herbivorous by nature, there was no reason to kill animals for their food.

Carnivores are those that can handle their prey with their body parts. A tiger kills its prey with its sharp canines and penetrating nails. Human being has no such morphology and cannot kill an animal without a weapon. A carnivorous animal eats the flesh of its prey and digests and assimilates the nutrients in the flesh. A human individual cannot digest flesh as such, because our digestive system does not secrete enzymes that could digest flesh. Therefore, man has to denature it by heat. Thus, humans should not have been meat-eaters before the discovery of fire. In some

V. Singh, *Textbook of Environment and Ecology*, https://doi.org/10.1007/978-981-99-8846-4_10

regions of the Earth, in ancient days, some people must have switched over to animal-based foods out of some natural compulsions, such as the nonavailability of plant-based foods. Due to the availability of fire to process flesh into meat, it must have been easier for them to digest animal-based foods.

In many areas of the world people, especially forest dwellers, are still dependent on uncultivated foods readily available in natural forest areas. Natural forests serve as a rich repository of a diversity of foods derivable from different parts of trees, shrubs, and herbs thriving in the environment. Foods available from natural forests and planted trees are far more palatable and nutritious than cultivated ones. They do not need heat to process before consumption and many of them are laden with medicinal values.

Forest dwellers almost everywhere rear animals in large numbers for different purposes: milk, meat, fiber, carting, riding, loading luggage, tillage, recreation, etc. These animals feed only on fodders derived from natural forests. Thus, people transform forest products into animal products and services for various purposes. Although forests in our times do not provide the bulk of the foods needed for humanity, their contribution to food and livelihood security for some sections of society is critical.

10.2 Sources of Foods

As we discussed above, primitive societies derived their foods from natural forests. They relished a variety of food items obtainable from immense varieties of plants. People living in traditional farming also derived their foods from a very large number of food crops and their numerous varieties (genotypes). The genetic base of the food crops shrunk in the Green Revolution era, and now our food habits count on a limited number of plants. In our times, nearly 67% of the foods are contributed by cultivated crops, mostly cereals. Pulses, vegetables, and fruits are also consumed throughout the world. About 17% of the foods are of animal origin. The meat of the animals reared in open rangelands is mostly eaten in developed countries. The rest of the food, about 7%, comes from aquatic resources with the bulk from fisheries (Fig. 10.1).

10.2.1 Food Crops

Living plants have been and continue to be the foundation of human foods and nutrition. About 5000 years ago, human beings derived foods from as many as 5000 plant species. There also existed quite a large number of plant species providing

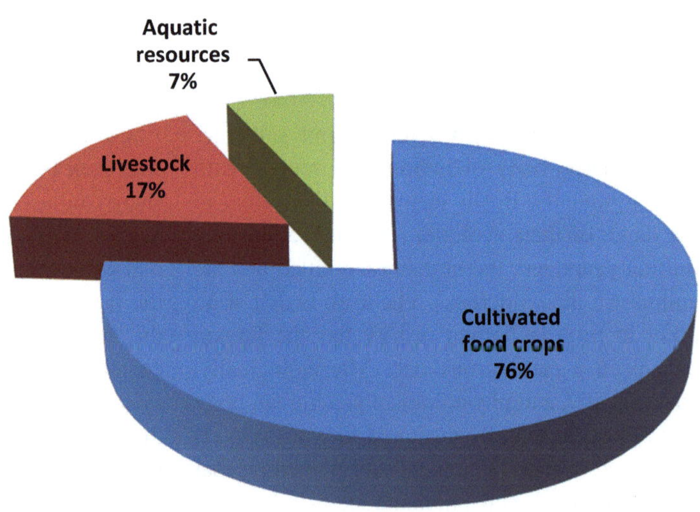

Fig. 10.1 Sources of the foods humanity depends on

food in traditional agricultural systems. The erosion of natural agro-biodiversity has gone intensified with the strengthening of agriculture-marketing links. Now only those foods grown that have their "market value." Not only plant species but also their genetic varieties have been rapidly squeezed out of cultivation practices. For example, Indian traditional agriculture once was blossoming with as many as 60,000 varieties of rice. But now only a few varieties having high market value are cultivated. Humanity the world over now derives food only from a few dozen kinds of plants and nearly three-fourths of the total energy needs of human beings are fulfilled by just three food crops: rice, wheat, and maize. Despite the shrinkage of agro-biodiversity in cultivated lands, food crops' contribution to total food demands and food security of human populations is phenomenal.

10.2.2 Livestock

Domesticated animals provide milk, eggs, and meat that make up a good part of the diet for many people inclined to nonvegetarian diets. Among all foods of animal origin, milk is in the highest demand all over the world. It is a product consumed by the majority of people, including vegetarians and nonvegetarians. Milk is also consumed through tea and coffee and as buttermilk, curd, butter, and refined butter (*ghee*). Cows are the most important dairy animals in the world. Buffaloes yield a significant amount of milk in India and many other Asian countries. India is the world's leading country in terms of cattle and buffalo population. About half of the world's total buffaloes are in India. India is also the world's top milk producer. Goats and sheep are generally reared for meat. However, as far as meat production is concerned, the scenario changes from country to country and it is also a matter of culture and religion.

10.2.3 Aquatic Foods

Foods from aquatic resources – marine as well as freshwater – constitute an important proportion of the foods availed to human beings. Fisheries and other foods obtainable from water habitats, in some countries, like Japan, where about 50% of animal proteins are met by fishery only, are crucial resources of their income. Rivers, streams, and rivulets are used for harvesting fisheries for consumption purposes. In several villages, people make up small ponds for aquaculture and incorporate fish into their diets.

10.3 Cultivated Food Crops

Almost entire humanity in our times depends on cultivated foods and domestic and aquatic animals. Food crops are most critical for global food security. Our survival and sustenance and our tactics for ushering in a sustainable and happy future depend on a limited number of food crops. Just three crops – rice, wheat, and corn – fulfill nearly three-fourths of our energy requirements, as also mentioned earlier. The other major food crops we encounter in day-to-day life include sugarcane, potato, banana, pulses, oilseeds, and vegetables. Although food crops vary from region to region, area to area, and from place to place, most people derive their foods from selected crops being grown extensively. The land area devoted to various crops is determined largely by market factors. Thus, many food crops people depended on years ago vanish from the cultivated land. And many other crops with high market value, such as cash crops, replaced the indigenous ones.

Horticultural crops, mainly fruits and vegetables, make up another important category of cultivated crops. A variety of condiments and spices are grown in many countries of the world, especially in South Asia. India is well known for the diversity of spices it produces and exports across the world. Tea and coffee are cultivated in many countries to provide beverages and also part of foods.

Globally, rice contributes most to the human food supply, which on average is equal to 541 kcal/person/day, as in 2013. In second place is wheat with 527 kcal/person/day, on average. The contribution of sugarcane is 200 and of corn 147 kcal/person/day, on average. However, these figures vary greatly from region to region. For example, rice makes important food grain in

South Asia, but not in Europe and the West; in Africa, root and tuber crops make the bulk of people's food providing 421 kcal/person/day. In some areas, such as the Himalayan mountains, pseudo-cereals – mainly amaranth, buckwheat, and chenopod – are consumed. The pseudo-cereals are used the same way as cereals but they do not belong to the Poaceae family.

10.4 Livestock Resources

The history of animal domestication is older than that of plants. Livestock plays a vital socioeconomic role and constitutes livelihood systems for millions of people worldwide. Mountain communities almost everywhere are especially livestock-dependent. Although milk is the most important product obtainable from dairy animals, domestic animals contribute in several ways – in terms of edible products as well as various services. Livestock provide about 17% of all the food to human beings, and for millions, they are the chief source of food security.

Bovine (buffalo and cattle), ovine (sheep and goat), equine (horses, mules, and donkeys), pigs, llamas, camels, yak, etc. are the farm animals that provide crucial products and services to mankind (Table 10.1). They provide not only edible products and services but also serve as the basis of livelihood security for farmers, herders, nomads, and several other societies.

Livestock are an integral part of farming systems being managed by small and marginal areas in a traditional setting (Singh et al. 2001). They serve as a bridge between cultivated land and forests. Livestock feed on forest fodders and, in this way, transfer nutrients of a more stable forest ecosystem to more fragile cultivated land via their manure. They also recycle the nutrients into croplands. Livestock-mediated linkages in the farming system are depicted in Fig. 10.2.

Draught animals, while engaged in land preparation, tillage, leveling, puddling, interculture, etc., contribute to in situ fertilization of the soil by urinating and voiding dung. Ruminant animals, in the process of fermentation of crop residues in the rumen, transform crop residues into useful products and work. Draught animals play a vital role in the diversification of cropping of the fields. They also help avoid the use of fossil fuels which are used in tractors and other machines operated in farming activities (Singh 1998). Thus, livestock helps prevent carbon emissions.

10.4.1 Draught Animal Power

The draught animal power (DAP)-based energy system is the most suited to small and marginal farmers and is of special importance in mountain agriculture. Why does animal power seem to be the best way of developing sustainable agriculture in the mountains and why is it much better than fossil fuel-powered machines? Many sound reasons can be given for this. Arguments in favor of the draught animals especially in the context of mountain agriculture largely characterized by small and marginal landholdings are as follows (after Singh 1998):

Table 10.1 Contributions of livestock in terms of products and services

Livestock	Milk	Meat	Fiber	Draught power	Carting	Riding	Manure	Carry-pack
Bovine (buffaloes and cattle)	√	√		√	√		√	
Ovine (sheep and goats)	√	√	√				√	
Equine (horses, mules, donkeys)				√	√	√	√	√
Llama				√	√		√	√
Camel	√				√	√	√	√
Yak	√	√	√	√	√	√	√	√
Pig		√					√	

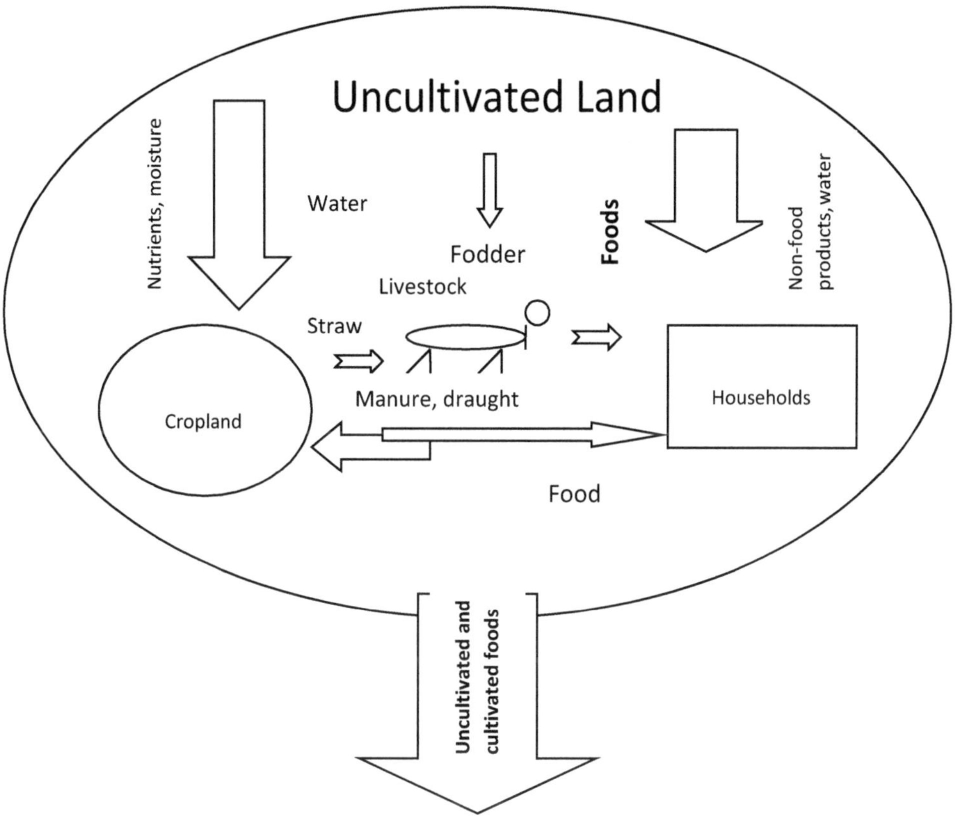

Fig. 10.2 Interconnections between uncultivated land (forest/rangeland/pasture), livestock, cultivated land, and households in traditional farming system

- The source of energy already exists in mountain areas. DAP does not have to be manufactured or bought at a high cost.
- The use of animals considerably increases a farmer's "workforce."
- It enables the farmer to diversify the crop planted, increase the cultivated area of the farm, and carry out agricultural work in time.
- Machine-based farm energy system causes the production activity to concentrate on a limited number of crops, thus reducing the diversity of the production system.
- Animal-drawn plows and other implements are cheaper and more affordable than a large tractor-drawn plow/power tiller. Animal-drawn implements can be made in the village itself and are more suitable for the small, often fragmented, and scattered farms found in the mountains.
- The use of draught animals does not entail any investment in expensive and nonrenewable fuel. Another enormous advantage of the ruminants used as draught animals is that they can be fed residues and by-products available on the farm, producing in return not only energy but food (milk from the female cattle), methane (biogas) for lighting and cooking, manure to fertilize the fields, and many other products obtainable after their death without competing with people for food.
- The use of draught animals enables the farmers to integrate livestock with crop production and permits the exploitation of the very large potential of cattle kept on settled, subsistence farms.
- Mechanization causes direct labor displacement in land preparation. If it does not also contribute directly to increasing cropping intensities and yields, or a switch to more

labor-intensive crops, there will be a net loss of employment opportunity in areas, e.g., the mountains, where alternative sources of income are extremely meager.

- Fossil energy to be used in the machines is a finite resource and its use has a considerable negative impact on the environment. Most farmers in the mountains cannot afford fossil energy-based technology.
- Where animals are used as draught power, it is possible for farmers to cultivate more land or they make time free for other activities.

The above-listed arguments reveal the DAP system's supremacy over the mechanization of farms. In the mountain areas, owing to specific resource base characteristics altogether different from those of the plains, improvement in DAP system efficiency and its sustainable use seem to be a pressing necessity.

Cattle and buffaloes serve the draught animals in Asia. Vital multiple contributions of the draught animals in terms of farm operations, products provided, income and employment gains, and social, cultural, and ecological gains are listed in Table 10.2. These attributes underline the importance of draught animals in the development of sustainable livelihood systems and ecological and sustainable agriculture.

Table 10.2 Key attributes of draught animals

	Cattle		Buffalo	
Attributes	Male	Female	Male	Female
Agricultural operations				
Ploughing	√			
Leveling	√			
Puddling	√			
Weeding and earthing up	√			
Threshing	√	√	√	√
Carting	√		√	
Products				
Milk		√		√
Meat*	√	√	√	√
Dung/manure	√	√	√	√
Income/employment				
Direct productivity improvement	√	√	√	√
Smaller gains through sale	√	√	√	√
Larger gains through sale	√	√	√	√
Income through hiring out	√		√	
Social, cultural, and ecological gains				
Cropping diversification	√			
In situ manuring of fields	√			
Renewable energy supply	√		√	
Religious, ethical values	√	√		
Festivity, fairs, rituals	√	√	√	√
Social status, prestige	√	√	√	√
Social cohesion encouragement	√	√		
Farming system sustainability enhancement	√	√	√	√

Source: Based on Singh (1998)

[a]Cow slaughter is banned in India, as the cow is considered a sacred animal by Hindus. Consuming beef is viewed as a religious transgression in India and among Hindus worldwide.

10.5 Agriculture, Foods, and Sustainable Future

Humanity's future depends on the availability of food. The foods are derived from different natural resources we have already discussed. Agriculture is the largest source of food humanity has exclusive dependence on. Agriculture is an umbrella term encompassing all the land-related activities managed by human beings, such as cropping, horticulture, animal production, dairying, poultry, fisheries, etc.

Glancing through the history of agriculture, we find that so far humanity has experienced four major eras of agriculture, viz., (i) primitive agriculture, (ii) traditional agriculture, (iii) Green Revolution agriculture, and (iv) LPG agriculture (Singh et al. 2014).

Primitive man depended largely on natural forests for gathering varieties of edible fruits, nuts, beans, pods, seeds, buds, flowers, leaves, stems, underground roots, honey, etc. At the same time, he had also begun plowing land after clearing wild vegetation. Cultivated land proportion in comparison to natural forests was extremely meager and most of the foods were uncultivated (of forest origin) ones. There was no irrigation system. Primitive agriculture encompassed as much biodiversity as could thrive in an area. This type of food production system was extremely productive, resilient, and sustainable.

With ever-increasing cultivated areas carved out of forests came into being what is categorized as traditional agriculture. This agriculture had a strong relationship with the forests. No external inputs were used. The forest-cultivated land ratio was quite larger. A very broad base of agrobiodiversity (food plant species and their genotypes) was a striking characteristic of this agriculture. In the traditional type of agriculture, the soil was regarded as a living system—an ecosystem. In India, traditional farmers held a reverential attitude toward the soil. Agriculture-market linkages were poor and the farmers were quite self-dependent in this system of food production.

Then, towards the 1960s, came what was counted as one of the most significant achievements of the twentieth century – the Green Revolution, a term first of all used by William S Gaud. This type of agriculture was dependent on high-yielding varieties (HYVs) of crop seeds, external inputs (chemical fertilizers, pesticides, mechanization, expertise, etc.), frequent irrigation, and banks and other financial sources for loans, debt, etc. The Green Revolution, in a few years, began revealing its side effects and after effects: pollution of soil, water, and atmosphere, squeezing out of biodiversity from croplands, monocultures of crops, ill health of people, socio-economic inequity, and suicides by farmers (for example, in India) trapped in loans for purchasing external inputs.

Advancement in biotechnology did not leave any sector unaffected, agriculture being no exception. In our contemporary times, agriculture is in a transition phase: from the Green Revolution to a more liberalized kind of agriculture. It is biotechnology-driven. New varieties of seeds are being developed by the private sector, such as seed companies and all its inputs and outputs are the products of the free market of the globalized world. Singh et al. (2014) refer to this emerging agriculture as LPG agriculture (liberalization, privatization, and globalization agriculture). With the rapid development of genetically modified organisms (GMOs), such as Bt crops, this system of food production has introduced a different kind of pollution, which is the worst of all kinds of pollution: genetic pollution. There is a danger even of wiping out pollinators due to the wide spread of Bt crops. This would be the biggest jolt to the very ecological integrity of the biosphere. All planning, all strategies, and all tactics of this agriculture are at the disposal of the corporate sector of the globalized world. Farmers in this agriculture have been reduced to mere free laborers to operate the practices instructed by multinational companies.

Among all these four development phases in the history of agriculture development, primitive agriculture was the most sustainable one and the

LPG agriculture is the most unsustainable one. Represented by a straight ecological pyramid, we find that the food production system lying at the base of the pyramid was the most sustainable. The sustainability of the traditional farming system above the base was comparatively of a lower degree. Green Revolution-type agriculture was comparatively more unsustainable and the agricultural system lying towards the tip of the ecological pyramid is the most unsustainable one (Fig. 10.3).

As mentioned above, the deterioration of the environment began with the inception of the Green Revolution and is further intensifying with the gradual strengthening of LPG agriculture. The environment-deteriorating features of these two agricultural systems are outlined in Table 10.3.

10.5.1 Agriculture and Sustainable Future

Whatever is out there in our present times need not be there in the future. The future has to be there in the lap of time. But whether with us or without us, that is the question. What is, then, a sustainable future? The sustainable future is the future that would be witnessed by healthy, vibrant, and happy humanity (Singh 2019). A sustainable future will not be out there on its own. It would be there for us to usher in if we are creating the same in our present times. As food is the source we are made up of, the source we derive energy for all kinds of activities, the source we derive our materialistic progress from, the source our peace depends on, and a source of our happiness evolves, agriculture would be a foundation on which a sustainable future

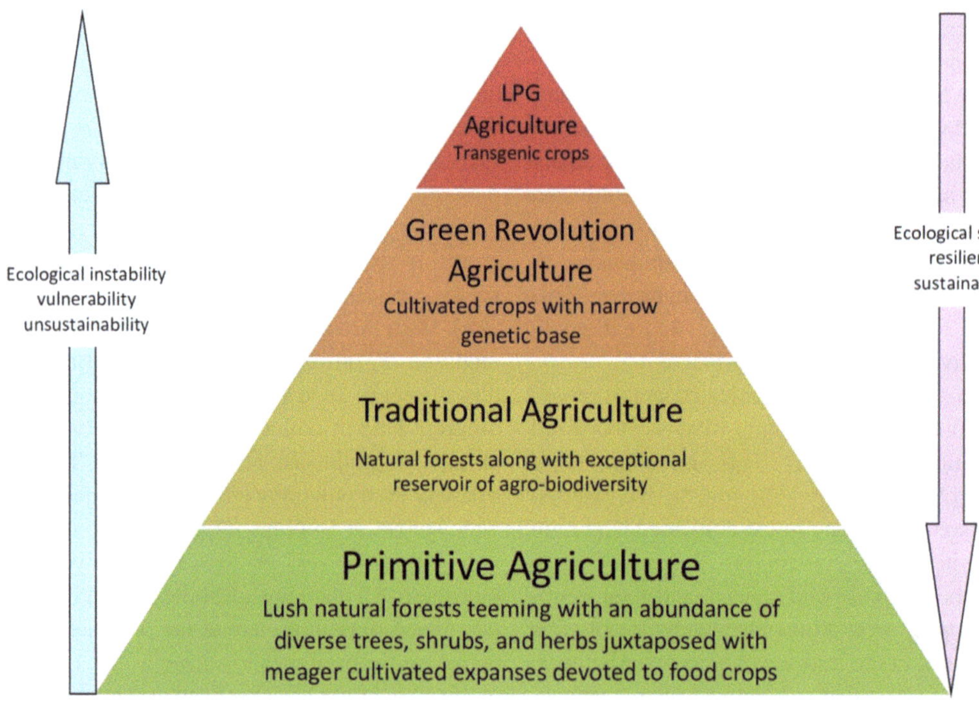

Fig. 10.3 Throughout the historical evolution of agriculture, the sustainability of agricultural systems tends to diminish as we approach the pinnacle of the ecological pyramid

Table 10.3 Negative environmental aspects of contemporary agricultural systems

Agricultural system	Major environment-deteriorating features
Green Revolution	• Volcano of pollution (contribution to about one-third of greenhouse gases) • Chemical warfare against life (indiscriminate use of pesticides and chemical fertilizers eventually mingling into soil, water, and air) • Extinction of seed varieties (trend of monocultures) • Human health hazards and shrinking life due to wiping out of soil organisms and erosion of agro-biodiversity • Agribusiness designs to put the farmers into a serfdom
LPG	• GMOs as weapons of genocide • Anthroposynthesis of poison • (cloning of Bt crystal protein [*cry*] gene; *Bacillus thuringiensis* [Bt]) • GMOs' impact on human health: toxicity, allergenicity, antibiotic resistance • Genetic contamination, or genetic pollution • Adaptation towards pest protection mechanism • Modification in soil ecology • Pollinator-annihilating • Ecological disintegration

would be constructed. But not the kind of agriculture we are operating in our present times, but the agriculture that fulfills ecological rules and is environmentally safe and regenerative. We call this type of agriculture ecological or sustainable agriculture.

10.6 Foods and Nutrition

Based on the nutrients – macro and micro – the foods are categorized into three groups: (i) energy foods, which have an abundance of carbohydrates and fat; (ii) body-building foods, which have a higher proportion of protein content; and (iii) protective foods, which are rich in vitamins and minerals. Although energy is there in all the organic compounds in foods, including vitamins, DNA, RNA, etc., we generally count only macronutrients, and among them only carbohydrates and fats for energy needed by the body. Proteins are the major body-building components; therefore, they are not generally counted for energy purposes. Vitamins and minerals are parts of some enzymes, and thus, some of them also participate in boosting the energy system and protecting the body from a variety of metabolic disorders.

The three macronutrients make the bulk of the food nourishing human beings. Carbohydrate-rich foods include cereals, potatoes, sugars, etc. Whereas potatoes and cereals have insoluble starch that acts as fuel for energy production, sugars make up soluble carbohydrates, the instant sources of energy. Fats are rich in energy and all fat-rich foods play a role in accumulated energy in adipose tissues of the body. All oils, butter, refined butter (*ghee*), and some seeds like sesame, mustard, rapeseed, castor, etc. are rich in fats. When there is a dearth of stored carbohydrates in the body (glycogen) due to hunger for a prolonged period, accumulated fats are used for providing energy to carry out metabolism. Proteins play a role in the repair of body parts and body-building. All the pulses are rich sources of proteins.

Micronutrients, vitamins, and minerals are generally met from fruits and vegetables. Vitamins are organic metabolic regulators. All fresh fruits are rich in one vitamin or the other. Minerals are the only inorganic substances functioning as growth regulators. All foods contain minerals in varying proportions. However, fruits and vegetables are especially rich in minerals.

There are certain ingredients of the macronutrients that have to be taken into consideration. These are essential fatty acids and essential amino acids. Essential fatty acids are the fatty acids that humans (as well as other animals) cannot synthesize in their bodies and, therefore, need to ingest for metabolic processes. There are two essential fatty acids, viz., alpha-linolenic acid, an omega-3 fatty acid, and linoleic acid, an omega-6 fatty acid. There are several sources of these essential fatty acids, such as walnuts, olive oil, soya oil, linseed (flaxseed), seaweed oil, sunflower seeds, canola oil, pumpkin seeds, leafy vegetables, etc.

Table 10.4 Composition of a healthy diet

Nutrient	Sources	Function	Percent of calories per day
Macronutrients			
Carbohydrates	Cereals (rice, wheat, maize), potato, yam, sugar	Energy	45–55
Fiber	Salad, peas, beans, oats, whole grains, seeds	Digestion, bowel function, regulation of blood sugar level	Part of the carbohydrates
Fats	Nuts, seeds, oils, butter, ghee, etc.	Energy, storage energy, hormone production	20–35
Proteins	Pulses, soybeans, nuts, etc.	Tissue growth, maintenance	10–35
Micronutrients			
Vitamins	Fresh fruits, green vegetables, etc.	Regulation of metabolism, biochemical functions	Trace
Minerals	Vegetables, fruits, nuts, etc.	Regulation of growth, metabolic functions	Trace

There are 20 amino acids which all are necessary for the synthesis of proteins in the body. However, 10 of them are essentially required to be consumed through food as these cannot be synthesized by the human body. These are called essential or dietary essential amino acids. These are tryptophan, threonine, histidine, arginine, lysine, leucine, isoleucine, methionine, valine, and phenylalanine. Among them, arginine is dietary essential only for the young, not for adults. Therefore, generally, only nine amino acids are considered essential.

A healthy diet must include a varied selection of foods (Table 10.4). As far as possible, foods must be derived from as many plant sources as possible. This is especially essential for vegetarians. Due to apprehensions of many contaminations and accumulation of all fat-soluble pollutants and toxic heavy metals in animal fats, an overwhelming human population the world over has vegetarian foods as their prime choice. Further, vegetarian diets provide everything essential, including some unidentified nutritional factors, for the physical, intellectual, ethical, aesthetic, emotional, and psychological development of human beings. Relishing the diversity of food plants is necessary as not all plants have all the nutrients in the required amounts. When foods are derived from a variety of plants, all the essential nutrients, as well as many unidentified factors, become available for perfect growth and holistic human development.

Water is not a nutrient, but essential for nutrition. Apart from many other roles of water in the body, it is necessary for the digestion, absorption, transportation, and assimilation of nutrients. There can be no food and nutritional security without water. All enzymes in the biochemical/metabolic reactions function in an aquatic medium. Therefore, water is of key importance in regulating metabolism, including in energy generation for various body functions. There are three types of water we depend upon – drinking water, water in the foods, and metabolic water. Nearly 20% of the water is contributed by the food we ingest. The rest must be met through drinking water. Metabolic water is produced in the cells in the process of biochemical reactions and is essential for keeping the metabolism going on.

10.7 Food Problems of the World

Food production and food supplies are not something static. They vary according to many circumstances. On the whole, our agriculture is in a state of unsustainability. The processes of unsustainability emerge from the mismanagement of the resource base, which results in poor production flows. The resource base goes on deteriorating and production flows go on decreasing. Slowly and slowly, the processes of unsustainability dynamically advance from the state of a climax towards the state of disclimax (Fig. 10.4).

Some crucial indicators of sustainability are rapid soil erosion, reduced level of agro-biodiversity in cultural practices, and side effects and after effects of chemical fertilizers and pesticide applications (environmental pollution, human health hazards, exacerbating global warming, etc.).

The current scenario relating to the food production system is that there are plentiful foods the world over. There is no dearth of fresh foods as well as processed (fast) foods. The food pyramid, ecologically speaking, is straight or upright. Food production is an ecosystem function. Every ecosystem, including an agroecosystem, and every species have a fixed productivity potential. The uppermost limit of production potential is sustained when an ecosystem is ecologically healthy. On account of rapid rates of soil erosion and reduced biodiversity levels, which are consequences of current management practices, the agroecosystems as of today are in poor ecological health and the ecosystem functions through which foods are produced are becoming increasingly paralyzed, and the ecological pyramid has assumed "inverted" position (Fig. 10.5).

The gloomy state of the inverted ecological pyramid and the straight food pyramid is the current state of the human-controlled biosphere. The straight food pyramid, indeed, is a result of the overexploitation of natural resources (forests, lands, soils, biodiversity, waters, etc.). With agricultural practices inducing unsustainability in agroecosystems, ecosystem functions go on getting impaired, and food production would not be enough to maintain the food pyramid in an upright position in the years to come.

10.7.1 Populations' Growth and Food Supplies

The land is limited, and cultivated area on land is limited. The finiteness of the resources used for food production is something we cannot do any-

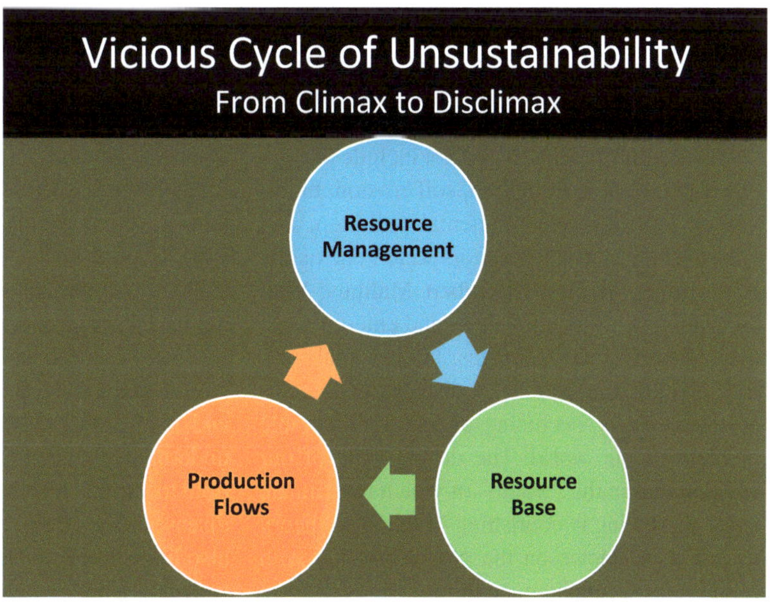

Fig. 10.4 The path to unsustainability: mismanagement of resources leads to resource degradation, resulting in reduced production, and the ecosystem transitions from a state of ecological climax towards disclimax

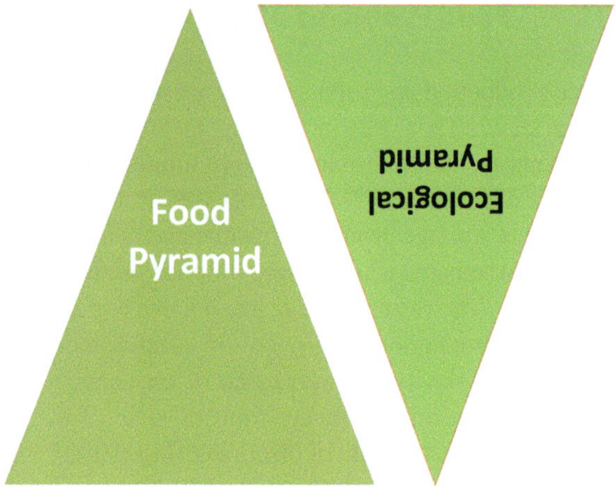

Fig. 10.5 The food pyramid and the ecological pyramid arising from contemporary agricultural practices: a portrait of unsustainability

thing to. This aspect of food production becomes the grimmest problem when populations dependent on finite resources go on multiplying. And this is the common scenario in our world. Barring a few developed countries, a trend of increasing human populations is a matter of great concern. One of the most talked about theories projecting population growth and food supply relationships – the Malthusian theory – is being criticized since its very inception at the threshold of the nineteenth century by sociologists.

Times seem to have arrived for the Malthusian theory to prove its "worth." A large number of countries in the world are unable to produce enough food for their populations. They are bound to import foods from food-surplus countries. The food-surplus (and food-exporting) countries are only those whose populations are not increasing faster than the food production potential of their lands.

The Malthusian theory did not include the elements like land deterioration, soil erosion, biodiversity depletion, environmental pollution, and climate change, as these factors were nonexistent in the times of Thomas Robert Malthus. With escalating environmental mess and emerging climate stresses, food shortages, hunger, malnutrition, and nutrient-deficiency diseases and deaths are increasingly becoming an almost incurable problem of our world. The dismal state of our environment in the "womb" of which the climate crisis is growing is nourishing Malthusian theory to cast its shadows on the planet infested with mushrooming human populations.

Adding fuel to the fire is another problem emanating from sociocultural and political reasons – that of economic inequity. There are situations of enough food grain storage, but large numbers of people of a poverty-ridden section of the populace have no access to food. As a result, they suffer from hunger and malnutrition and nutrient-deficiency diseases. In some cultures females of a population are discriminated against; whatever nutritive foods are available with a family nourish the males in the first place. Thus, the less preferred poor females become a victim of nutrient deficiencies.

10.7.2 Undernourishment

A person receiving less than 90% of minimum dietary intake on a long-term basis is considered undernourished. Lack of sufficient calorie intake for a prolonged period causes stored fats and proteins to break down, leaving people unable to work or even move. Undernourishment may not lead to death but an undernourished person does not have enough energy for an active and productive life. Undernourished children suffer from permanently stunted growth, mental retardation, and many other disorders. People living under undernourished conditions are extremely susceptible to various infectious and nutrient-deficiency diseases. The life expectancy of such people is also likely to be much shorter than those who are well-nourished.

Undernourishment is a global problem but is more prevalent in underdeveloped countries. A gloomy global scenario of undernourishment is reflected in a joint report of FAO, IFAD, UNICEF, WFP, and WHO (2020). This scenario persists even though there is more than enough food produced in the world that no one would go to bed hungry. Some salient points picked up from this report are as follows:

- Nearly 960 million people worldwide are undernourished.
- In 2000–2004, the proportion of undernourished people in the world was 15%; in 2019, the proportion was 8.9% (i.e., despite target-oriented global hunger-alleviation programs, the reduction in the total number of people suffering from undernourishment has not come down up to an appreciable level).
- About 70% of the world's total foods are produced by small farmers, herders, and fishermen, and yet these are the people who are especially vulnerable to food insecurity.
- Rural populations of the world more acutely suffer from poverty and hunger.
- The conflict between and within countries are major drivers of hunger: Out of 144 million stunted children, as many as 122 million are those who live in these countries.

10.7.3 Malnourishment

Malnourishment is a condition when a specific component of the food – for example, proteins, minerals, vitamins, etc. – is required for sound health and proper growth of a person is deficient or absent in the diet. Dependence on such diets, even if they are in required bulk, for a prolonged period causes certain health-related problems, such as lack of appetite, indigestion, and reduced absorption of nutrients. Reduced body growth, the decline in working capacity, lowering of reproduction capacity, and many other well-pronounced disorders are the major health problems malnourished people have to face. Some striking malnutrition problems and their specific causes are presented in Table 10.5.

Table 10.5 Issues of malnutrition and their underlying factors

Malnutrition problem	Cause of the malnutrition problem
Kwashiorkor	Lack of required amount of protein causing failure of neural development in infants
Marasmus	Lack of protein and calories causing progressive body emaciation
Anemia	Lack of iron in diets or inability to absorb iron causing reduced level of hemoglobin in blood
Goiter and hypothyroidism	Iodine deficiency in diet causes thyroid gland not to produce enough thyroid hormone resulting in disorders and abnormalities such as mental retardation and deep mutism among children
Pellagra	Lack of niacin (vitamin B-3) in diet causing diarrhea, dementia, and dermatitis – "three Ds")
Chronic hunger	Diets persistently inadequate in terms of quantity and/or poor quality causing people to stay alive but at the cost of properly productive life

Malnutrition-related problems are more prevalent in poverty-ridden masses in underdeveloped countries. According to a report by FAO, IFAD, UNICEF, WFP, and WHO (2020):

- An estimated 14 million children all over the world under the age of five suffer from severe acute malnutrition – also known as severe wasting.
- Among all the children suffering from severe acute malnutrition, only 25% have access to life-saving treatment.
- The rate of stunting – that is, children too short for their age due to chronic malnutrition – was 21.3% in 2019 and 33% in 2000.

10.7.4 Overnutrition

Overnutrition is also one of the global problems in no way less serious than undernutrition and malnutrition. While the latter two are the problems of poor people, especially in developing countries, the former is of rich people and developed countries. Global markets are overstocked by fast foods, vitamins, nutraceuticals, and several types of industrially processed foods and a variety of drinks. The problem of overnutrition arises when energy consumption exceeds energy expenditure. The problem of overnutrition multiplies when people do not exercise enough to expend an excess of energy.

People in the habit of overnutrition suffer from several health-related problems. The high content of fat-soluble vitamins and minerals in the diet cause toxicity in the body. Obesity is increasingly becoming a global health problem. It is thanks to the habit of overnutrition. Obesity itself is a disease that results in so many other health problems, notably high blood pressure, arthritis, cardiovascular diseases, hypertension, respiratory diseases, diabetes, etc.

10.7.5 Micronutrient Deficiency Diseases

Deficiency diseases caused by the lack or deficiency of certain vitamins and minerals are spelled out in Table 10.6. On the whole, a micronutrient deficiency leads to metabolic disorders, consequently leading to various other health-related problems.

Table 10.6 Deficiency diseases caused by micronutrients

Micronutrient	Deficiency diseases
Vitamins	
Vitamin A	Nyctalopia (night blindness), xeropthalmia (dryness of eyes), keratomalacia (eye disorder), and complete blindness; dermatosis (skin disease) and infertility
Thiamine	Loss of appetite, or anorexia, fatigue, irritability, stomach problem, beriberi
Riboflavin	Ariboflavinosis (stomatitis – painful red tongue, sore throat, chapped and fissured lips, and angular stomatitis – inflammation of the mouth corners
Niacin	Pellagra (rashes on skin exposed to sun), swollen mouth and bright red tongue, vomiting and diarrhea
Cobalamin	Pernicious anemia (stomach cannot create enough cobalamin-absorbing intrinsic factor leading to inadequate supply of healthy red blood cells)
Vitamin C	Scurvy (depresses immune system, weakens muscles and bones, makes people feel fatigued
Vitamin D	Rickets (skeletal disorder in children) and osteomalacia (marked softening of bones)
Vitamin E	Infertility, muscle weakness, coordination and walking difficulties, and numbness and tingling
Vitamin K	Nonclotting of blood: increased bleeding and bruising, blood in vomiting, heavy blood losses in menstruation, presence of blood in urine and stools
Mineral nutrients	
Calcium and phosphorus	Weak bones and teeth, retarded growth, metabolic disorders
Sodium	Hyponatremia (low blood pressure and loss of body weight)
Potassium	Hypokalemia (muscle weakness, muscle cramp, fatigue, fast or irregular heartbeat, frequent urination, constipation
Iron	Anemia (hemoglobin deficiency in blood)
Manganese	Osteoporosis, anemia, poor immunity, chronic fatigue, hormonal imbalance, worsened premenstrual syndrome (PMS), infertility, impaired glucose sensitivity
Zinc	Weak immunity, diarrhea, motor disorders, hair loss, allergies, loss of appetite, poor night vision, mood swings, joint and hip pain, hypogonadism (testicles cannot produce enough testosterone, can perform damaged reproductive functioning), impaired brain functioning
Iodine	Goiter (no synthesis of thyroxin hormone in thyroid gland), cretinism (physical deformity and learning disabilities due to congenital thyroid deficiency), slow heartbeat, weight gain, depression, fatigue, hair loss, memory and mental ability impairment, increased sensitivity to cold

10.8 Sustainable Agriculture

Sustainable agriculture ensures that the current population's food and nutritional requirements are met while safeguarding the ability of future generations to produce enough food to meet their own needs.

There can be no sustainability without ecological sustainability of its own. Ecological sustainability is the first precondition for a socioeconomic system to attain sustainability. Sustainability, thus, is rooted in ecosystems. Sustainability, in essence, is a phenomenon emanating from ecosystem functions. Ecological agriculture and sustainable agriculture are the same.

Sustainable agriculture is essentially founded on a farming system or an agroecosystem. An agroecosystem is a multicomponent system (e.g., forest/rangeland, cultivated land, livestock), all components being in organic linkages with each other (see Fig. 10.2). Naturally empowered to transform solar energy into food energy, an agroecosystem nourishes all the processes of food production. With a considerable exchange of materials through photosynthesis, nature's nutrient cycles and recycles moisture circulation, and biodiversity conservation, an agroecosystem, is a self-sustaining unit of nature.

Sustainable agriculture operates with ecological principles. There are three basic principles of ecological or sustainable agriculture:

1. Biodiversity
2. Living soil
3. Cyclic nutrient flows

10.8.1 Biodiversity

Biodiversity infuses resilience in agricultural systems. The higher the level of biodiversity, the higher the degree of sustainability. A biodiversity-laden agroecosystem is more productive than one based on monocultures and provides a variety of foods of varying tastes, flavors, nutritive, and often medicinal values (Singh 2023). If a crop provides food deficient in a certain nutrient, the other crop would be richer in the same. If food derived from a crop is rich in carbohydrates, the same derived from the other would be rich in proteins. An agroecosystem should be stocked with as many species and as many varieties of the species of all food crops as have been thriving and can thrive in an ecological region. Biodiversity must flourish in every component of an agroecosystem: in forests, cultivated lands, and livestock. A variety of foods and other essentials would be accrued from all components of a biodiversity-based agroecosystem.

10.8.2 Living Soil

As also discussed elsewhere, the soil is not a physical substratum; it is an ecosystem in itself and one of the largest and most wonderful terrestrial ecosystems on Earth (Singh 2020). In ecological agriculture, the soil has to be treated and taken care of as an ecosystem. In conventional agriculture, the emphasis is on feeding the plants with nutrients. But in ecological agriculture soil is fed with nutrients, and the soil, then, feeds the nutrients to the plants. Traditional Indian farmers also believed in feeding the soil, as it is the "duty" of the soil to feed the plants. The above-soil biodiversity helps nourish and sustain the soil biodiversity or biodiversity. Deep-rooted (trees and shrubs) crops help draw nutrients from deeper soil layers and help nourish the shallow-rooted (cereals, millets, vegetables, and other annuals) crops that depend on the fertility of topsoil. Thus, a combination of deep-rooted and shallow-rooted crops would be vital for taking care of soil fertility. A variety of crops, including those that boost nitrogen fixation in the soil, helps maintain soil fertility. Minimum tillage, frequent applications of manure, use of mulch material, and effective means of soil and water conservation are vital for maintaining the fertility status of the soils.

10.8.3 Cyclic Nutrient Flows

Soil nutrients are exhaustible. Soil microflora and fauna as well as soil fertility are all attributes of soil nutrients. Every crop depletes the soil of its nutrients to a certain extent. Those soil nutrients coming through the crop reach consumers of the foods and fodders – humans and livestock, respectively. The consumers retain certain proportions of the nutrients in their body and the rest is voided as excreta. The first principle of soil fertility management is that its nutrient loss be recouped to some extent by recycling the lost nutrients. Human and livestock excreta should reach the cultivated soils after processing – as slurry, compost, and/or vermicompost. Since this nutrient loss is recovered to a small extent only, the bulk of the lost nutrients is made up of cropland linkages with forests and trees and shrubs (deep-rooted crops) in the fields. When the ratio between forest and cultivated land is wider enough, as it is a precondition for a sustainable agroecosystem, there is always nutrient increment in the soils – employing biogeochemical cycles, nutrient flows by water and air, and through grazing animals and human management like transfer of forest floor litter to the croplands via organic manure and/or mulching.

Sustainability is not a static phenomenon, but a dynamic one. Sustainability is also not an outcome. It is a cyclic process. Traditional farmers manage the soil in such a way that it continues to be replenished by nutrients through manure, recycling, in situ fertilization, mixed cropping, mulching, and other management practices.

Farmers cultivate as much agrobiodiversity as could be possible in a particular area. They also manage the natural biodiversity in uncultivated areas (forests, grasslands, rangelands, etc.). This biodiversity is a key to sustainability. The higher the degree of biodiversity, the higher the level of sustainability (Singh et al. 2014). Farmers also manage cyclic flows of nutrients. Whatever nutrients are extracted from croplands are recycled into the same soil through manure. The soil fertility is further enhanced by supplementing the nutrients from forest soil. This wonderful practice of traditional farming is an example of farmers' management of sustainability in traditional agriculture (Fig. 10.6).

Thus, the sustainability of agriculture is not something on its own. It is not a state of things. It is not a certain level of production. It is a phenomenon: a dynamic phenomenon. A phenomenon that completes its cycle fed at each of the four steps, viz., biodiversity, nutrient flows, a nutrient increment from forests, and living soil.

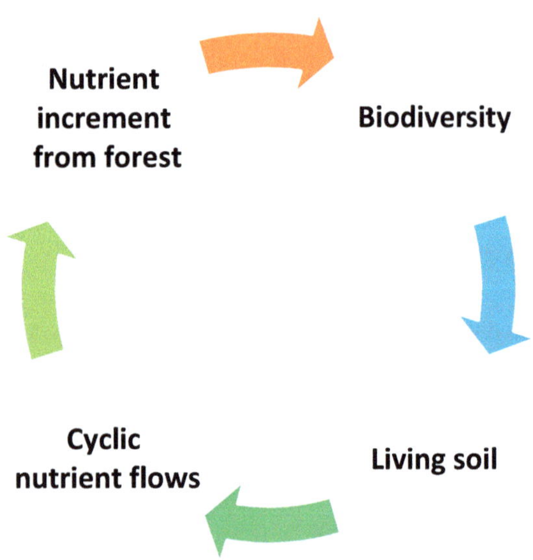

Fig. 10.6 Ecological or sustainable agriculture life cycle

10.9 Summary

Foods are the most essential and critical need of all heterotrophs on Earth. Nutrients and energy are the two components of food. While the nutrients take care of body structure, the energy makes the organisms accomplish their metabolism and all other functions and activities. Primitive man derived their foods from the bounty of natural forests which provided several types of edible fruits, nuts, pods, flowers, buds, leaves, stems, underground stems, roots, mushrooms, honey, etc. in abundance. Foods available from natural forests and planted trees are far more palatable and nutritious than cultivated ones. People living in traditional farming also derived their foods from a very large number of food crops and their numerous varieties (genotypes). Following the Green Revolution, not only plant species but also their genetic varieties have been rapidly squeezed out of cultivation practices. Humanity the world over now derives food only from a few dozen kinds of plants and nearly three-fourths of the total energy needs of human beings are fulfilled by just three food crops: rice, wheat, and maize. Domesticated animals provide milk, eggs, and meat that make up a good part of the diet for many people inclined to nonvegetarian diets. Among all foods of animal origin, milk is in the highest demand all over the world. It is the product consumed by the majority of people. Foods from aquatic resources – marine as well as freshwater – constitute an important proportion of the foods availed to human beings.

Almost entire humanity in our times depends on cultivated foods and domestic and aquatic animals. Food crops are most critical for global food security. Our survival and sustenance and our tactics for ushering in a sustainable and happy future depend on a limited number of food crops. Horticultural crops, mainly fruits and vegetables, make up another important category of cultivated crops. India is well known for the diversity of spices it produces and exports across the world. Tea and coffee are cultivated in many countries to provide beverages and also part of foods.

Livestock play a vital socioeconomic role and constitute livelihood systems for millions of people worldwide. Although milk is the most important product obtainable from dairy animals, domestic animals contribute in several ways – in terms of edible products as well as various services. Bovine (buffalo and cattle), ovine (sheep and goat), equine (horses, mules, and donkeys), pigs, llamas, camels, yak, etc. are the farm animals that provide crucial products and services to mankind. Cattle and buffaloes serve the draught animals in Asia. Vital multiple contributions of the draught animals in terms of farm operations, products provided, income and employment gains, and social, cultural, and ecological gains are phenomenal. Humanity's future depends on the availability of food. Agriculture is the largest source of food humanity has exclusive dependence on. The deterioration of the environment began with the inception of the Green Revolution and is further intensifying with the gradual strengthening of LPG (liberalization, privatization, and globalization) agriculture.

The three macronutrients make the bulk of the food nourishing human beings. Carbohydrate-rich foods include all cereals, potatoes, sugars, etc. Micronutrients (vitamins and minerals) are generally met from fruits and vegetables. Water is not a nutrient, but essential for nutrition and, therefore, for nutritional security.

The current scenario relating to the food production system is that there are plentiful foods the world over. The gloomy state of the inverted ecological pyramid and the straight food pyramid is the current state of the human-controlled biosphere. The straight food pyramid, indeed, is a result of the overexploitation of natural resources (forests, lands, soils, biodiversity, waters, etc.). The major issues concerning food production, food availability, and nutrition are population growth, undernourishment, malnourishment, overnutrition (by rich class people), and micronutrient deficiency diseases.

Sustainable agriculture ensures that the current population's food and nutritional requirements are met while safeguarding the ability of future generations to produce enough food to meet their own needs. Sustainable agriculture is essentially founded on a farming system or an agroecosystem. Sustainable agriculture operates with ecological principles. There are three basic principles of ecological or sustainable agriculture: biodiversity, living soil, and cyclic nutrient flows.

10.10 Exercises

10.10.1 Multiple-Choice Questions

1. From morphological, anatomical, and physiological angles, the human species is
   (a) herbivorous or vegetarian
   (b) carnivorous
   (c) omnivorous
   (d) scavenger
2. Which of the following sources provides the largest proportion of food for humanity in our contemporary world?
   (a) Food crops
   (b) Livestock
   (c) Fisheries
   (d) Insects
3. What proportion of the total food does humanity derive from cultivated food crops?
   (a) 86%
   (b) 76%
   (c) 17%
   (d) 7%
4. Which of the following is a pseudo-cereal?
   (a) Amaranth
   (b) Buckwheat
   (c) Chenopod
   (d) All of the above
5. Globally, which food crop contributes the most to human food supplies?
   (a) Corn
   (b) Wheat
   (c) Rice
   (d) Yam
6. Most of the human energy needs worldwide are fulfilled by
   (a) sugar
   (b) rice, wheat, and maize
   (c) potato
   (d) nuts
7. Most of the food to humanity on Earth is contributed by
   (a) C_3 plants
   (b) C_4 plants
   (c) CAM plants
   (d) Fisheries
8. What C_4 plant provides most of the energy to human beings?
   (a) Sugarcane
   (b) Maize
   (c) Sorghum
   (d) Amaranthus
9. Which country is the world's top milk producer?
   (a) India
   (b) Canada
   (c) USA
   (d) France
10. Which of the following countries has the largest number of descript cattle breeds?
    (a) India
    (b) China
    (c) Russia
    (d) Philippines
11. The term Green Revolution was first used by
    (a) Norman E. Borlaug
    (b) M.S. Swaminathan
    (c) William S. Gaud
    (d) Lal Bahadur Shastri
12. Foods rich in proteins are categorized as
    (a) Energy foods
    (b) Body-building foods
    (c) Protective foods
    (d) Essential amino acid foods
13. Which of the following are essential amino acids?
    (a) Tryptophan
    (b) Lysine
    (c) Phenylalanine
    (d) All of the above
14. Which food grains are rich sources of protein?
    (a) Cereals
    (b) Pseudo-cereals
    (c) Pulses
    (d) Vegetables

15. Lack of niacin (nicotinamide) in diet causes
    (a) Goiter
    (b) Pellagra
    (c) Marasmus
    (d) Kwashiorkor
16. Lack of this vitamin in the diet leads to infertility, muscle weakness, coordination and walking difficulties, and numbness and tingling. This vitamin is
    (a) Vitamin A
    (b) Vitamin C
    (c) Vitamin D
    (d) Vitamin E
17. Which of the following pairs is a mismatch?
    (a) Vitamin C – scurvy
    (b) Thiamine – night blindness
    (c) Cobalamin – pernicious anemia
    (d) Vitamin D – rickets
18. The amino acid essential for the young, not for the adults, is
    (a) Proline
    (b) Lysine
    (c) Arginine
    (d) Tryptophan
19. Which of the following is a correct matching?
    (a) Iodine – goiter
    (b) Sodium – hyponatremia
    (c) Potassium – hypokalemia
    (d) All of the above
20. To be sustainable, an agriculture system must depend on
    (a) crop monocultures
    (b) biodiversity, living soil, and cyclic flows of nutrients
    (c) chemical fertilizers
    (d) pesticides

Answers: 1-a, 2-a, 3-b, 4-d, 5-c, 6-b, 7-a, 8-a, 9-a, 10-a, 11-c, 12-b, 13-d, 14-c, 15-b, 16-d, 17-b, 18-c, 19-d, 20-b

10.10.2 Short-Answer Questions

1. Name the foods that can be found in natural forests.
2. What are three crops human beings derive most of their foods from?
3. Why are pseudo-cereals called so? Name three pseudo-cereals.
4. Which three cereals provide most of the foods to human beings?
5. Name the four agriculture development phases humanity has witnessed in the history of agriculture.
6. Name some examples of energy foods, body-building foods, and protective foods.
7. What do you mean by malnourishment?
8. What is kwashiorkor?
9. What is the deficiency disease caused by the deficiency of iron and calcium in diet?
10. What cultivated crops serve as rich sources of protein?

10.10.3 Long-Answer Questions

1. Describe different sources of food available for the humankind.
2. Write an essay on various food problems that our contemporary world encounters.
3. Write an account of the effects of modern agriculture, viz., the Green Revolution and LPG agriculture.
4. What is sustainable agriculture? Explain the principles of sustainable agriculture.
5. Write short notes on:
    (a) Draught animal power
    (b) Agriculture, foods, and sustainable agriculture
    (c) The Indian food and nutrition scenario

References

FAO, IFAD, UNICEF, WFP and WHO (2020) The state of food security and nutrition in the World 2020: transforming food systems for affordable healthy diets. FAO, Rome. https://doi.org/10.4060/ca9692en

Singh V (1998) Draught animal power in mountain agriculture: a study of perspectives and issues in Central Himalayas, India. International Centre for Integrated Mountain Development (ICIMOD), Kathmandu. (ISSN: 1024–7548)

Singh V (2019) Fertilizing the universe: a new chapter of unfolding evolution. Cambridge Scholars Publishing, London, 285 pp

Singh V (2020) Environmental plant physiology: botanical strategies for a climate smart planet. CRC Press (Taylor and Francis), Boca Raton, 216 pp

Singh V (2023) Biodiversity: concept, crises, and conservation. NIPA, New Delhi, 190 pp

Singh V, Tulachan PM, Partap T (2001) Livestock feeding management at smallholder dairy farms in uttaranchal hills. Ind J Animal Sci 71(12):1172–1177

Singh V, Shiva V, Bhatt VK (2014) Agroecology: principles and operationalisation of sustainable mountain agriculture. Navdanya, New Delhi. 64p + viii pp

11 Mineral Resources

Minerals, the exhaustible and nonrenewable resources found in Earth's crust, rule over the global economy. Our life is dependent on minerals to the extent that we cannot do anything without them. They are the most precious inputs of the Industrial Age we are living in. Minerals play a critical role in the life of a person from his/her birth to death. The socioeconomic status of a society or a country is determined to a great extent by the mineral resources in its possession. The distribution of mineral resources varies from one geographical area to the other. Their availability, types, quantities, proportions, properties, etc. are attributable to the geological features of a geographical area.

11.1 Mineral Proportions

The mineral deposits of economic significance are known as ores. Elements used in a variety of ways are extracted from these ores through metallurgical processes. The Earth's crust, the aquatic ecosystems – mainly the oceans and the seas – and the atmosphere are the major sources accommodating mineral resources. Some 88 elements are occurring in Earth's crust. About 99% of the Earth's crust is constituted of oxygen, silicon, nickel, sodium, potassium, and uranium. The rest compose only about 1% of the crust.

11.2 Mineral Resource Formation Processes

Formation of the mineral resources, unlike renewable ones, is rather a very slow process. It does often take millions of years for the mineral deposits to come into being. The following could be the processes of the formation of mineral resources:

1. The concentration of minerals during the cooling of molten rock material
2. The concentration of minerals due to weathering, transport, and sedimentation
3. Intense heat and pressure
4. Evaporation of water from water bodies, especially from lakes, seas, and oceans
5. Microbial actions

11.3 Types of Minerals

Minerals are categorized into metals and nonmetals. The metallic minerals include gold, silver, copper, iron, aluminum, nickel, etc. The nonmetallic minerals include asbestos, calcium carbonate, granite, sand, gravel, phosphates, carbonates, etc. The metallic minerals are originally found in and extracted from their respective ores. Nonmetallic resources generally occur as minerals.

V. Singh, *Textbook of Environment and Ecology*, https://doi.org/10.1007/978-981-99-8846-4_11

Coal and crude oil widely used as fuels for transportation, industries, food processing, etc. are also categorized as mineral resources. These are the organic mineral resources constituting about 88% of the total value of the world's mineral production with 42% coal, 33% petroleum, 10.5% natural gas, and 2.5% lignite. The metallic and nonmetallic minerals, on the other hand, constitute only some 6–7% (Misra and Pandey 2018).

11.4 Uses of Important Mineral Elements

Uses of minerals and metals date back to the prehistoric period, about 4000 BC. The extraction of metals from their respective ores dates back to 1100 BC. Copper is said to be the first metal to have been used followed by gold and silver. It was possible because of the development of furnaces capable to raise very high temperatures at which ores could be reduced to metals. This discovery led to the extraction of iron the rate of which went on increasing with time. With the inception of the Industrial Age, iron extraction rapidly increased. A little amount of charcoal was mixed with iron to develop it into steel, a much harder metal than iron. As of today, iron makes 95% of all the metals consumed (Misra and Pandey 2018) although its use is comparatively less than many nonmetallic minerals used as, for example, building materials. Further, the nonmetallic minerals are consumed at a much faster rate and in much greater amounts than all the metallic minerals. It is because the former are generally readily accessible and their extraction is comparatively much more cost-effective than that of the latter.

Based on their various uses, the mineral resources can be categorized into some groups, for example, minerals used in agriculture, industry, building construction, and road construction and elements used in metal production, transport, defense equipment, building material, etc. Some important minerals – metallic elements, liquid-metal elements, and nonmetallic minerals – and their uses are presented in Table 11.1.

11.5 Mining

The process of extracting out required minerals or their ores from the Earth is known as mining. Over 100 minerals are mined. India produces as many as 86 metals and minerals. Nearly 80% of all kinds of mining in India is carried out for coal alone (Rajagopalan 2017). Some minerals are easily accessible for their extraction or mining, whereas there are many which lie deep beneath the Earth's surface and their mining needs many advanced techniques and improved methods. The minerals found at or close to the Earth's surface and those occurring at deeper layers are mined using surface mining and subsurface mining techniques, respectively.

11.5.1 Surface Mining

The surface mining of the mineral resources at or near Earth's surface is carried out in the following three ways:

11.5.1.1 Open-Pit Mining

Minerals or mineral ores are taken out by removing the overburden, that is, the materials from above the mineral deposits. Marble granite, sandstone, gravel, limestone, copper (copper pyrites), iron ores, etc. are extracted using the open-pit mining technique.

11.5.1.2 Dredging

When the mineral deposits lie in the water sources, they are extracted using chained buckets and draglines.

11.5.1.3 Strip Mining

Power shovels, bulldozers, and stripping wheels are used in this technique to extract the minerals. The striking examples of the minerals stripped

Table 11.1 Important minerals and their applications

S. No.	Mineral	Uses
I. Metal elements		
1.	Aluminum	Rockets, aircraft, utensils, electric wiring, building material, packaging
2.	Beryllium	Copper alloys, refractory
3.	Chromium	Textile and tanning industries, metallurgy, chemicals, refractory
4.	Cobalt	Alloys, therapeutics, catalysts, radiography
5.	Columbium	Stainless steel, nuclear reactor
6.	Copper	Alloys, brass and bronze, gold jewelry, silverware, electric wire, utensils, pipes, electronic goods
7.	Gold	Jewelry, alloys, dentistry, monetary purposes
8.	Iron	Primary component of steel, building material, transport vehicles, several industrial uses, lead pipes, battery electrodes, pigments
9.	Lead	Batteries, gasoline, ammunition, alloys, paints
10.	Magnesium	Alloy, structural refractoriness
11.	Manganese	Alloy steels, disinfectant
12.	Molybdenum	Alloy steels
13.	Nickel	Metal plating, coins, alloys, chemical industry
14.	Platinum	Jewelry, industrial catalyst, equipment, automobile, medicinal use
15.	Silver	Jewelry, alloy, vessels, photography
16.	Thorium	Electricity generation, nuclear bombs
17.	Tin	Cans, containers, alloys, tin plates, soldering, house construction (tin roofs)
18.	Tungsten	Alloys, chemicals
19.	Titanium	Alloys, pigments, aircrafts
20	Uranium	Nuclear bombs, nuclear reactor (electricity generation), tincting glass
21.	Vanadium	Alloys
22.	Zinc	Brass, electrodes, galvanizing, soldering, die-casting, chemicals, medicine
I. Liquid-metal element		
23.	Mercury	Thermometer, dental inlays, electric switches
II. Nonmetal minerals		
24.	Asbestos	Roofing, insulation, textile, ceramics, gasoline, solid propellants
25.	Feldspar	Artificial teeth, ceramic flux
26.	Fluorspar	Refrigerants, propellants, acid, flux
27.	Limestone	Building material, cement, reclamation of acidic soils, whitewashing
28.	Nitrates	Fertilizers, chemicals
29.	Phosphorus	Fertilizers, chemicals, detergents, medicine
30.	Potassium	Fertilizers, chemicals
31.	Salt	Chemicals, metallurgy, medicine
32.	Sulfur	Fertilizers, medicine, insecticides, acid, rubber tires, iron and steel industry

off employing this technique include the phosphate rocks.

11.5.2 Subsurface Mining

The mineral resources in deeper layers of the Earth have to be extracted horizontally. To make access to these resources, big holes are dug in the Earth's surface and then horizontal channels are made to reach the mineral deposits.

11.6 Environmental and Socioeconomic Effects of Mining

The environmental impacts of mining operations are disastrous. Disposal and processing of the mineral resources/ores further add to environmental disasters. Mining operations cause extreme harm to the environment especially when these are carried out carelessly and in an unthinkable and unethical manner. Let us know how the mining operations for bringing the minerals into human use adversely affect our environment and socioeconomic resources.

- Mining operations inevitably begin with the removal of topsoil from above the mineral deposits. Topsoil is the most important natural resource on which depend our food production systems as well as our sustainable and happy future. The topsoil constitutes the largest ecosystem on the land portion of the Earth and ecological disruptions leading eventually to the desertification of the land are triggered by its loss (Singh 2019). Its removal from the mining site costs not only our present but also our secure, sustainable, and happy future.
- Biodiversity erosion with the loss of flora and fauna from mining sites is one of the most disastrous consequences of mining operations. Large-scale deforestation takes place when many important minerals are extracted from forest areas. All the adverse consequences of deforestation come into effect when mining operations cost a forest. Wildlife is also disturbed due to forest loss. Blasting and bulldozing often carried out during mining processes drive away wildlife and also create dust clouds playing havoc with life in the surrounding area.
- Mining operations result in defacing the landscape. A land defaced and spoilt by mining is called derelict land or mine spoil. Natural picturesque values of such land are lost and the land is rendered uninhabitable.
- Mining often leads to ground subsidence that can result in the buckling of roads, bending of rail tracks, tilting of buildings, and cracks in houses. Leakage of gases from the pipes cracked due to the machinery of mining can also take place which might cause an environmental disaster.
- Mining operations severely affect the hydrological cycle and are hazardous to water resources. They may change the natural routes of rivers and streams and break down hydrological processes vital for generating springs and falls in mountain ecosystems. Mining pollutes groundwater resources as well as surface water resources. Mine wastes such as those containing radioactive substances, like uranium, may also contaminate water resources causing widespread health hazards to the people and animals dependent on these water resources.
- Mining activities in fragile ecosystems, such as in the mountains and hills, lead to even more damaging and more extensive ecological repercussions. Catastrophic landslides, rampant soil erosion, disruption of the hydrological cycle, and disappearance of small forest-fed rivers, etc. have been experienced during the limestone quarrying in the Himalayan mountains. Mining operations in the fragile ecosystems cast their shadows even up to distant plain areas due to highland-lowland ecological linkages.
- Mining operations and mineral ore processing cause air pollution by throwing particulate matter, soot, and metal particles into the atmosphere. In many cases, hazardous gases are also released into the atmosphere.

- Mining causes direct and indirect health hazards to life, including human beings. The indirect effects are related to air, water, and radioactive pollution contributed by mining. The direct effects are to be experienced by those engaged in mining operations. For example, respiratory diseases like asbestosis, silicosis, and black lung diseases are caused to the miners' exposure to suspended particulate matter (SPM) and toxic substances during mining operations. Skin diseases, headaches, vomiting sensation, reduced appetite, diminishing eyesight, etc. are the other health problems often caused to workers engaged in mining operations for a longer duration.
- There have been numerous cases of the displacement of people from the mining areas. Such a situation brings a flood of miseries to their lives. Their precious local cultures begin disintegrating and their livelihood base is destroyed. The worst hit is the tribal communities who lose their traditions, culture, livelihoods, right to lands, and unique lifestyles. They are often not given adequate compensation. Of course, no compensation can heal the wounds of displacement the victims of the mining are inflicted with.

11.7 Reclaiming the Derelict Lands

The derelict lands or mine spoils can be reclaimed or restored to a certain extent – to seminatural conditions – in the following ways:

1. **Revegetation of the mine spoils** should be taken up first by planting herbs, shrubs, and tree saplings of the species that can thrive on a mine-damaged site, or a mine spoil. As the primary vegetation establishes, intensive plantation of the site should be carried out involving native plant species. Vegetative cover on the derelict lands would be phenomenal for the gradual restoration of native flora and fauna. Fertilization of the land should be done with organic manures that would help conserve moisture and plant nutrients necessary for plant growth.
2. **Prevention of drainage discharge** would contribute to the retention of soil and moisture on the site. The natural vegetation on the mine spoils flourishes amid appropriate environmental conditions. Prevention of drainage discharge would contribute to building up appropriate environmental conditions for the ecological regeneration of the mine spoil.
3. **Adoption of eco-friendly mining technology** would phenomenally help minimize the adverse effects of mining in some cases. For instance, the microbial leaching technique for extraction of gold with the help of the bacterium *Thiobacillus ferrooxidans* is highly useful for minimizing ecological damage at goldmines.

11.8 Conservation of Minerals

As minerals are exhaustible or nonrenewable resources, once their natural deposits are exhausted, they cannot be replenished or regenerated. Therefore, in situ mineral conservation should be taken up as a task of immediate attention and top priority. The following steps would be vital for mineral conservation.

11.8.1 Decreased Consumption

Decreased consumption is an ethical obligation in a low-waste society we must strive to evolve into one. The lowest possible rates of mineral consumption are the minimum requirement of their extraction, processing, and marketing. This, as a consequence, would translate into the maintenance of the ecological integrity of the biosphere. Instead of discarding waste, durable and repairable nonrenewable products should be used and reused again and again. This would help reduce repeated requirements for fresh products.

11.8.2 Use of Mineral Waste

The waste generated by discarded nonrenewable things is nonbiodegradable and is, in fact, no waste at all. This "waste" could be used as a raw material for an industry manufacturing certain products. Use of the discarded minerals or waste of the minerals as raw material would help reduce the pressure of extraction on natural mineral deposits.

11.8.3 Reuse of Mineral Products

The use of certain items manufactured from minerals over and over again would help conserve exhaustible natural resources. For example, glass bottles and tin and aluminum containers can be used repeatedly for a very long time.

11.8.4 Recycling

Recycling is the process in which discarded mineral items (often referred to as waste) are collected, melted, and reprocessed into new usable products. For example, metallic minerals like iron, steel, silver, gold, copper, nickel, zinc, etc. can be easily recycled. A bit more expensive than other ways of mineral conservation, recycling is a promising technology applicable to most minerals, especially metallic ones.

11.8.5 Substitution

The scarce minerals in nature, if extracted at a rapid rate, would get exhausted in the near future. Such minerals can be substituted with minerals occurring in abundance. For example, gold jewelry is substituted by relatively less expensive and more abundant minerals. Scarcer materials like steel, tin, and copper in many industries have been substituted with plastics, high-strength glass fibers, alloys, ceramics, etc.

11.8.6 Alternative Energy Sources

Alternative sources of energy developed from inexhaustible energy such as solar energy, wind energy, hydro-energy, etc. would greatly help avoid mining activities necessitated for extraction of energy-generating minerals, e.g., lignite. Alternative energy sources would be instrumental in the in situ conservation of the minerals for which most of the mining operations are carried out.

11.9 Summary

Minerals are exhaustible and nonrenewable resources found in Earth's crust. The socioeconomic status of a society or a country is determined to a great extent by the mineral resources in its possession. The distribution of mineral resources varies from one geographical area to the other. The mineral deposits of economic significance are known as ores. Some 88 elements are occurring in Earth's crust. About 99% of the Earth's crust is constituted of oxygen, silicon, nickel, sodium, potassium, and uranium. The rest compose only about 1% of the crust.

The processes of the formation of mineral resources are concentration of minerals during cooling of molten rock material, the concentration of minerals due to weathering, transport and sedimentation, intense heat and pressure, evaporation of water from water bodies, and microbial actions.

Minerals are categorized into metals and nonmetals. The metallic minerals include gold, silver, copper, iron, aluminum, nickel, etc. The nonmetallic minerals include asbestos, calcium carbonate, granite, sand, gravel, phosphates, carbonates, etc. Coal and crude oil widely used as fuels for transportation, industries, food processing, etc. are also categorized as mineral resources. Copper is said to be the first metal to have been used followed by gold and silver. The nonmetallic minerals are consumed at a much faster rate and in much greater amounts than all the metallic minerals. Based on their various uses, the mineral resources can be categorized into some groups, for example, minerals used in agriculture, industry, building construction, and road construction and elements used in metal production, transport, defense equipment, building material, etc.

The process of extracting out required minerals or their ores from the Earth is known as mining. Mining is categorized into surface mining (open pit mining, dredging, and strip mining) and subsurface mining. The environmental impacts of mining operations are disastrous. Disposal and processing of the mineral resources/ores further add to environmental disasters.

Mining operations begin with the removal of topsoil, the most important natural resource on which our food production systems as well as our sustainable and happy future depend. Biodiversity erosion with the loss of flora and fauna, defacing the landscape (called derelict land or mine spoil), ground subsidence (resulting in buckling of roads, bending of rail tracks, tilting of buildings, and cracks in houses, leakage of gases from the pipes cracked due to the machinery of mining), disruption of the hydrological cycle, change in natural routes of rivers and streams, breakdown hydrological processes vital for generating springs and falls in mountain ecosystems, and air, water and soil pollution, etc. are the major consequences of mining operations. Mining causes direct and indirect health hazards to life, including human beings.

The derelict lands or mine spoils can be reclaimed or restored to a certain extent – to seminatural conditions – by revegetation of the mine spoils (by planting herbs, shrubs, and tree saplings of the native species), prevention of drainage discharge (for retention of soil and moisture on the site), and the adoption of eco-friendly mining technology (for instance, the microbial leaching technique for extraction of gold with the help of bacterium *Thiobacillus ferrooxidans* for minimizing ecological damage at goldmines).

Conservation of mineral resources is very important and it can be ensured employing decreased consumption, use of mineral wastes, reuse of mineral products, recycling, substitution, and alternative energy sources (for conserving energy-yielding minerals).

11.10 Exercises

11.10.1 Multiple-Choice Questions

1. Which of these minerals is metallic?
   (a) Sand
   (b) Stone
   (c) Phosphates
   (d) None of the above
2. Which is a nonmetallic mineral?
   (a) Mercury
   (b) Copper
   (c) Silver
   (d) None of the above
3. Which of these is a metal?
   (a) Mercury
   (b) Tungsten
   (c) Manganese
   (d) All of the above
4. Which metal is used in dentistry?
   (a) Gold
   (b) Copper
   (c) Uranium
   (d) Zinc
5. Which of these minerals is used in detergents?
   (a) Phosphorus
   (b) Sulfur
   (c) Chromium
   (d) Calcium
6. Which of the following metals is used in dental inlays and electric switches?
   (a) Nickel
   (b) Mercury
   (c) Silver
   (d) Zinc
7. Brass is an alloy made up of
   (a) silver and gold
   (b) copper and zinc
   (c) aluminum and nickel
   (d) iron and platinum
8. Which of the following is a metal alloy?

(a) Copper
(b) Zinc
(c) Brass
(d) Gold

9. Which mineral is used for electricity generation?
(a) Mercury
(b) Tungsten
(c) Thorium
(d) Mica

10. India is one of the world's largest producers of
(a) gold
(b) uranium
(c) bauxite
(d) mica

11. Which is an energy-generating mineral?
(a) Lignite
(b) Bauxite
(c) Mica
(d) All of the above

12. Bauxite ore is a source of
(a) copper
(b) aluminum
(c) gold
(d) silver

13. Uranium is obtained from
(a) magnetite ore
(b) hematite ore
(c) pitchblende or uranite ore
(d) copper pyrites

14. Which of the metals is primarily used in the production of stainless steel?
(a) Magnesium
(b) Manganese
(c) Copper
(d) Iron

15. Which country is the largest producer of crude oil in the world?
(a) Saudi Arabia
(b) Indonesia
(c) Kuwait
(d) Russia

16. When the mineral deposits lie in the water sources, they are extracted using
(a) dredging
(b) open-pit mining
(c) subsurface mining
(d) strip mining

17. What terms in column A match with the appropriate terms in column B?

Column A	Column B
(a) Derelict land	(i) Magnetite and hematite
(b) Mercury	(ii) Mine spoil
(c) Nickel	(iii) *Thiobacillus ferrooxidans*
(d) Iron	(iv) Coins
(e) Gold	(v) Liquid metal

(a) a-i, b-ii, c-iii, d-iv, e-v
(b) a-ii, b-v, c-iv, d-i, e-iii
(c) a-v, b-iv, c-iii, d-ii, e-i
(d) a-i, b-iii, c-v, d-ii, e-iv

18. The metallic mineral that can be extracted using an eco-friendly microbial leaching technique is
(a) silver
(b) gold
(c) platinum
(d) tungsten

19. The bacterium used for the extraction of gold is
(a) *Escherichia coli*
(b) *Bacillus ferrooxidans*
(c) *Thiobacillus ferrooxidans*
(d) *Clostridium* spp.

20. The dredging technique of mining is applied to the minerals found
(a) deep beneath the Earth's surface
(b) just above topsoil
(c) in the Himalayan mountains
(d) underwater mineral deposits

Answers: 1-d, 2-d, 3-d, 4-a, 5-a, 6-b, 7-b, 8-c, 9-c, 10-d, 11-a, 12-b, 13-c, 14-d, 15-a, 16-a, 17-b, 18-b, 19-c, 20-d

11.10.2 Short-Answer Questions

1. What is the difference between metallic and nonmetallic minerals?
2. Give examples of some metallic and nonmetallic minerals.

3. Name the minerals found in abundance in India.
4. What are the uses of mercury, tungsten, and copper?
5. Which region in India is famous for diamonds?
6. Mention the area in India where the gold ores are found.
7. Name the two ores of iron.
8. Where is the iron ore in India concentrated?
9. What is mining?
10. What do you mean by derelict land?

11.10.3 Long-Question Answers

1. Write a brief account of the types of minerals found in India and their distribution.
2. Mineral resources are vital for the economy of a country. Discuss explaining the socioeconomic uses of important minerals.
3. Discuss the environmental impacts of mining.
4. Explain various ways of mineral conservation.
5. Write short notes on the following:
   (a) Formation of mineral deposits
   (b) Mineral wealth of India
   (c) Eco-friendly mining

References

Misra SP, Pandey SN (2018) Essential environmental studies. Ane Books Pvt. Ltd, New Delhi. 790 pp

Rajagopalan R (2017) Environmental studies: from crisis to cure. Oxford University Press, New Delhi. 333 pp

Singh V (2019) Fertilizing the universe: a new chapter of unfolding evolution. Cambridge Scholars, London. 285 pp

12 Energy Resources

Energy is a non-thing. But, in our contemporary world, it is everything. All socioeconomic activities and all politics of the world revolve around energy resources. Energy is power. The world dances to the tune of the countries that possess energy resources in abundance. Energy is a Devil's Bargain. It is a bone of contention among the nations. At the same time, it is also a source of peace.

Energy, the term coined by Thomas Young (1773–1829), is defined as the capacity to do work. It is the energy that works. Matter does not work. Energy makes the matter work. A machine, an engine, a motor, a rocket, or a spacecraft does not work on its own. It is energy that operates them and makes them work. Everything in our economic systems, our world, and the whole universe is accomplished by energy. The human race uses energy as an input in all socioeconomic activities, as well as in improving the quality of life. Energy consumption in the world increases from year to year. In 2018, for example, the world's energy consumption rates increased by 2.9% over those of 2017. In this chapter, our focus will be only on the energy sources applicable in the socioeconomic arena of the world.

12.1 Energy as the Basis of Socioeconomic Development

The state of socioeconomic development of a defined society or a country rests on how they use the energy resources they have at their disposal. Energy, in essence, is a cornerstone of socioeconomic development. The world is dynamic due to energy. The world is in movement thanks to the overwhelming usage of energy resources. The world is progressive on account of energy. The essence is that all domestic chores, like cooking, heating, cooling, and lighting, and all socioeconomic activities are operated by the use of energy. The energy we are talking about is inanimate, not the animate energy generated by human and farm animals' muscles. However, there are many countries and many societies that are overwhelmingly dependent on animate energy. Animate energy is of significant value in agriculture in many underdeveloped countries, especially in mountain areas. In our contemporary world, it is the inanimate energy that matters.

Energy use globally as well as in individual countries has increased over time in close association with GDP (Stern 2011) which implies

V. Singh, *Textbook of Environment and Ecology*, https://doi.org/10.1007/978-981-99-8846-4_12

that energy plays a central role in economic growth. So far, conventional sources of energy comprising fossil fuels are the major determinant of the socioeconomic status of a country or a society. Generally, the higher the amount of energy resources consumed by a country, the higher the socioeconomic status it enjoys. It is the industry sector that uses the maximum proportion of the available energy resources – to the extent of 40%. With consumption of 30% of the total energy consumption, domestic and commercial sectors stand in second place. The transportation sector consumes about 24% of energy. Transportation, solely an energy-dependent sector, maintains the dynamism of the human-controlled world. Air travel has turned this world into a global village. If you want to visit a country in just one or two days, it is possible because transportation and aviation industry exhaust a pretty large proportion of global energy resources – to the extent of 24%. The rest 6% of the energy is used for other purposes, including agriculture.

Industries, factories, instruments, commercial establishments, cities, towns, and villages all consume considerable quantities of electric energy to accomplish all the necessary activities and realize desired economic goals. Dependence on energy for socioeconomic development is as inevitable as food for survival. It is the electricity generated by different energy sources (Fig. 12.1) that makes the core of the energy in its direct usable form.

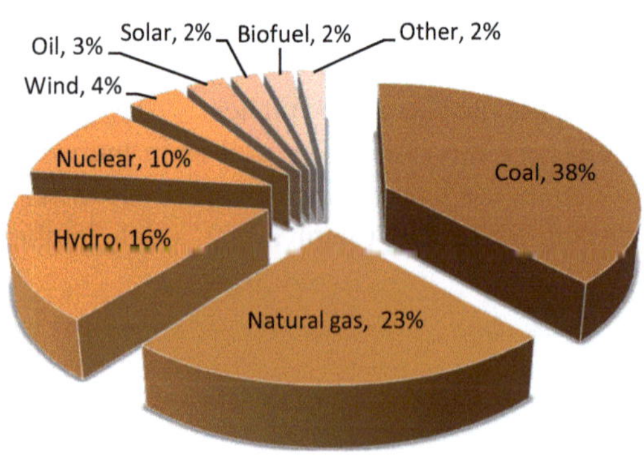

Fig. 12.1 Power generation from various energy sources worldwide in 2017

12.2 Energy as a Key to Sustainability

As the economic sectors and domestic chores operate exclusively with the help of energy, energy has to be central to sustainability processes. Further, sustainability is a dynamic process, so it will be operationalized, maintained, and sustained by constant energy inputs. Energy availability, energy usage pattern, and energy efficiency of conventional energy (fossil fuel-based) sources are counted as various indicators of the sustainability of a system. The processes leading to sustainability can only be attained by ensuring environmental safety. Thus, a system based on conventional energy sources causing environmental pollution would hardly be a sustainable one. If the source of energy is nonconventional (inexhaustible and renewable), a system would most likely be a sustainable one. Since the use of conventional energy and carbon emissions go hand in hand, consequently causing environmental pollution, global warming, and climate change, in our times, nonconventional energy sources that generate no, or minimum, environmental pollution, variously known as renewable, green, or clean energy sources, are crucial to meet the criterion of sustainability. Green energy generation to reduce emissions of greenhouse gases in energy generation, according to Agovino et al. (2018), has become necessary for achieving sustainability in the energy sector, and various international and national bodies are recommending and encouraging policies that lead to green energy generation for sustainability.

Exhaustion of conventional energy sources in due course of time, in addition to their pollution-creating behavior, is the main concern of the human race. There is much emphasis on developing and increasing the use of inexhaustible and pollution-free sources of energy, like solar energy, wind energy, tidal energy, etc. However, such energy sources have not come into use to a considerable extent. Once conventional energy sources get replaced by clean energy sources, the processes leading to sustainability would be strengthened. There can be no sustainability without ecological sustainability of its own

(Singh 2019). Clean energy is environmentally friendly and contributes to ameliorating ecological processes. Thus, clean energy is central to the processes leading to sustainability. Of course, other principles of sustainability nurturing the processes of ecological well-being of a production system are also necessary to be operationalized.

12.3 Renewable and Nonrenewable Energy Sources

Energy resources, generally, are categorized as renewable and nonrenewable ones. The rate at which we are using available energy resources affects their long-term availability. Renewable resources are those that get regenerated or replenished through natural processes within a reasonable length of time. These resources can further be classed into inexhaustible and exhaustible resources. Inexhaustible energy can be used indefinitely irrespective of the rate it is consumed at. Solar energy, wind energy, geothermal energy, tidal energy, etc. are examples of inexhaustible energy. Renewable energy can also be exhaustible if is frequently overused over a long period. For example, fuel wood or biomass energy is renewable energy, but the indiscriminate cutting of trees or the continuance of deforestation for a long time leads to the exhaustion of even renewable resources emanating from forests/trees.

Nonrenewable energy resources, on the other hand, cannot be regenerated in a reasonable length of time and stay in limited stock. Coal, peat, and petroleum are examples of nonrenewable energy resources. Another form of nonrenewable energy is nuclear energy, which is sometimes also referred to as sustainable energy. Uranium and plutonium, the sources of nuclear energy, are also limited and will get exhausted over time with the continuous use of these resources for the generation of nuclear energy.

The inexhaustible form of renewable energy creates almost no environmental pollution and is also referred to as clean energy. This energy is environment-friendly and not harmful to human health. Since this form of energy has not so far been in intensive use, this is also called new and nonconventional energy.

Nonrenewable energy sources, on the other hand, are linked with environmental pollution as they release greenhouse and various toxic gases. The overwhelming consumption of nonrenewable energy sources is largely responsible for global warming and climate change.

The generation of nonconventional and renewable energy is pretty expensive due to technical reasons. Therefore, this form of energy is not so far much affordable for the majority of people. Nonrenewable energy sources such as coal, petrol, diesel, natural gas, etc. are readily available and comparatively less expensive.

12.4 Fossil Fuels

Much of our world today owes to fossil fuels. Fossil fuels have come into being as a result of the anaerobic decomposition of buried dead organisms. These have a high percentage of carbon and, upon combustion, yield energy that is used for several purposes. Fossil fuels are millions of years old “products” of photosynthesis. These include coal, petroleum, and natural gas. Sometimes peat is also included in this category of fuels. As per the figures of 2018, the share of fossil fuels among all the energy sources was as high as 85%, of which petroleum, coal, and natural gas comprised 34%, 27%, and 24%, respectively (Agovino et al. 2018).

Fossil fuels have been generated by natural processes. However, these fuels are not considered renewable because it took millions of years for them to build up and come into existence. Further, their rates of depletion are extremely faster than the rates of their regeneration.

12.4.1 Coal

Coal is the most abundant fossil fuel in the world. World coal reserves, as of 2019, stood at 1070 billion tons. Most of the coal reserves are concentrated in the USA (23%), Russia (15%), Australia

(14%), and China (13%). Coal comprises carbon, ash (minerals), water, nitrogen, and sulfur. The average composition of coal as shown in Fig. 12.2 may vary according to the type of coal. There are three types of coal, namely, (i) anthracite or hard coal with 90% carbon and calorific value of 8500 kcal per kg, (ii) bituminous or soft coal with 80% carbon and calorific value of 7600 kcal per kg, and (iii) lignite or brown coal with 70% carbon and calorific value 5000 kcal per kg. The calorific values of different types of coal are not fixed but will vary according to the composition of the coals.

Anthracite and bituminous comprise as large as 70% proportion of the world's total coal reserves. With current consumption rates and a predicted increase in consumption rates in the future, it is estimated that the coal reserves are likely to be exhausted by the end of the twenty-first century.

On burning, coal yields heat energy which is used for many purposes, especially for electricity generation. About 40% of the world's electricity is generated in coal-fired power plants. The most coal-reliant countries in the world include China, India, the USA, Australia, Indonesia, Russia, South Africa, Germany, Poland, and Kazakhstan. China, where about 70% of the energy consumed is obtained by using coal, is the largest producer of coal (3411 million tons per annum), followed by India (692 million tons) and the USA (661 million tons). Australia, Indonesia, and Russia stand at fourth, fifth, and sixth places with annual coal production of 493, 434, and 385 million tons, respectively.

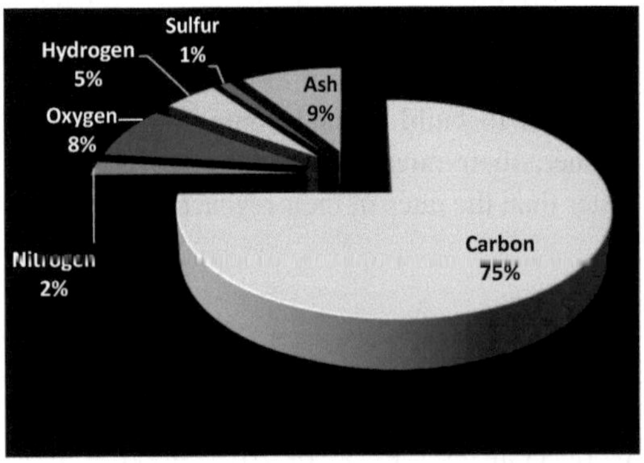

Fig. 12.2 Coal composition, with values subject to variation based on the type of coal

Global coal capacity between 2000 and 2019 increased worldwide, from 1066 gigawatts (GW) to 2045 GW. This is largely attributable to explosive growth in China and India. The use of coal for electricity generation and other purposes is notorious for creating environmental pollution, including carbon emissions contributing to global warming. A coal-fired power plant using low-quality lignite can emit as much as 1200 tons of carbon dioxide per gigawatt hour (GWh) of electricity generation. There are as many as 2425 coal-fired plants in the world, as of 2019, which are emitting about 15 billion tons of CO_2. All coal-fired plants would have to be abandoned by 2040 if the target of global atmospheric temperature rise has to be held well below 1.5 °C.

12.4.2 Petroleum

Petroleum, or crude oil, often called oil, is a yellowish-black liquid occurring in the geological formations below the Earth's surface. It is believed to have been formed millions of years ago upon burial in sedimentary rocks of excessive amounts of dead organisms, especially algae and zooplankton which were subsequently subjected to high temperature and pressure. Crude oil is divided into various fractions using the fractional distillation technique. Although natural gas and solid bitumen are also parts of petroleum, they are generally counted separately. Crude oil is a much-needed fuel by the world and is the lifeline of the global economy. This is the fuel over which many wars have been fought in the world.

Petroleum resources are concentrated in 11 countries that have formulated what is known as OPEC – Organization of the Petroleum Exporting Countries. Nearly 70% of the global petroleum reserves are found in OPEC of which Saudi Arabia alone possesses one-fourth of the reserves. Formed in 1960, to regulate production and control prices of exported petroleum, OPEC has member countries, namely, Saudi Arabia, Iran, Iraq, Algeria, Kuwait, Nigeria, Qatar, Libya, UAE, Venezuela, and Indonesia.

12.4.3 Natural Gas

Natural gas, sometimes also called fossil gas, is a naturally occurring gas found in geological formations, often in association with petroleum, and also in coal beds. It consists of a hydrocarbon gas mixture, primarily methane with varying proportions of higher alkanes. Nearly 40% of the natural gas is found in Kazakhstan and Russia.

Natural gas is widely used for cooking, heating, transport, and electricity generation. It has a high calorific value, and compared to coal and petroleum, it is a cleaner energy source. It burns without releasing smoke and is also safe for human health. LPG, that is, liquefied petroleum gas, with butane as the main content, is a domestic fuel for cooking. Ethyl mercaptan is added to LPG to give a foul smell to identify gas leakage from a safety point of view. CNG, that is, compressed natural gas, with methane as the main gas, is an alternative to diesel and petrol in transport vehicles.

12.5 Nuclear Energy

Nuclear power is generally generated by nuclear fission reactions using uranium and plutonium. A huge amount of heat is generated which then is used in steam turbines to generate electricity in a nuclear power plant or nuclear reactor (Fig. 12.3). Nuclear power can also be obtained by employing nuclear fusion and nuclear decay reactions. In nuclear fission, the nucleus of a certain isotope with large atomic mass – generally uranium-235 (U-235) – is split into lighter nuclei upon bombardment of neutrons. In the nuclear fission process, a huge amount of energy through chain reactions takes place.

Nuclear fission is the most common, almost exclusive, method to harness nuclear energy for electricity production. However, electricity generation using nuclear fusion remains an agenda in nuclear research. In this process, two isotopes of a light element are fused at extremely high temperatures to form heavier nuclei yielding enormous amounts of energy. Nuclear decay processes are applied in, for instance, isotope thermoelectric reactors.

There are 455 nuclear power plants in the world with a net capacity of 392,779 MWe. These are operating in 31 countries, as of December 2019. The USA with 95 nuclear reactors is the world's largest producer of nuclear power with 809,359 GWh of nuclear electricity followed by France with 56 nuclear reactors with 382,403 GWh. China, Russia, Japan, the Republic of Korea, and India have 49, 38, 33, 24, and 22 nuclear reactors, respectively. However, only three countries – France, Slovakia, and Ukraine – depend on the electricity produced by nuclear reactors for use by more than 50% of their population. France is on top of nuclear power generation for use by 70.6% of its population. As many as 54 nuclear reactors with a net capacity of 57,441 MWe are under construction, of which 7 reactors with the capacity of 4824 MWe are in India and 11 of 10,564 MWe in China.

The brightest side of nuclear power is that no, or minimum, emission of greenhouse gases takes place. A huge amount of power is generated by using very less amount of energy sources, uranium or plutonium. For example, only one atomic mass unit (amu) of U-235 yields energy that is equal to the combustion of 15 metric tons of coal or about 14 barrels of crude oil. Power production using nuclear reactions is quite inexpensive. The initial cost of power plant establishment might be high, but running costs are substantially low. A nuclear reactor functions well for 40–60 years. Unlike solar and wind power, nuclear power production takes place irrespective of weather conditions.

There is a negative side to nuclear power production also. Uranium is a finite resource. Dependence on such a source for electricity production would go on becoming expensive with the decreasing supplies and increasing demands in the future. The process of uranium refinement is somewhat complex and transportation of nuclear fuel is pretty risky. The disposal of radioactive waste generated in the process of electricity generation is flawed by the transmission of harmful radiation. Nuclear accidents may not be common, but the world has had a very bitter experience of two major nuclear accidents – the Chornobyl nuclear disaster in

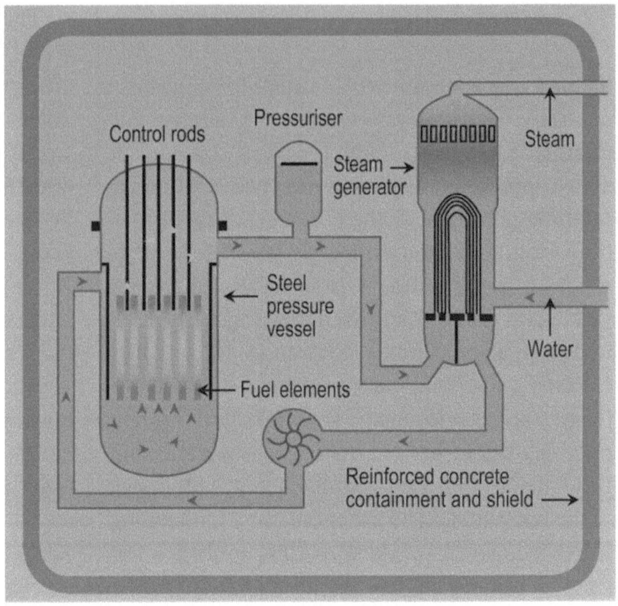

Fig. 12.3 Structural framework of a nuclear reactor

the erstwhile USSR (Union of Soviet Socialist Republics) on April 26, 1986, and the Fukushima Daiichi nuclear disaster in Japan on March 11, 2011. Accidents of this nature can result in severe and long-term consequences for both the environment and human life. The running cost of electricity generation using nuclear fuel might be quite inexpensive, but the maintenance cost involving the nuclear waste disposal process is pretty high.

12.6 Solar Energy

The sun serves as the primary source of energy, and all the various forms of energy we utilize on Earth can be traced back to solar energy, including the energy harnessed from fossil fuels – which are the products of photosynthesis in ancient times. All biomass energy is solar energy we utilize indirectly. Even the animate energy obtainable from work animals, such as bullocks, horses, camels, donkeys, etc., which emanates from human labor, is a solar energy that is derived from the plants which trap it during photosynthesis (Singh 2020). Our concern in this chapter, however, is the direct solar energy we can use for electricity generation and various other purposes.

12.6.1 Direct Solar Energy

Solar radiation propagates through space in the form of waves. As soon as the radiation strikes an object, it transforms into heat or thermal energy. This form of energy matters as it can be used for heating water, warming houses, drying food grains, cooking food, lighting houses and streets, running motor vehicles, and generating electricity for running motors for pumping out groundwater for irrigation of crops and for operating very many gadgets we use in our daily life.

Photovoltaic (PV) cells convert direct solar energy into electricity – a process known as the thermal or photovoltaic conversion of solar energy. But what would happen when sunlight is not available, for example, at night or on cloudy days? Uninterrupted electricity supply under such conditions is ensured by connecting the photovoltaic conversion device with a backup system to store and generate electricity. Extra power so generated can be transported to a grid or utility center from where it can be supplied back to the user when needed (Fig. 12.4).

Till the 1980s, photovoltaics were used for generating electricity at a very small scale, for example, for charging calculators or for availing hot water in the bathrooms through an off-grid rooftop

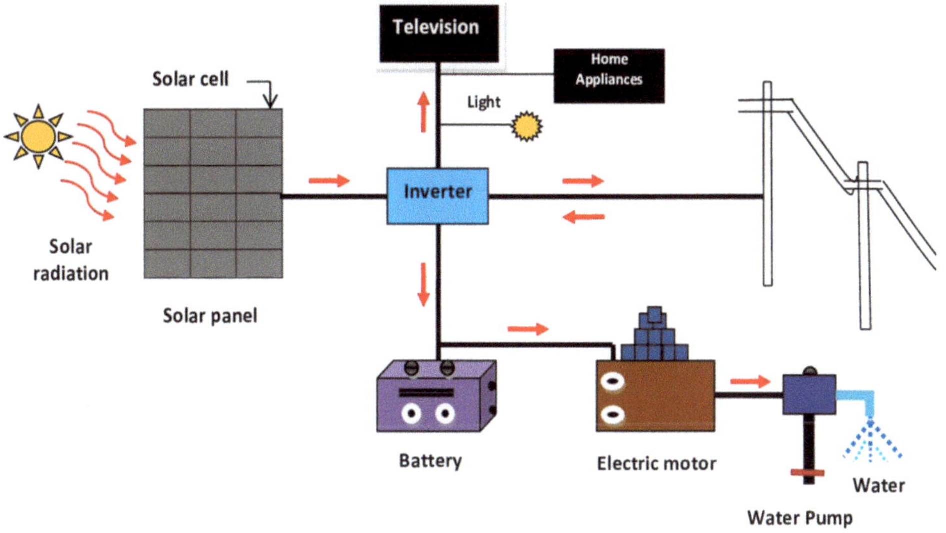

Fig. 12.4 Solar cell-generated electricity: a schematic illustration

PV system. But now grid-connected PV systems and photovoltaic power stations are emerging worldwide, and humanity seems to be heading toward a solar power revolution. PV systems and solar-based electricity are very inexpensive and people are being encouraged and given incentives to install PV systems for electricity generation at the local level. New transport vehicles operated by solar power are in the process of running on the roads, and solar energy-based airplanes are ready to take off. Ever-increasing generation of electricity employing solar energy would be a boon for the world caught in climate chaos.

Solar power is a clean source of energy. Solar energy, according to the International Energy Agency, would be the world's largest source of electricity by 2050, and most of the solar installations will be in India and China. In 2004, the world produced only 2.6 TWh of solar electricity which increased to 31.4 TWh in 2010 and 442.6 TWh in 2017 (Fig. 12.5). Thus there has been a whopping 170 times increase in solar electricity generation in the world in just 14 years. However, solar electricity share is still very low. In 2017, solar power contributed only 1.7% of the world's total electricity production. By 2050, solar electricity consumption will go up to 27%, of which 16% would be generated by PV systems and 11% through concentrated solar power. Of the ten largest PV stations, as of February 2020, five with a capacity of 5963 MW are in India. China with a capacity of 3397 MW and the USA with a capacity of 1129 MW are in second and third place, respectively. Pavagada Solar Park in the Karnataka State of India with a capacity of 2050 MW is the world's largest PV station (Fig. 12.6).

12.6.2 Biomass Energy: The Indirect Solar Energy

The human race has been utilizing biomass energy since time immemorial. Biomass energy is the solar energy that exists in the bonds of biomolecules via photosynthesis. Forest residues (e.g., fuel wood), crop residues, marine or freshwater algae, biodegradable organic wastes, animal dung, etc. that we use for various purposes all are the products of photosynthesis. The sources we derive biomass energy from are (i) solid (e.g., firewood, charcoal, animal dung), (ii) liquid (e.g., ethanol and methanol used in internal combustion engines of automobiles), and (iii) gas (e.g., methane along with mixtures of some other gases produced in digesters using animal dung and crop residues and utilized especially for cooking purpose).

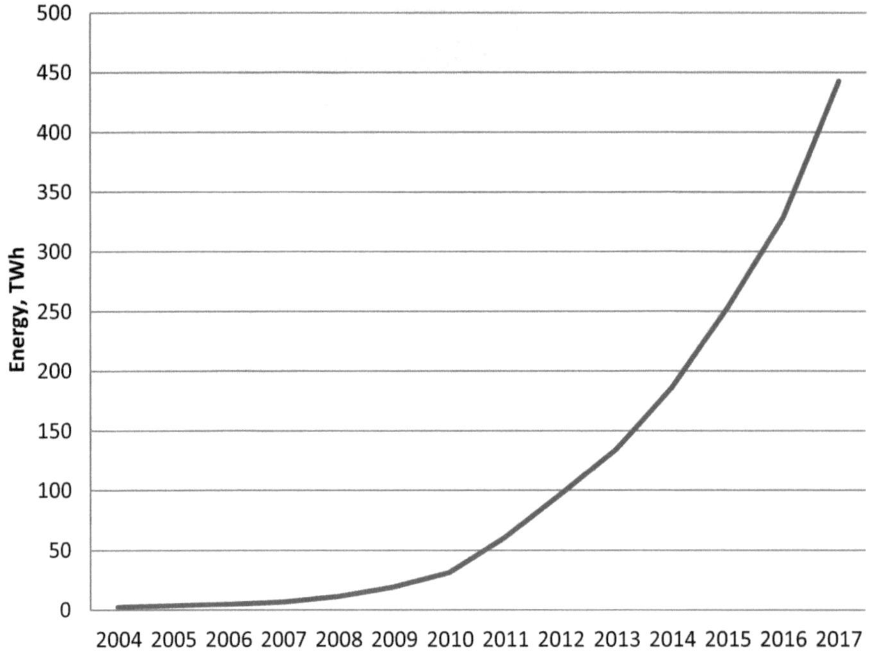

Fig. 12.5 Global solar power generation from 2004 to 2017. (Data collected from https://en.wikipedia.org/wiki/Solar_power)

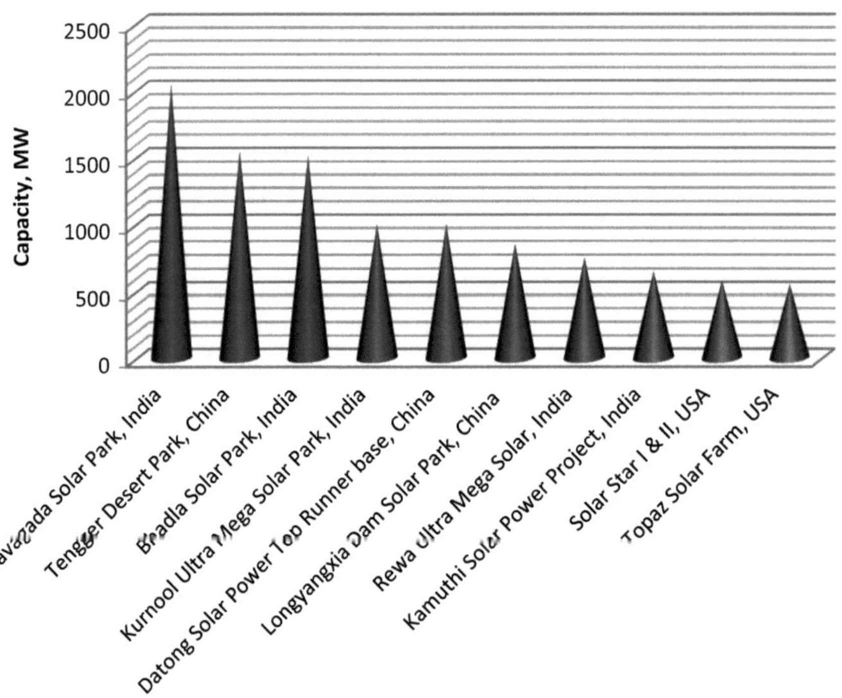

Fig. 12.6 The top ten largest PV power stations worldwide as of February 2020. (Information gathered from https://en.wikipedia.org/wiki/Solar_power)

12.7 Wind Power

The atmosphere of the planet is dynamic. The fluidity of the air – or air in motion – at a certain speed can generate electricity. The use of the kinetic energy of wind to provide mechanical power and/or generate electricity is what we call wind power or wind energy. In the process, the wind passes through wind turbines to turn electric generators to produce electricity. The use of wind power by human beings is not something new. People in the medieval period sailed by using the direction of the wind currents. People in the early days used wind power for grinding grains and pumping water. However, the use of the kinetic energy of the wind in generating electricity is what is of key significance in our times.

Wind power is an environmentally safe, clean, and sustainable form of energy. Wind power, in a sense, is driven by solar radiation. Thus, it depends on natural solar input that energizes atmospheric gases (air) to turn into winds. Wind power is an intermittent energy source. It cannot be used for energy provision uniformly or at our will. It would depend on atmospheric behavior that varies from region to region, area to area, place to place, and following seasons and weather conditions varying from time to time. Thus, the harnessing of wind power is feasible and cost-effective only in the areas or places receiving constant winds with markedly high speeds. Mountain passes, islands, coastal areas, and desert areas are some of the naturally demarcated areas suitable for running wind farms.

Wind energy in our world generally operates through what we call wind farms, that is, a group of wind turbines, for electricity generation. The area wind turbine installation covers are about 0.3 hectares per MW of electricity production (Denholm et al. 2009), but the land area between turbines can be used for crop cultivation or other purposes. A wind turbine has an upwind rotor with three blades attached to a nacelle that lies on top of a tall tubular tower.

There has been phenomenal growth in wind power installed capacity all over the world. In 1996, the cumulative wind power capacity was only 6.1 gigawatts (GW) which increased to 591 GW in 2018, registering as many as 97 times increase in the power over 23 years. The trend of wind power growth is shown in Fig. 12.7.

The Gansu Wind Farm in China is the world's largest farm with a capacity of 7965 MW followed by the Muppandal Wind Farm in India with a capacity of 1500 MW. The USA's Alta (Oak Creek – Mojave) and India's Jaisalmer Wind Park, with capacities of 1320 and 1064 MW, respectively, are in third and fourth order among the onshore wind farms.

Among the offshore wind power farms, that is, the farms installed in large water bodies, the Walney Extension in the United Kingdom is the largest one in the world at 659 MW, as of September 2018. The London Array in the United Kingdom and Gemini Wind Farm in the

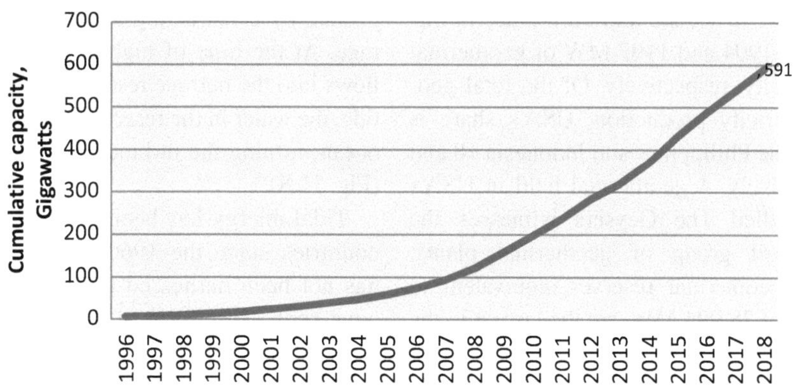

Fig. 12.7 Accumulative worldwide expansion of wind power capacity. (Data sourced from https://en.wikipedia.org/wiki/Wind_power)

Netherlands at 630 and 600 MW, respectively, are in second and third place, respectively.

12.8 Geothermal Energy

The source of geothermal energy lies deep inside the Earth. The slow decaying of radioactive particles in the Earth's core generates heat, the original form of geothermal energy. The temperature inside the Earth, at the core-mantle boundary may, goes high up to 4000 °C. Very high temperatures as well as high pressure cause some rocks to melt. In the process, the mantle portion, lighter than rock, comes upward. In the core, rock and water are heated to the extent that sometimes the temperature goes high up to 370 °C.

Water heated with geothermal energy when comes out of the Earth's surface is utilized for various purposes. People from ancient times have been using water from the hot springs or natural geysers for domestic uses, such as for bathing, cooking, and warming up houses. Hot water from the springs is also said to carry some medicinal value, especially for the treatment of skin ailments. But in our times, when we call it a source of energy, we generally refer to it to be used for electricity generation.

There was 11,700 MW of geothermal energy available worldwide, as of 2013. The USA is the leading country in respect of harnessing geothermal energy. The installed geothermal electric capacity of the USA was 3086 MW from 77 power plants, as of 2010. The Philippines and Indonesia were in second and third place in this respect, with 1904 and 1197 MW of geothermal electric capacity, respectively. Of the total geothermal electricity production, USA's share is 29% and of the Philippines and Indonesia 18 and 11%, respectively. A geothermal field in USA's California called The Geysers witnesses the world's largest group of geothermal plants. Indonesia's geothermal reserves, equivalent to the capacity of 28,994 MW, are the largest in the world. However, in terms of harnessing geothermal energy, it is far behind the USA. In India, natural geysers are common in Kullu and Manali in Himachal Pradesh, Jammu and Kashmir, and Uttarakhand.

Geothermal energy is eco-friendly, cost-effective, reliable, and sustainable. Although greenhouse gas emission takes place along with hot water coming from the Earth's interior, it is in small measure compared to that taking place upon burning of conventional fuels. Installation of geothermal plants to harness energy is quite expensive. The problem is that bound with the conditions of geological development history, geothermal energy is available only near tectonic plate boundaries.

12.9 Tidal Energy

Tidal energy, or tidal power, is artificially tapped from tidal motions in the oceans and the seas. It is a form of hydropower derived from the tides generated in marine water bodies. The heights of the tides are not uniform. The difference between low and high tides gives rise to tidal currents that can drive turbines to produce electricity. This is the form of energy that is attributable to celestial bodies, largely to the Moon and to some extent to the Sun. Gravitational forces exerted by the Moon and the Sun attract ocean waters creating tidal forces. Changing positions of these celestial bodies relative to the Earth's position, as well as the Earth's rotation and geographical location of the ocean floors and coastlines, are the factors that influence the creation of tidal forces.

Tidal waves have the power to turn a turbine and generate electricity. This can be accomplished by constructing a barrage, the tidal barrage. At the time of high tide, the ocean water flows into the barrage reservoir. When it is a low tide, the water in the reservoir flows back into the ocean, turning the turbine to produce electricity (Fig. 12.8).

Tidal energy has been in operation in some countries since the 1960s. However, so far it has not been harnessed to the extent it should have been. This form of inexhaustible, renewable, dependable, and sustainable energy has enormous potential for generating electricity and fulfilling most of the human energy needs

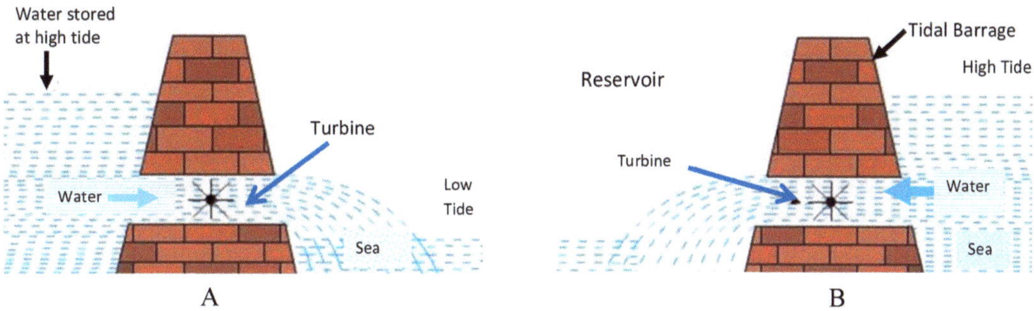

Fig. 12.8 Utilizing tidal barrage for harnessing tidal power: Turbine rotation during high tide (**a**), turbine rotation again as water flows from tidal basin to the sea (**b**)

on the planet. Although relatively more expensive to install, appropriate technologies for generating electricity out of ocean tides are also available.

The Rance Tidal Power Station with a 240 MW installed capacity in operation in France since 1966 was the world's first and the largest tidal power station before the Sihwa Lake Tidal Power Station with an installed capacity of 254 MW came into existence in South Korea in 2011. MeyGen, the largest tidal power project with an electric capacity of 398 MW in the Pentland Firth in Scotland, is under construction. South Asia's first commercial-scale tidal power plant has been planned in the Gulf of Kutch in the Gujarat state of India.

12.10 Hydroelectric Energy

Running water assumes kinetic energy. Hydroelectric energy is produced from the kinetic energy of water. The faster the speed of the running water, the greater the measure of the kinetic energy. Thus, the rivers flowing in mountain areas where voluminous amounts of water flow downward from high slopes are more suitable for hydroelectricity generation. About 17% of the world's total electricity is generated by hydropower. Again, nearly 70% of the world's renewable energy is contributed by hydroelectricity. Since hydroelectricity generation needs tapping of the lotic aquatic systems, only the countries that have such water resources in favorable geographical locations can go in for producing hydroelectricity. In total 150 countries of the world depend on hydroelectricity to some extent. The Asia-Pacific region is leading in hydropower, contributing to about one-third of the electricity produced in the world.

The methods of generating electricity using hydropower are constructing dams, running turbines on the river, constructing tidal power stations, and utilizing pumped storage. Among them, damming the rivers and creating a huge reservoir are the conventional and the most talked about way of generating electricity. Most of the hydroelectricity in the world is generated in this way. When a river is dammed in the highlands, hills, or mountains, a huge water reservoir is created behind the dam. The potential energy of the water in the reservoir is used to drive the water turbine and generator. The height difference between the source and outflow and the volume of the reservoir water determine the extent to which potential water energy can be used in generating electricity.

China is the leading country producing hydroelectricity with 1064 terra-watt hour (TWh) annual hydroelectricity production, followed by Canada and Brazil with 383 and 373 TWh annual production. Norway produces as much as 96% of its electricity from hydropower. Next in order are Venezuela, Brazil, and Canada producing 68.3, 63.2, and 58.3% of their electricity using hydropower. The proportion of hydroelectricity to the electricity from all other sources is quite small in other countries.

12.10.1 Pros and Cons of Hydroelectric Power

Hydroelectricity production is a boon as well as a bane in many respects. Wherever a large dam damming a river comes into existence, it becomes a matter of controversy. Hydroelectricity is the most controversial form of usable energy that changes the socioecological and environmental scenario of the area where it is generated and the socioeconomic scenario where it is supplied and used. Table 12.1 presents a brief account of the pros and cons of hydroelectric power.

12.11 Biogas Energy

Organic matter originates with photosynthesis and in the process accumulates energy which gets released during its degradation. If this energy is tapped, it can be used for various purposes. If the organic matter is subjected to microbial degradation, the gases highly useful for producing energy are produced. Organic matter is biodegradable. Subjected to the actions of anaerobic organisms, or methanogen bacteria, the biodegradable organic matter results in the production of biogas, mainly methane along with other gases. Burning of methane yields energy that can be used for various purposes, for example, for cooking, heating, lighting, and electricity generation. Production of the fuel gases employing anaerobic bacterial decomposition is what the biogas energy is.

To produce biogas, the biodegradable organic wastes are fermented under anaerobic conditions in a closed system variously known as a digester, anaerobic digester, biodigester, or bioreactor. Food wastes, industrial wastes, municipal wastes, plant biomass, manure, sewage, and any other

Table 12.1 Pros and cons of hydroelectric power

Pros	Cons
1. **A Long-lasting energy asset** Hydroelectric projects might last over 100 years and is quite reliable.	1. **Ecological ruin** Very large area in the valleys is inundated, valuable flora and fauna get destroyed, and many native species might be pushed toward extinction.
2. **Low operating cost** Once project begins working, the operation costs incurred are quite low.	2. **En masse displacement** A large proportion of the population inhabiting a valley area to be inundated behind dam have to be displaced. This process cuts the cultural roots of people bringing them unending miseries.
3. **Clean energy** Use of electricity does not lead to greenhouse gas emissions.	3. **Water pollution** The running water has power of purification, but when it becomes stagnant in the dam, it gets polluted and becomes a source of many waterborne diseases.
4. **A multipurpose project** Hydropower projects, in addition to electricity generation, provide other benefits, such as irrigation, drinking water supplies to large population, navigation, water spots, fish production, etc.	4. **Greenhouse gas emissions** Methane emission from the submerged area takes place contributing to global warming.
	5. **Risk of large-scale failure** Catastrophic events, such as excessive soil erosion, landslides, etc., may cause damage to a dam. Weight of the artificial water body might induce what is called as reservoir-induced seismicity (RIS). A large dam in a seismologically prone area, such the Hindu Kush-Himalayas, might be extremely risky.

biodegradable organic material may be used for biogas production upon fermentation in a digester under anaerobic conditions. In India, biogas is more popular as gobar gas, as it is the gobar (dung) which is the main, often the sole, input meant for fermentation in the digesters and produces methane mainly used for cooking purposes.

Biogas is primarily a methane gas which is the energy-yielding output, along with carbon dioxide, water vapor, and traces of hydrogen sulfide and other gases. Generally, biogas is used for cooking food at a mass scale. However, it can be brought into use for many other purposes. After removing CO_2 gas, the biogas so produced can also be compressed and used as fuel to drive motor vehicles and may also be brought in industrial use as well as for the generation of electricity. The residue left in the digesters, called slurry, serves as a very good nitrogen-rich organic fertilizer for ameliorating soil fertility.

Eventually as a "byproduct" of photosynthesis, biogas is a renewable gas as its production-and-use cycle continues. As an indirect solar energy, biogas is a clean gas in the sense that there is no net production of CO_2 and CH_4 is utilized as fuel. Biogas production is most appropriate in rural areas where dung voided by livestock can be efficiently utilized in this process.

Biogas plants used in India are typical of two types, namely, floating gas holder type and fixed dome type. The floating gas holder type consists of a well-shaped digester with a partition wall. One portion of the digester is fed with a dung-water mixture through an inlet pipe. The other portion is connected with an outlet pine for discharging spent slurry. An inverted steel drum floats to hold the biogas generated through fermentation by methanogen bacteria (Fig. 12.9).

The fixed dome type consists of a single unit in the digester along with inlet and outlet chambers. The roof, generally made of bricks and cement, is dome-shaped holding the biogas generated in the digester (Fig. 12.10).

12.12 Energy from Urban Waste

The generation of waste is proportional to an increase in populations, changing lifestyles, consumption patterns, and degree of industrialization. Cities and towns of the world generally produce solid wastes disproportionate to their management capacities. According to an estimate, the world's cities produced about 18 billion tons of waste in 2020. These wastes chiefly serve as a source of environmental pollution and pollution-related diseases. The environmental pollution emanating from huge amounts of urban waste includes greenhouse gas emissions contributing to global warming and climate

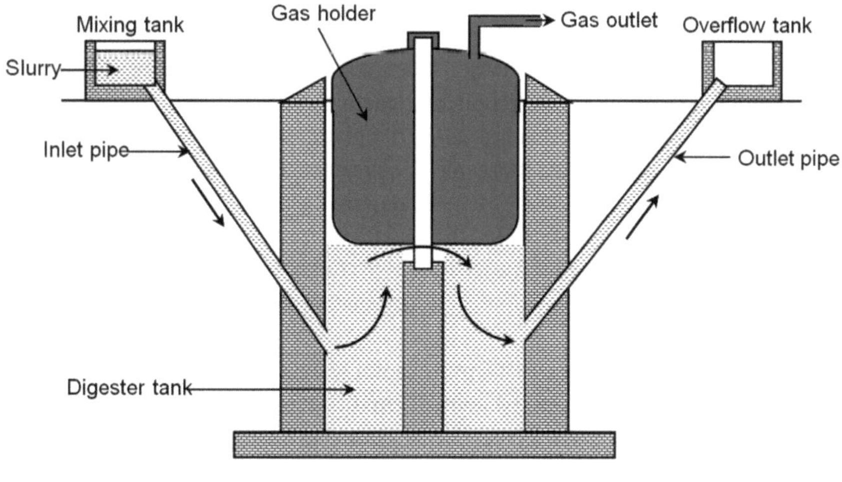

Fig. 12.9 Floating gas holder type biogas plant

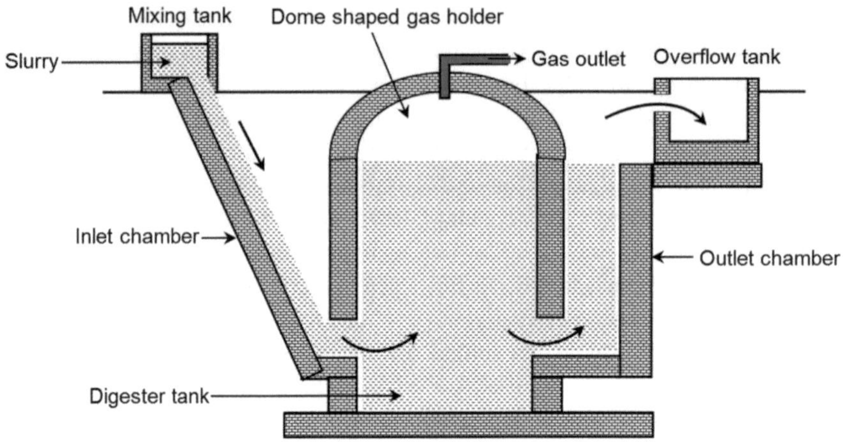

Fig. 12.10 Fixed dome type biogas plant

change. Disposal sites in urban areas are too limited to accommodate "mountains" of waste unabatedly generated by the domestic, commercial, and industrial establishments in the towns and cities.

A ray of hope is that the wastes are also a promising source of energy that is so valuable for socioeconomic development. If recovery of energy from the wastes – or what is popularly known as waste-to-energy – is done, it offers to address pollution and pollution-related diseases as well as the paucity of renewable energy often faced by urbanites. Another ray of hope is that technologies – physical, thermal, and biological types – necessary for implementing waste-to-energy processes are available.

- Physical technologies are used to make the waste more useful as a source of fuel energy.
- Thermal technologies are used to produce heat energy, fuel oil, or synthetic natural gas (syngas) from organic as well as inorganic wastes.
- Biological technologies are usable in the anaerobic biodegradation of organic wastes to yield energy.

The thermochemical technologies extracting energy out of wastes are of three types: (i) combustion, (ii) gasification, and (iii) pyrolysis. The combustion process is carried out in the excess of air, gasification in reduced air, and pyrolysis in the absence of air.

12.12.1 Pyrolysis of Municipal Wastes

Thermal decomposition of organic wastes in the absence of oxygen involves both simultaneous and successive reactions producing a variety of energy-rich compounds and yielding energy for various purposes including electricity generation. The temperature at the start of the reactions under nonreactive conditions – that is, in the absence of oxygen – is between 350 and 550 °C and goes as high as 700–800 °C. Carbon-rich wastes are especially suitable for use in pyrolysis. From a mix of municipal solid wastes (MSW), wastes of inorganic origin, such as glasses, metallic items, nonreactive wastes, etc., are separated, and the remaining organic matter is subjected to a pyrolysis reactor. To maintain the very high temperature required for pyrolysis, the reactor is connected to an external heating source (Fig. 12.11).

Pyrolysis of MSW yields a range of useful products, namely, syngas (a high calorific value gas), bio-oil or pyrolysis oil (a biofuel), and char (a solid residue). Syngas, or synthesis gas, is a mixture of energy-rich gases with CO, H_2, CH_4, water vapor, and a range of VOCs (volatile organic compounds) as combustible constituents of it. Net cal-

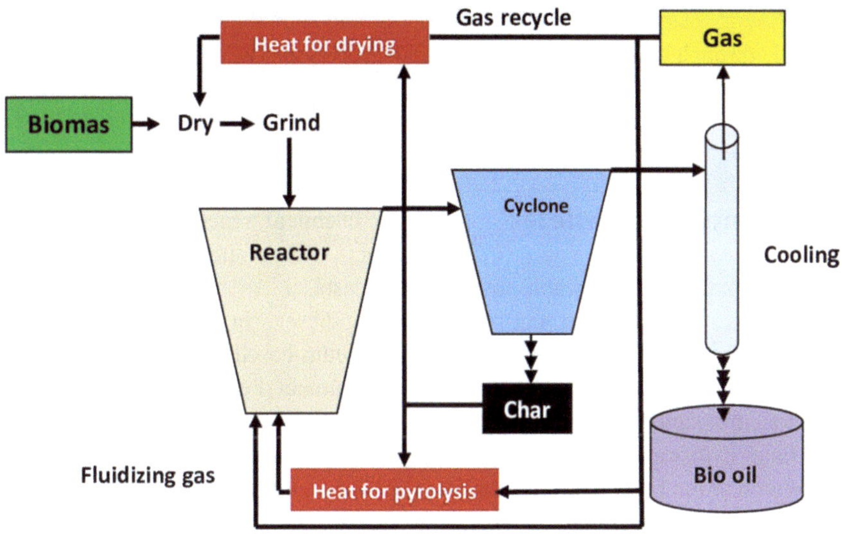

Fig. 12.11 An outline diagram of pyrolysis

orific values of syngas range between 10 and 28 MJ/Nm^3 varying according to the type of organic matter and temperature inside the pyrolysis reactor. Impurities of particulate matter, soluble matter, and hydrocarbons are removed to clean the syngas, which, then, can be combusted to generate energy and turn the same into electricity. Syngas, as a basic chemical, can also be used in the petrochemical and petroleum refining industries.

Bio-oil, the other product of pyrolysis, is a dark brown liquid that can be used in diesel engines and in a gas turbine to produce electricity. It is also used in a wide range of chemicals and the manufacturing of many organic compounds.

The solid residual product, char, obtainable during pyrolysis of MSW contains noncombustible matter in addition to carbon. It can be used in the manufacture of activated carbon and soil amendment.

12.13 Dendrothermal Energy

The energy emanating from plants, or dendrothermal energy, is a renewable form of energy but not inexhaustible. Once the plant biomass is burnt to release energy, it is used up and cannot be replaced. We need to be dependent on growing plants, again and again, to keep pace with the continuous supply of energy sources. Dendrothermal energy is achievable at some cost of environmental pollution, which is a negative side of the energy source. However, plantations for achieving the energy source compensate for greenhouse gas release by employing carbon sequestration following photosynthesis.

A dendrothermal energy base is built up by an energy plantation. Energy plantation especially for fuel wood and charcoal is an ambitious project in the Philippines and many other countries. In India, energy plantation on a large scale has been tried in Rajasthan and Madhya Pradesh.

Although achieving self-sufficiency in energy at local levels is the main objective of the energy plantation, many other advantages – tangible, as well as intangible – are achievable through energy plantation. There are several indigenous plants that, in addition to energy-yielding biomass, also offer fruits and other edible products, fodder, fiber, fertilizer, ayurvedic medicines, etc. They ameliorate soil fertility by pulling nutrients from deeper layers of land and by adding soil organic matter. Energy plantation enhances landscape value and, through photosynthesis, helps absorb atmospheric CO_2, thus contributing to climate regulation.

12.14 Petroplants and Biodiesel

Petroplants are alternative and renewable sources of liquid fuel that can replace diesel and petrol. They provide hydrocarbons in a liquid form worth substituting conventional fuels. The hydrocarbons these specific plants contain can be easily converted into hydrocarbons that occur in petroleum. A search of the petroplants as alternative sources of the fuels conventionally used in motor vehicles began during the days of World War II when parts of Europe experienced a petroleum crisis. *Euphorbia abyssinica*, a plant species yielding latex, was successfully used in a gasoline refinery. Later on, many other plants, especially *Euphorbia lathyris*, *E. tirucalli*, and *Calotropis procera*, were used for the same purpose. By now nearly 400 latex-yielding plants as alternative fuel sources belonging to Euphorbiaceae, Urticaceae, Convolvulaceae, Asclepiadaceae, Sapotaceae, and Apocynaceae have been screened for their hydrocarbons convertible into petroleum hydrocarbons. Some 15 of them have been found more promising.

One interesting aspect about the petroplants belonging to the family Euphorbiaceae can grow in the wasteland and normally do not need irrigation. *Jatropha* as a source of biofuel has especially attracted attention in recent years and has been planted intensively in many areas of the world. Its oil, popular as biodiesel, can be used in diesel engines and can also be refined into gasoline. Many studies to look into the effects of *Jatropha* plantation on food crops have also been carried out (Rastogi et al. 2019; Rastogi and Singh 2019, 2020), in which it has been concluded that this petroplant does not provide a healthy environment for the cultivable food crops. Therefore, *Jatropha* should be cultivated on degraded wastelands or on land not used for growing food crops.

Biodiesel is a diesel consisting of esters of long-chain fatty acids derivable from vegetable oils and animal fats and can be used in propelling diesel engines as pure or as a blend. According to the US Biodiesel Board, biodiesel is a mono-alkyl ester. Biodiesel is prepared by transesterification, a chemical process separating glycerin from vegetable oil or fat. Of the two products of the chemical reaction – glycerin and methyl esters – the latter is the chemical name of biodiesel.

Biodiesel, in comparison to conventional petroleum-based diesel, is much better for the environment, because it:

(a) Enhances diesel engine life as it is a better lubricant than fossil-fuel-based diesel
(b) Reduces CO_2 lifecycle to the extent of about 86%
(c) Decreases SO_2 emission by up to about 100%
(d) Decreases hydrocarbon emissions by about 70%
(e) Reduces particulate matter pollution by about 50%
(f) Reduces the formation of smog

12.15 Energy Sources for the Future

As energy plays a quintessential role in socioeconomic growth and the happiness of the humanity, its availability in its clean form is one of the most important needs to usher in an environmentally safe, healthy, and vibrant future. Environmental quality and climate pattern of our biosphere critically depend on the source of energy we overwhelmingly depend upon. Concern about climate change is growing, and the energy sources we use today will have a definite impact – positive or negative – on global climate. And there is a set of energy sources that would certainly have a positive bearing on the environment, climate, human health, and, of course, our socioeconomic progress. All the forms of nonconventional, inexhaustible, and renewable energy sources – solar power, wind power, tidal power, and geothermal and hydroelectric energy – that we have discussed in this chapter are very promising for the future. The following innovative and clean energy sources can also be developed and brought into use at a mass scale.

12.15.1 Space-Based Solar Power

So far we have harnessed solar energy as it falls on Earth's surface. However, the space-based solar power system can also be developed and technology for the same is available. Much of the solar radiation is absorbed as it enters into the outermost atmospheric layer of the Earth, and before striking Earth's surface and passing through the four atmospheric zones, it "cools down" considerably. High-intensity solar radiation flows uninterruptedly in outer space that can be harnessed there itself and transmitted to Earth for use.

Installation of giant solar farms in space would be instrumental in collecting huge amounts of high-intensity solar radiation with the help of giant mirrors and subsequently reflecting it to smaller solar collectors and then transmitting it to Earth in the form of laser beams or microwaves. Harnessing solar power in space for generating electricity and for other purposes on Earth would avert all sorts of energy crises humanity on Earth faces. Initiatives on the solar-based power system are being taken in a few countries, but the world is yet to be practically dependent on it.

12.15.2 Hydrogen Energy

Hydrogen is one of the promising energy sources for the future. Hydrogen on the planet is available in combination with many elements, for example, with oxygen in the water. In its independent form (H_2), it serves as an energy source, yielding large amounts of energy, at the rate of 150 kJ per gram. Taking into consideration such a huge amount of energy, hydrogen is regarded as an ideal source of energy we can be dependent on. Hydrogen from water can be readily separated through photolysis, electrolysis, or thermal dissociation.

Hydrogen in the environment is inflammable and highly explosive. Its handling, therefore, is pretty difficult. Further, owing to its lightness, its storage is difficult, and its transport is also quite tedious. Hydrogen is a clean source of energy with zero pollution. It can be used as a fuel in hydrogen-burning engines. NASA is making use of hydrogen energy, in the form of liquid hydrogen, for charging its space shuttles for years.

12.15.3 Fuel Cells

A fuel cell is an electrochemical cell enabled to convert the chemical energy of a fuel into electricity. Conversion of chemical energy into electrical energy involves a pair of redox reactions generally using hydrogen as a fuel and oxidizing agent, oxygen. There are various types of fuel cells, but they all in common consist of an anode, a cathode, and an electrolyte, allowing H^+ ions, or protons, to move between two sides of a fuel cell. At the anode, a catalyst causes oxidation of the fuel generating H+ ions and electrons. Movement of the ions takes place from the anode to the cathode through the electrolyte. Simultaneously, the movement of electrons from the anode to the cathode takes place via an external circuit, generating electricity (direct current). At the cathode, another catalyst induces reactions of ions, electrons, and oxygen, forming water (Fig. 12.12).

So far, the most common chemical reaction in a fuel cell is based on hydrogen and oxygen:

$$2H_2 + O_2 \rightarrow 2H_2O$$
$$E = 1.23\,V \text{ under standard conditions}$$

A fuel cell, generally, does not work under standard, or even optimal, conditions. Therefore, it generates power of about 0.6–0.7 V. It provides clean energy and other useful product, the water.

A fuel cell is distinguishable from most batteries in the sense that, to sustain, it requires a continuous supply of fuel (H_2) and O_2 (from the air). Most batteries, on the other hand, generate energy from metals and their ions or oxides. Fuel cells generate electricity continuously subject to the uninterrupted supplies of fuel and oxygen. Apart from hydrogen, fuel cells can also work on carbon-rich materials and other energy sources. A variety of fuel cells – for example, PDMFC (passive direct methanol fuel cells), DEFC (direct ethanol fuel cells), PEMFC (proton exchange membrane fuel cells), MCFC (molten carbonate fuel cells), SOFC (solid oxide fuel cells), and PAFC (phosphoric acid fuel cells) – have been developed and are in operation.

Hydrogen-run fuel cells are being used by NASA to provide energy to launch manned space

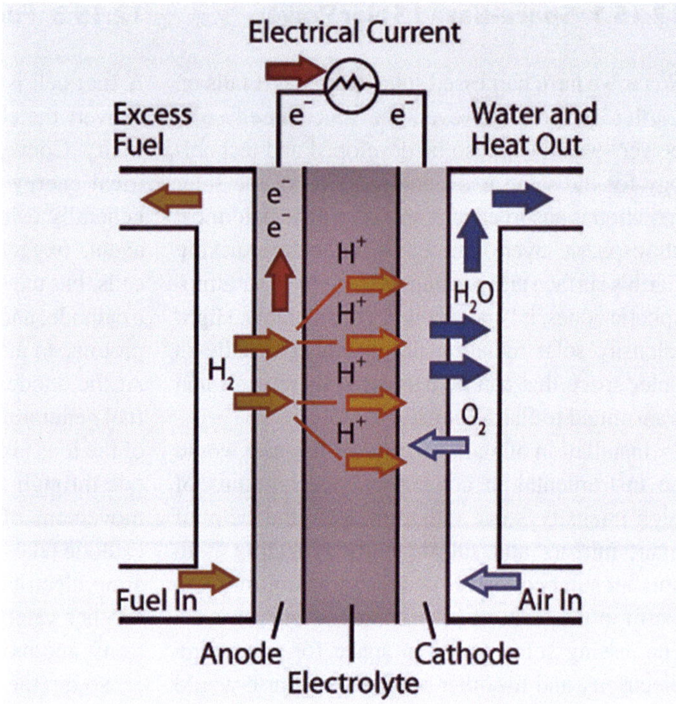

Fig. 12.12 Functioning of a fuel cell with hydrogen as a fuel and oxygen an oxidizer

vehicles. Production of water, in addition to electricity, is a very important advantage accrued to astronauts. They are also being increasingly used in cars. In the years to come, most transport vehicles are likely to operate by fuel cells.

12.16 Summary

Energy is defined as the capacity to do work. The human race uses energy as an input in all socioeconomic activities, as well as in improving the quality of life. Energy plays a central role in the economic growth of a country. So far, conventional sources of energy comprising fossil fuels are the major determinant of the socioeconomic status of a country or a society. Generally, the higher the amount of energy resources consumed by a country, the higher the socioeconomic status it enjoys. If the source of energy is nonconventional (inexhaustible and renewable), a system would most likely be a sustainable one. Once conventional energy sources get replaced by clean energy sources, the processes leading to sustainability would be strengthened. Energy resources, generally, are categorized as renewable and nonrenewable ones. Renewable resources are those that get regenerated or replenished through natural processes within a reasonable length of time. These resources can further be classed into inexhaustible and exhaustible resources. Inexhaustible energy can be used indefinitely irrespective of the rate it is consumed at. Solar energy, wind energy, geothermal energy, tidal energy, etc. are examples of inexhaustible energy. Nonrenewable energy resources, on the other hand, cannot be regenerated in a reasonable length of time and stay in limited stock. The inexhaustible form of renewable energy creates almost no environmental pollution and is also referred to as clean energy. Clean energy is environment-friendly and not harmful to human health.

Among fossil fuels, coal is the most abundant fossil fuel in the world. There are three types of coal, namely, anthracite or hard coal, bituminous or soft coal, and lignite or brown coal. Anthracite and bituminous comprise as large as 70% proportion of the world's total coal reserves. About 40% of the world's electricity is generated in coal-fired power plants. Petroleum, or crude oil, is the much-needed fuel by the world today and is the lifeline of the global economy. Natural gas is a naturally occurring gas found in geological formations, often in association with petroleum, and also in coal beds. Nearly 40% of the natural gas is found in Kazakhstan and Russia.

Nuclear power is generally generated by nuclear fission reactions using uranium and plutonium. However, electricity generation using nuclear fusion remains an agenda in nuclear research. There are some 455 nuclear power plants in the world operating in 31 countries with a net capacity of 392,779 MWe. A huge amount of power is generated by using very less amount of energy sources, uranium or plutonium. The disposal of radioactive waste generated in the process of electricity generation is flawed by the transmission of harmful radiation.

Solar power is a clean source of energy. This energy source is expanding worldwide at a very fast rate. Biomass energy is the solar energy that exists in the bonds of biomolecules via photosynthesis. The use of the kinetic energy of wind to provide mechanical power and/or generate electricity is what we call wind power or wind energy. Wind energy in our world generally operates through wind farms – that is, a group of wind turbines, for electricity generation. The source of geothermal energy lies deep inside the Earth. Slow decaying of radioactive particles in the Earth's core generates heat, the original form of geothermal energy. Tidal energy, or tidal power, is a form of hydropower derived from the tides generated in marine water bodies. Hydroelectric energy is produced from the kinetic energy of water. The methods of generating electricity using hydropower are constructing dams, running turbines on the river, constructing tidal power stations, and utilizing pumped storage. Among them damming the rivers and creating a huge reservoir are the conventional and the most talked about the way of generating electricity. This way of generating hydroelectricity has many serious ecological, environmental, and social repercussions. Production of the fuel gases employing anaerobic bacterial decomposition is what the biogas energy is. If recovery of energy from the wastes – or what is popularly known as waste-to-energy – is done, it offers to address pollution and pollution-related diseases as well as the paucity of renewable energy often faced by urbanites. Thermal decomposition of organic wastes in the absence of oxygen involves both simultaneous and successive reactions producing a variety of energy-rich compounds and yielding energy for various purposes including electricity generation. Pyrolysis of MSW yields a range of useful products, namely, syngas (a high calorific value gas), bio-oil or pyrolysis oil (a biofuel), and char (a solid residue). The energy emanating from plants, or dendrothermal energy, is a renewable form of energy, but not inexhaustible. Petroplants are alternative and renewable sources of liquid fuel that can replace diesel and petrol. They provide hydrocarbons in a liquid form worth substituting conventional fuels.

Environmental quality and climate pattern of our biosphere critically depend on the source of energy. There is a set of energy sources that would certainly have a positive bearing on the environment, climate, human health, and, of course, our socioeconomic progress. Space-based solar power, hydrogen energy, and fuel cells open new gates for usable and sustainable energy for the future.

12.17 Exercises

12.17.1 Multiple-Choice Questions

1. Which of the following invertebrate animals depends on chemosynthesis for its energy requirements?
   (a) *Riftia pachyptila*
   (b) *Octopus vulgaris*
   (c) *Pheretima posthuma*
   (d) None of the above

2. What sector does consume the maximum proportion of available energy in the world?
   (a) Industry
   (b) Agriculture
   (c) Domestic
   (d) Transport
3. At present most electricity worldwide is generated using
   (a) Coal
   (b) Natural gas
   (c) Bio-fuel
   (d) Solar power
4. What is the most abundant fossil fuel in the world?
   (a) Petroleum
   (b) Natural gas
   (c) Coal
   (d) Peat
5. What is the correct descending order of carbon percentage and calorific value in three types of coal?
   (a) Bituminous > anthracite > lignite
   (b) Bituminous > lignite > anthracite
   (c) Anthracite > bituminous > lignite
   (d) Lignite > bituminous > anthracite
6. Which of the following is a correct match?
   (a) Anthracite – hard coal
   (b) Bituminous – soft coal
   (c) Lignite – brown coal
   (d) All of the above
7. Which of the following is the world's largest coal-producing country?
   (a) India
   (b) China
   (c) USA
   (d) Russia
8. What is added to liquid petroleum gas (LPG) used for cooking food to give a foul smell to identify gas leakage from a safety point of view?
   (a) Ethyl mercaptan
   (b) Acetyl acid
   (c) Methyl gas
   (d) Ammonia
9. Which country is on top of nuclear power generation for use by 70.6% of its population?
   (a) Ukraine
   (b) Russia
   (c) France
   (d) Norway
10. With an electric capacity of 2050 MW, which is the world's largest PV station at present?
    (a) Pavagada Solar Park, India
    (b) Tengger Desert Park, China
    (c) Rewa Ultra Mega Solar, India
    (d) Topaz Solar Farm, USA
11. Gansu Wind Farm, the world's largest wind farm, is located in
    (a) Romania
    (b) Ukraine
    (c) China
    (d) Brazil
12. Which is the leading country in respect of harnessing geothermal energy?
    (a) India
    (b) China
    (c) Indonesia
    (d) USA
13. Which energy source on Earth is determined by the gravitational force of the Moon and to some extent by that of the Sun?
    (a) Geothermal energy
    (b) Tidal energy
    (c) Fossil fuel energy
    (d) Solar energy
14. The world's largest tidal power station, the Sihwa Lake Tidal Power Station, with an installed capacity of 254 MW, came into existence in 2011 in
    (a) South Korea
    (b) Vietnam
    (c) Thailand
    (d) Cambodia
15. Which country produces as much as 96% of its electricity from hydropower?
    (a) Romania
    (b) Russia
    (c) Norway
    (d) Costa Rica
16. Floating gas holder and fixed dome are the types of
    (a) Nuclear reactors
    (b) Biogas plants
    (c) Tidal barrage
    (d) Alcohol-manufacturing plants

17. The energy generation processes taking place in the excess air, reduced air, and absence of air, respectively, are
    (a) Combustion, gasification, and pyrolysis, respectively
    (b) Gasification, combustion, and pyrolysis, respectively
    (c) Pyrolysis, gasification, and combustion, respectively
    (d) Combustion, pyrolysis, and gasification, respectively
18. Dendrothermal energy is the energy obtainable from
    (a) Refined petroleum
    (b) Plants
    (c) Natural geysers
    (d) Soft coal
19. What is special about the plants belonging to Euphorbiaceae, Urticaceae, Convolvulaceae, Asclepiadaceae, Sapotaceae, and Apocynaceae families?
    (a) They provide methyl alcohol.
    (b) They are latex-yielding plants of high medicinal value.
    (c) They are specially used for biogas production.
    (d) They have hydrocarbons convertible into petroleum hydrocarbons.
20. What energy source has hydrogen as a fuel and oxygen as an oxidizing agent?
    (a) PV cell
    (b) Geothermal energy
    (c) Biogas energy
    (d) Fuel cell

Answers: 1a, 2a, 3a, 4c, 5c, 6d, 7b, 8a, 9c, 10a, 11c, 12d, 13b, 14a, 15c, 16b, 17a, 18b, 19d, 20d

12.17.2 Short-Answer Questions

1. What is energy?
2. Write some examples of conventional and nonconventional energy sources.
3. What is the difference between renewable and nonrenewable energy sources?
4. What energy sources are responsible for global warming?
5. What are the three types of coal?
6. What do you mean by nuclear fission?
7. What is geothermal energy?
8. What is the difference between gasification and pyrolysis?
9. What are petroplants?
10. What are the advantages of biodiesel?

12.17.3 Long-Answer Questions

1. How are the conventional energy sources causing environmental disruption and climate change? Discuss citing facts and figures.
2. How can solar power solve energy crises facing the world?
3. Discuss the pros and cons of hydroelectricity.
4. What are the energy sources you would like to be evolved for the future? Discuss from the viewpoint of environmental health, climate change mitigation, and sustainability.
5. Write short notes on the following:
    (a) Wind power
    (b) Tidal power
    (c) Fuel cells

References

Agovino M, Garofalo A, Romano AA, Scandurra G (2018) Explanatory analysis of the key factors in an energy sustainability index. Qual Quant 52:2597–2632. https://doi.org/10.1007/s11135-017-0679-0

Denholm P, Hand M, Jackson M, Ong S (2009) Land–use requirements of modern wind power plants in the United States. Technical report NREL/TP-6A2–45,834. National Renewable Energy Laboratory, Colorado. 40pp

Rastogi A, Singh V (2019) Energy budget of Jatropha-based cropping systems in Tarai area of Central Himalayas, India. Ind J Agri Sci 89(8):1298–1302

Rastogi A, Singh V (2020) Allelopathic effects of Jatropha curcas stem extracts on food crops in Tarai area. Ind J Agri Sci 90(7):1277–1281

Rastogi A, Singh V, Arunachalam A (2019) Allelopathic effects of aqueous leaf extract of Jatropha curcas L. on food crops in the Himalayan foothills. J Emerg Technol Innovative Res 6(6):866–875

Singh V (2019) Fertilizing the universe: a new chapter of unfolding evolution. Cambridge Scholars Publishing, London. 285 pp

Singh V (2020) Environmental plant physiology: botanical strategies for a climate smart planet. CRC Press (Taylor and Francis), Boca Raton. 214 pp

Stern DI (2011) The role of energy in economic growth. CCEP Working Paper 3.10. Centre for Climate Economics & Policy, Crawford School of Economics and Government, The Australian National University, Canberra.

Section III

Biodiversity

Biodiversity: Concepts and Values 13

Biodiversity is a fundamental concept in ecology and conservation biology, encompassing the variety of life on Earth at various levels of organization. It is a measure of the richness and abundance of different species, ecosystems, and genetic variations. Understanding the concepts and values of biodiversity is crucial for recognizing its importance and formulating effective conservation strategies. In this chapter, we will explore the definition of biodiversity, the levels at which it operates, the patterns it forms along gradients, and the diverse values it holds for human societies and the planet.

13.1 Definition of Biodiversity

Biodiversity refers to the variety of life forms, including plants, animals, microorganisms, and the ecosystems in which they occur. It encompasses three main components: genetic diversity, species diversity, and ecosystem diversity.

(a) **Genetic diversity:** Genetic diversity refers to the variation within species, both within and between populations. It includes the diversity of genes, alleles, and genetic traits that enable species to adapt and evolve over time.
(b) **Species diversity:** Species diversity represents the variety of different species present in a particular area or on the planet as a whole. It includes species richness (the number of species) and species evenness (the relative abundance of different species).
(c) **Ecosystem diversity:** Ecosystem diversity refers to the variety of habitats, communities, and ecological processes occurring in a given region. It encompasses different types of ecosystems, such as forests, wetlands, grasslands, coral reefs, and deserts, each with their unique assemblages of species and ecological interactions.

13.2 Levels of Biodiversity

Biodiversity operates at multiple levels, each with its own unique patterns and processes. The levels of biodiversity encompass genetic, species, and ecosystem diversity.

13.2.1 Genetic Diversity

Genetic diversity operates within populations of a species, as well as between different populations. It is essential for species' adaptability, resilience, and ability to respond to environmental changes.

13.2.2 Species Diversity

Species diversity operates at the scale of different species occurring within a given area or globally. It

V. Singh, *Textbook of Environment and Ecology*, https://doi.org/10.1007/978-981-99-8846-4_13

provides the basis for ecological interactions, such as competition, predation, and mutualism, shaping the structure and functioning of ecosystems.

13.2.3 Ecosystem/Community Diversity

Ecosystem diversity operates at the scale of entire ecosystems, encompassing the variety of habitats and ecological communities present in a particular region. It influences the flow of energy and nutrients, the stability of ecosystems, and the provision of ecosystem services. These are many factors that affect ecosystem diversity, as shown in Fig. 13.1.

A community is the biotic content of an ecosystem. Alpha, beta, and gamma diversity are three components of community biodiversity, each representing a different aspect of species richness and composition within a given area. These measures help us understand the distribution and diversity of species at different scales.

Alpha Diversity Alpha diversity refers to the diversity within a specific location or habitat, such as a local ecosystem or a single community. It represents the number of species present and their relative abundances within that particular area. Alpha diversity measures the richness and evenness of species within a local scale and provides insights into the biodiversity of a specific site.

Beta Diversity Beta diversity measures the turnover or change in species composition between different habitats or locations. It quantifies the differences in species composition among different ecosystems within a region. Beta diversity is a measure of species turnover or replacement and reflects the heterogeneity of habitats or environmental conditions across the landscape. It helps us understand how species composition changes as we move from one habitat or location to another.

Gamma Diversity Gamma diversity represents the total species diversity within a large geographic region or entire landscape. It is a measure

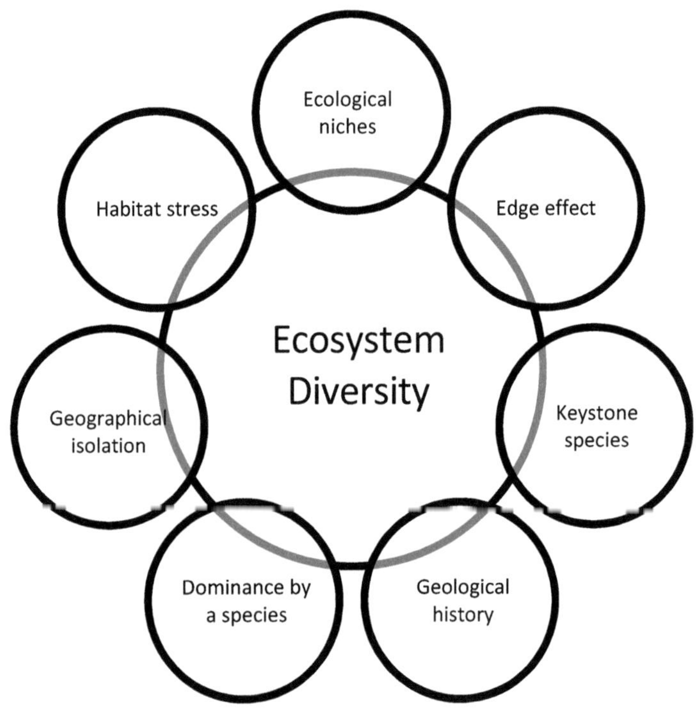

Fig. 13.1 The factors affecting ecosystem diversity

of biodiversity at a regional or global scale, encompassing multiple habitats, ecosystems, and communities. Gamma diversity takes into account the combined species richness and composition of all the local habitats within the larger area. It provides an overview of the overall biodiversity within a region and can help identify areas of high species richness and conservation priority.

These three measures (Fig. 13.2) together provide a comprehensive understanding of biodiversity within and between habitats, contributing to our knowledge of the overall diversity patterns in a given region.

13.3 Biodiversity Gradients

Biodiversity exhibits patterns along spatial, temporal, and environmental gradients. These gradients reflect variations in species richness, genetic diversity, and ecosystem complexity across different geographical regions or environmental conditions.

(a) **Latitudinal gradients:** Latitudinal gradients refer to the changes in biodiversity from the poles to the equator. Generally, species richness increases toward the tropics, with the highest diversity found in tropical rainforests. This pattern is influenced by factors

Fig. 13.2 Alpha, beta, and gamma diversity: the three dimensions of community diversity (based on Singh 2023)

such as climate stability, evolutionary history, and energy availability.

(b) **Altitudinal gradients:** Altitudinal gradients describe changes in biodiversity with increasing elevation. As we move up mountains, the number of species tends to decrease due to changes in temperature, precipitation, and habitat availability. However, unique species adapted to specific altitudinal zones can also be found.

(c) **Habitat gradients:** Habitat gradients refer to biodiversity variations across different types of habitats or ecosystems. For example, coral reefs exhibit high species diversity due to their complex structure, while deserts have lower species diversity due to harsh environmental conditions.

13.4 Values of Biodiversity

Biodiversity holds immense value for both human societies and the planet as a whole. It provides a wide range of ecosystem services and cultural and aesthetic benefits and has intrinsic value as a part of Earth's natural heritage.

13.4.1 Ecosystem Services

Biodiversity plays a critical role in providing ecosystem services such as pollination, nutrient cycling, water purification, climate regulation, and soil formation. These services are essential for human well-being, food production, and the functioning of Earth's systems.

13.4.2 Cultural and Aesthetic Values

Biodiversity contributes to the cultural identity and well-being of human societies. It provides inspiration for art, music, literature, and spiritual beliefs. Natural landscapes, diverse species, and wildlife also have aesthetic values that enrich our lives.

13.4.3 Scientific and Educational Values

Biodiversity serves as a vast source of scientific knowledge and discovery. Studying different species and ecosystems helps us understand ecological processes and evolutionary mechanisms and develop new technologies inspired by nature.

13.4.4 Intrinsic Value

Biodiversity has intrinsic value, meaning it possesses worth and rights independent of its usefulness to humans. Each species has its unique evolutionary history and contributes to the complexity and beauty of life on Earth.

Understanding the concepts and values of biodiversity is vital for promoting its conservation and sustainable use (Singh 2020, 2023). By recognizing the various levels of biodiversity, the patterns it forms along gradients, and the multitude of values it holds, we can make informed decisions to safeguard the intricate web of life upon which we depend. Preserving biodiversity is not only an ethical imperative but also crucial for maintaining the resilience and long-term sustainability of our planet.

13.5 Summary

The chapter explores the concept of biodiversity, which refers to the variety of life forms and ecosystems on Earth. Biodiversity is comprised of genetic diversity, species diversity, and ecosystem diversity. Genetic diversity refers to variation within and between species, allowing for adaptation and evolution. Species diversity refers to the number and abundance of different species in an area. Ecosystem diversity encompasses different habitats and ecological communities.

Biodiversity operates at multiple levels, including genetic, species, and ecosystem diversity, each with its own patterns and processes. Biodiversity exhibits gradients along spatial, temporal, and environmental factors. Latitudinal gradients show higher diversity toward the tropics, while altitudinal gradients decrease with elevation. Habitat gradients also influence biodiversity, with more diverse habitats exhibiting higher species richness. Biodiversity holds immense value, providing ecosystem services like pollination, nutrient cycling, and climate regulation. It also has cultural, aesthetic, scientific, educational, and intrinsic values. Recognizing the importance of biodiversity is crucial for effective conservation and sustainable practices. Preserving biodiversity is essential for the long-term sustainability of the planet and our own well-being.

13.6 Exercises

13.6.1 Multiple-Choice Questions

1. Which of the following best defines biodiversity?
   (a) The total number of species in an ecosystem
   (b) The variety of life forms on Earth
   (c) The genetic diversity within a species
   (d) The distribution of species across different habitats forms on Earth
2. What is the primary cause of biodiversity loss worldwide?
   (a) Climate change
   (b) Pollution
   (c) Habitat destruction
   (d) Invasive species
3. Which of the following is an example of genetic biodiversity?
   (a) A rainforest containing various plant species
   (b) A coral reef with a diverse array of marine organisms
   (c) A population of cheetahs with different coat patterns
   (d) A wetland supporting different bird species
4. Which level of biodiversity encompasses the variety of ecosystems on Earth?
   (a) Species diversity
   (b) Genetic diversity
   (c) Ecosystem diversity
   (d) Functional diversity
5. Which of the following is an example of an ecosystem service provided by biodiversity?
   (a) Carbon sequestration
   (b) Soil erosion
   (c) Acid rain
   (d) Noise pollution
6. What is the main reason for the conservation of biodiversity?
   (a) Aesthetic value
   (b) Economic value
   (c) Cultural value
   (d) Ecological value
7. Which of the following is a direct threat to marine biodiversity?
   (a) Deforestation
   (b) Desertification
   (c) Overfishing
   (d) Urbanization
8. Which of the following is an example of an endangered species?
   (a) White-tailed deer
   (b) Bald eagle
   (c) House cat
   (d) Domestic cow
9. What is the term used for complete disappearance of a particular species?
   (a) Extinction
   (b) Endemism
   (c) Migration
   (d) Speciation
10. Which of the following is the largest contributor to the loss of tropical rainforest biodiversity?
    (a) Logging
    (b) Agriculture
    (c) Climate change
    (d) Mining
11. Which group of organisms is considered the most diverse in terms of species?
    (a) Mammals
    (b) Birds

(c) Insects
(d) Reptiles

12. What is the term used to describe the variety of different species within a given area?
(a) Species richness
(b) Species evenness
(c) Species abundance
(d) Species dominance
13. Which of the following would give highest measure of biodiversity?
(a) Sahara Desert
(b) Arctic tundra
(c) Amazon rainforest
(d) Australian outback
14. Which factor is not a major threat to freshwater biodiversity?
(a) Water pollution
(b) Habitat destruction
(c) Overfishing
(d) Desertification
15. What is the term used to describe the introduction of exotic species into a new ecosystem?
(a) Inbreeding
(b) Hybridization
(c) Natural selection
(d) Biological invasion
16. Which of the following is an example of a keystone species?
(a) Polar bear
(b) Honeybee
(c) Oak tree
(d) Grasshopper
17. What is the primary factor responsible for the loss of coral reef biodiversity?
(a) Pollution
(b) Overfishing
(c) Ocean acidification
(d) Habitat destruction
18. Which international agreement aims to conserve biodiversity and promote sustainable use of natural resources?
(a) Kyoto Protocol
(b) Paris Agreement
(c) Montreal Protocol
(d) Convention on Biological Diversity
19. Which of the following is an example of an ecosystem engineer?
(a) Earthworm
(b) Blue whale
(c) Bald eagle
(d) Tiger
20. What is the term used to describe the process by which new species evolve from existing species?
(a) Evolution
(b) Speciation
(c) Extinction
(d) Adaptation

Answers: 1-b, 2-c, 3-c, 4-c, 5-a, 6-d, 7-c, 8-b, 9-a. 10-b, 11-c, 12-a, 13-c, 14-d, 15-d, 16-b, 17-c, 18-d, 19-a, 20-b

13.6.2 Short-Answer Questions

1. What is biodiversity?
2. What are the three levels of biodiversity?
3. What is genetic diversity?
4. What is species diversity?
5. What is ecosystem diversity?
6. What is the difference between ecosystem diversity and community diversity?
7. What do you mean by the diversity within a community?
8. What factors influence ecosystem diversity?
9. What is species richness?
10. What is the value of biodiversity?

13.6.3 Long-Answer Questions

1. What is the concept of biodiversity, and why is it important?
2. What are the different types of biodiversity and their significance?
3. What are the levels of biodiversity, and how do they interact?
4. How can the value of biodiversity be recognized and protected?

5. Write short notes on the following:
   (a) Intrinsic value of biodiversity
   (b) Alpha, beta, and gamma diversity
   (c) Cultural and aesthetic values of biodiversity

References

Singh V (2020) Environmental plant physiology: botanical strategies for a climate smart planet. Taylor and Francis (CRC Press), Boca Raton/London. https://doi.org/10.1201/9781003014997. 230 pp

Singh V (2023) Biodiversity: concepts, crises, and conservation. NIPA, New Delhi. 190 pp

14 Threats to Biodiversity

Biodiversity is the foundation of life on Earth, encompassing the incredible variety of living organisms and their interactions within ecosystems. However, the world is currently facing a biodiversity crisis, with species disappearing at an alarming rate. This chapter explores some of the major threats to biodiversity and their impact on the delicate balance of life on our planet.

14.1 Habitat Destruction

One of the primary threats to biodiversity is habitat destruction. Human activities such as deforestation, urbanization, and agriculture have significantly altered or destroyed natural habitats, pushing many species to the brink of extinction (WWF 2020). The loss of habitats directly affects species that rely on specific environments, disrupting their life cycles, feeding patterns, and reproductive behaviors.

14.1.1 Habitat Fragmentation

Division of a large intact and integrated habitat into two or multiple segments is what we mean by habitat fragmentation. In our contemporary world desperate for economic development, encountering an integrated habitat would be a rare case. An extensive network of roads throughout the world has led to the fragmentation of habitats. Railway tracks, powerlines, canals, tourist spots, etc. are the other causes of habitat fragmentation. Habitat fragmentation threatens biodiversity in many ways, such as by reducing species' dispersal and colonization potential, disturbing ecological niches, and declining foraging ability of wild animals.

14.1.2 Introduction of Exotic Species

Introduction of exotic species poses a significant threat to biodiversity. When non-native species are introduced into new ecosystems, they often lack natural predators or competitors, allowing them to multiply rapidly and outcompete native species for resources such as food, water, and habitat. This can lead to the displacement or even extinction of native species, disrupting the delicate balance of the ecosystems. Exotic species can also introduce new diseases, parasites, or pathogens that native species have no natural defenses against, further compromising their survival. The loss of native species can have far-reaching consequences, affecting not only the ecological balance but also the functioning of ecosystems and the services they provide to humans. Therefore, the careful management and control of exotic species introductions are crucial to preserving biodiversity and maintaining healthy ecosystems.

V. Singh, *Textbook of Environment and Ecology*, https://doi.org/10.1007/978-981-99-8846-4_14

14.1.3 Deforestation

The clearing of forests for timber extraction, agriculture, and urban expansion has led to the destruction of vast areas of critical habitat worldwide. Forest ecosystems are home to a significant proportion of Earth's terrestrial biodiversity, including numerous plant and animal species. Deforestation disrupts the delicate balance of these ecosystems, resulting in the loss of biodiversity and contributing to climate change.

14.1.4 Urbanization

The rapid growth of cities and towns has led to the conversion of natural habitats into concrete landscapes. Urban areas fragment ecosystems, isolating populations and reducing gene flow between them. This fragmentation can lead to reduced genetic diversity and increased vulnerability to environmental changes, making species more susceptible to extinction.

14.1.5 Agriculture

The expansion of agricultural land, particularly for monoculture crops, has resulted in the destruction of natural habitats and the loss of biodiversity. Large-scale agriculture often involves the use of pesticides, herbicides, and chemical fertilizers, which can have adverse effects on both target and nontarget species, including pollinators and soil organisms.

14.1.6 Environmental Pollution

Environmental pollution plays a significant role in the loss of biodiversity (Parmar et al. 2023). Various forms of pollution, such as air, water, and soil pollution, have detrimental effects on ecosystems and the species that inhabit them. Air pollution, resulting from the emission of harmful gases and particulate matter from industrial activities and vehicle exhaust, can lead to acid rain and smog, affecting plants, animals, and their habitats. Water pollution, caused by the discharge of toxic chemicals, pesticides, and waste materials into water bodies, not only harms aquatic organisms directly but also disrupts entire food chains and ecosystems. Additionally, soil pollution, caused by the accumulation of pollutants like heavy metals and pesticides, can adversely impact plant growth, impairing the availability of food and shelter for various species. These forms of pollution contribute to habitat degradation, reduced reproductive success, genetic mutations, and overall population declines. As a result, the loss of biodiversity occurs as species struggle to adapt or face extinction due to the adverse impacts of environmental pollution. Therefore, addressing and mitigating pollution is crucial for the preservation of biodiversity and the long-term sustainability of our planet.

14.2 Man-Wildlife Conflicts

As human populations continue to expand and encroach upon natural habitats, conflicts between humans and wildlife have become more common. These conflicts arise when human activities negatively impact wildlife populations or when wildlife poses threats to human lives, livelihoods, or property. Man-wildlife conflicts can have severe consequences for biodiversity conservation.

14.2.1 Habitat Encroachment

When humans expand their settlements into wildlife habitats, it disrupts the natural behavior and movement patterns of animals. As a result, animals may venture into human-occupied areas in search of food or shelter, leading to conflicts with humans. This situation often ends with the persecution or displacement of wildlife, contributing to population declines and biodiversity loss.

14.2.2 Poaching and Illegal Wildlife Trade

The illegal trade in wildlife and their parts remains a significant threat to many species. Poaching for bushmeat, ivory, skins, and traditional medicines has driven numerous species to the brink of extinction. The demand for exotic pets and ornamental plants has also led to the illegal capture and trade of various species, further depleting biodiversity.

14.3 Susceptibility to Extinction

While all species face some level of risk, certain factors make some more susceptible to extinction. Understanding these factors is crucial for effective conservation planning and prioritization (Fig. 14.1).

14.3.1 Feeding at High Trophic Level

The most stable populations in the biosphere are those of the photosynthesizers followed by those closest to photosynthesis, that is, of herbivores. As the species deviate away from photosynthesis, they go on becoming vulnerable, the top carnivores being the most vulnerable one. The reason is that the species feeding at high trophic level often face scarcity of the preys they feed on. The flow of energy and nutrients to them via food chains is further reduced if the producers (photosynthesizers) in an ecosystem are extracted away due to human's socioeconomic activities. Lions, cheetahs, and other top carnivores, thus, are more vulnerable to extinction than the species feeding at lower trophic levels.

14.3.2 Small Population Size and Genetic Factors

Species with small population sizes are particularly vulnerable to extinction due to the increased effects of genetic drift, low fecundity, inbreeding, and reduced genetic diversity. These factors reduce a population's ability to adapt to changing environments and increase the risk of genetic disorders.

14.3.3 Specialist Species and Habitat Specificity

Specialist species that rely on specific habitats or have narrow ecological niches are more susceptible to extinction. If their habitats are destroyed

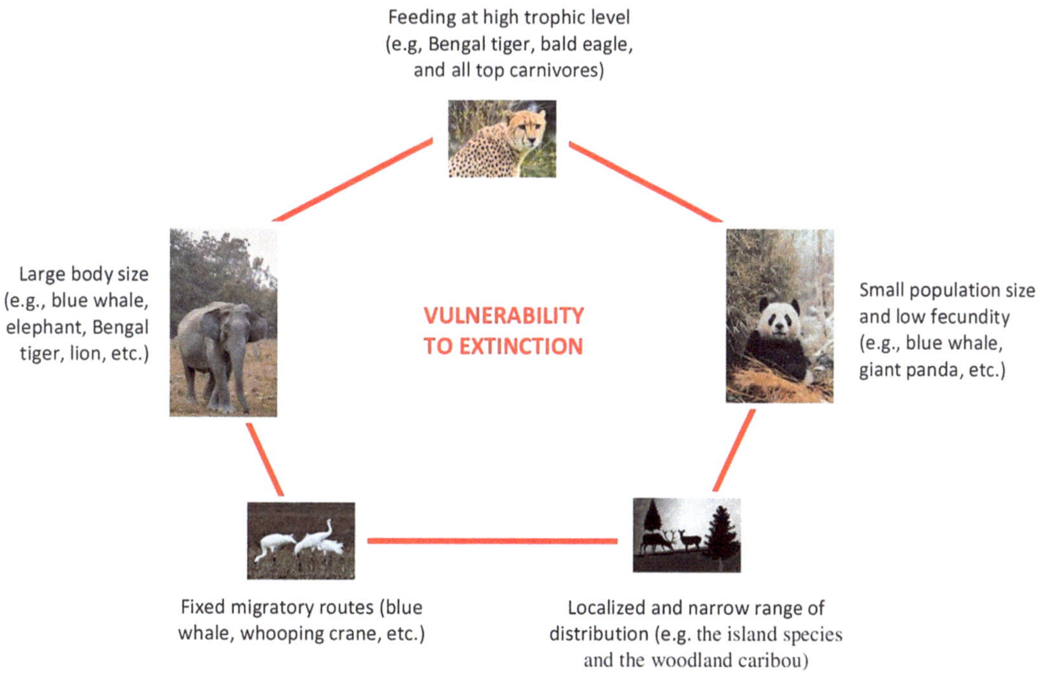

Fig. 14.1 Some factors contributing to enhancing species' vulnerability to extinction. (Source: Based on Singh (2023))

or significantly altered, specialist species may struggle to find suitable alternatives, leading to population declines and potential extinction.

14.3.4 Endemic Species and Geographic Isolation

Endemic species, found only in specific geographic regions, face a higher risk of extinction. Their limited distribution makes them more vulnerable to habitat loss, climate change, and other threats. Additionally, geographic isolation can lead to reduced genetic diversity and restrict the ability of populations to recover from disturbances.

14.3.5 Fixed Migratory Routes

Some species, such as blue whale and whooping crane, follow fixed migratory routes. Such species may be threatened by certain obstructions, for example, adverse environmental conditions, habitat destruction, lack of food, increased chances of encountering predatory species, etc. during their migration through fixed routes.

14.3.6 Large Body Size

Some species with large size, notably blue whale, elephant, Bengal tiger, lion, etc., are more likely to encounter critical problems like ecological space and sustainably meeting food requirements corresponding to their body size.

14.4 IUCN Red List

The International Union for Conservation of Nature (IUCN) Red List plays a vital role in assessing and categorizing the conservation status of species worldwide. The Red List categories provide a standardized framework for identifying species at different levels of risk (IUCN 2019).

Extinct, extinct in the wild, and critically endangered: Species categorized as extinct no longer exist, while those listed as extinct in the wild survive only in captivity. Critically endangered species face an extremely high risk of extinction in the near future if immediate conservation measures are not implemented.

Endangered and vulnerable: Endangered species face a high risk of extinction, while vulnerable species are at a lower but still significant risk. Both categories highlight the urgent need for conservation actions to prevent further population declines and potential extinctions.

Near threatened and least concern: Near-threatened species are close to qualifying for a threatened category, indicating that they may become vulnerable to extinction in the future. Species categorized as least concern are considered to have a lower risk of extinction, although ongoing monitoring is essential to ensure their long-term survival.

The threats to biodiversity outlined in this chapter highlight the urgent need for collective action to preserve and protect Earth's ecosystems and the myriad species they support. By addressing habitat destruction, mitigating man-wildlife conflicts, understanding susceptibility to extinction, and utilizing tools like the IUCN Red List, we can strive to reverse the current trajectory of biodiversity loss and secure a sustainable future for all life on our planet.

14.5 Mass Extinction of Species

Once upon a time, on a planet teeming with diverse and wondrous life, a series of cataclysmic events unfolded, forever altering the course of evolution (Singh 2023). These events would come to be known as the five mass extinction events. Each event left its mark, reshaping the Earth's inhabitants and reshuffling the tapestry of existence (Table 14.1).

The Ordovician-Silurian Extinction, occurring approximately 443–416 million years ago, witnessed a decline in marine life. Trilobites, brachiopods, and corals suffered greatly. The possible causes were linked to climate change, widespread oceanic anoxia, and intense volcanic activity.

Table 14.1 The five mass extinction events, impacted species affected, and their potential causes

Event	Time period	Species affected	Possible reason
The First Mass Extinction: Ordovician-Silurian	443–416 million	Marine life, trilobites, brachiopods, and corals	Climate change, oceanic anoxia, and volcanic activity
The Second Mass Extinction: Late Devonian	359–299 million	Fish, trilobites, corals, and early amphibians	Global cooling, marine anoxia, and asteroid impact
The Third Mass Extinction: Permian-Triassic	252 million	Marine life, trilobites, insects, and amphibians	Massive volcanic eruptions and global warming
The Fourth Mass Extinction: Triassic-Jurassic	201 million	Marine life, large amphibians, and reptiles	Climate change, volcanic activity, and asteroid impact
The Fifth Mass Extinction: Cretaceous-Paleogene	66 million	Dinosaurs, marine reptiles, ammonites, and plants	Asteroid impact, climate change, and volcanic activity

During the Late Devonian Extinction, between 359 and 299 million years ago, fish, trilobites, corals, and early amphibians experienced significant losses. Global cooling, marine anoxia, and possibly an asteroid impact were believed to be the contributing factors.

The Permian-Triassic Extinction, occurring 252 million years ago, marked the most severe mass extinction event in Earth's history. Marine life, trilobites, insects, and amphibians were among the affected species. Massive volcanic eruptions and subsequent global warming were considered the primary causes.

The Triassic-Jurassic Extinction, around 201 million years ago, witnessed the decline of marine life, large amphibians, and reptiles. Climate change, volcanic activity, and a potential asteroid impact were the probable reasons behind this catastrophic event.

Finally, the Cretaceous-Paleogene Extinction, occurring 66 million years ago, resulted in the extinction of dinosaurs, marine reptiles, ammonites, and numerous plant species. The event was triggered by an asteroid impact, which caused a global catastrophe, including widespread wildfires, climate disruption, and the subsequent collapse of ecosystems.

The stories of these mass extinction events serve as a stark reminder of the fragility of life on our planet. They emphasize the importance of preserving biodiversity and actively working to mitigate the impacts of environmental changes, ensuring the survival and flourishing of countless species for generations to come.

Every species, like the individuals of a species, has also its own age. At a certain point of time, a species has also to go extinct. More than 99.9% of the living species that ever lived on Earth have vanished forever. According to an estimate as many as five billion species have gone extinct during different epochs over the evolutionary history of the living Earth.

14.6 The Sixth Mass Extinction

The Sixth Mass Extinction, often referred to as the Holocene extinction or Anthropocene extinction, is an ongoing global phenomenon characterized by a significant loss of biodiversity. It is named as such because it represents the sixth major extinction event in the Earth's history, with previous events including the well-known extinction of the dinosaurs. What sets the Sixth Mass Extinction apart from previous ones is that it is primarily driven by human activities, specifically the impacts of habitat destruction, overexploitation of resources, introduction of invasive species, climate change, and environmental pollution.

Human-induced habitat destruction is one of the primary drivers of species extinction. Activities such as deforestation, urbanization, and conversion of natural habitats for agriculture or infrastructure result in the loss, fragmentation, and degradation of ecosystems. This directly impacts numerous plant and animal species that rely on these habitats for survival, disrupting their natural life cycles and pushing them toward extinction.

The Sixth Mass Extinction represents a grave threat to global biodiversity. It is estimated that species extinction rates are currently 1000–10,000 times higher than natural background rates. The loss of species has far-reaching consequences, affecting ecosystem functioning, food security, human well-being, and the stability of the planet's ecosystems as a whole.

Addressing the Sixth Mass Extinction requires urgent and collective action. Conservation efforts, habitat restoration, sustainable resource management, and measures to mitigate climate change and pollution are crucial for halting the loss of biodiversity. Recognizing the importance of biodiversity and implementing sustainable practices are key steps toward ensuring a healthy and resilient planet for future generations.

14.7 Summary

This chapter explores the threats to biodiversity and their impact on the delicate balance of life on Earth. The primary threats include habitat destruction due to habitat fragmentation, introduction of exotic species, deforestation, urbanization, agriculture, and environmental pollution. Man-wildlife conflicts are sharply pronounced into biodiversity depletion. Many species flourishing in an ecosystem might be susceptible to extinction. Their susceptibility pushes them toward extinction due to slight changes in the environmental/climatic conditions. Human activities, notably deforestation, urbanization, and agriculture, have significantly altered or destroyed natural habitats, pushing many species to the brink of extinction. The introduction of non-native species disrupts ecosystems and can lead to the displacement or extinction of native species. Environmental pollution, including air, water, and soil pollution, negatively affects ecosystems and species. Man-wildlife conflicts arise when human activities negatively impact wildlife populations or when wildlife poses threats to human lives or property. Species with small population sizes, specialist species, and endemic species are particularly vulnerable to extinction. The International Union for Conservation of Nature (IUCN) Red List categorizes species based on their conservation status. The five mass extinction events that have taken place on the living planet wiped out numerous species from the planet. The ongoing Sixth Mass Extinction, primarily driven by human species, poses a significant threat to global biodiversity.

14.8 Exercises

14.8.1 Multiple-Choice Questions

1. Which of the following is considered a major driver of biodiversity loss?
   (a) Habitat destruction
   (b) Genetic modification
   (c) Ecotourism
   (d) Climate change
2. Invasive species pose a threat to biodiversity because they:
   (a) Increase genetic diversity
   (b) Enhance ecosystem stability
   (c) Compete with native species for resources
   (d) Promote species conservation
3. What is the primary reason for overexploitation of species?
   (a) Preservation efforts
   (b) Economic demand
   (c) Habitat fragmentation
   (d) Climate change
4. Which of the following is a consequence of pollution on biodiversity?
   (a) Increased species diversity
   (b) Enhanced ecosystem productivity
   (c) Habitat degradation
   (d) Promotion of ecological balance
5. What is the main factor contributing to climate change-induced threats to biodiversity?
   (a) Deforestation
   (b) Ocean acidification
   (c) Soil erosion
   (d) Carbon sequestration

6. How does habitat fragmentation affect biodiversity?
   (a) Increases connectivity between habitats
   (b) Boosts species migration
   (c) Reduces gene flow and increases extinction risk
   (d) Enhances ecosystem resilience
7. Which of the following is an example of direct exploitation of biodiversity?
   (a) Pollution
   (b) Habitat destruction
   (c) Overfishing
   (d) Climate change
8. Which factor is driving the loss of coral reefs worldwide?
   (a) Desertification
   (b) Water pollution
   (c) Soil erosion
   (d) Ozone depletion
9. How does deforestation contribute to biodiversity loss?
   (a) Increases forest cover and enhances wildlife habitats
   (b) Provides economic opportunities for local communities
   (c) Disrupts ecosystems and destroys habitats
   (d) Boosts carbon sequestration and mitigates climate change
10. What is the main reason for the decline of pollinator populations?
   (a) Habitat conservation efforts
   (b) Genetic modification of crops
   (c) Climate change adaptation
   (d) Pesticide use
11. Which of the following characteristics makes a species more vulnerable to extinction?
   (a) Wide geographic range
   (b) High reproductive rate
   (c) Specialized diet or habitat requirements
   (d) Large population size
12. Which of the following factors contributes to an increased risk of species extinction?
   (a) High genetic diversity
   (b) Effective conservation measures
   (c) Fragmentation of habitat
   (d) Abundant food resources
13. Which of the following types of species is considered more vulnerable to extinction?
   (a) Generalist species with broad ecological niches
   (b) Specialist species with specific ecological requirements
   (c) Keystone species with significant influence on ecosystem dynamics
   (d) Indicator species with sensitivity to environmental changes
14. What is the purpose of the IUCN Red List?
   (a) To identify endangered species for commercial exploitation
   (b) To track the population sizes of all known species
   (c) To assess the conservation status of species and highlight those at risk of extinction
   (d) To categorize species based on their economic value
15. Which criteria are used to categorize species on the IUCN Red List?
   (a) Rarity and uniqueness
   (b) Economic importance and market demand
   (c) Geographic range and population size
   (d) Cultural significance and historical value
16. What does the "Critically Endangered" category on the IUCN Red List indicate?
   (a) The species is extinct in the wild.
   (b) The species is at high risk of extinction in the near future.
   (c) The species has a stable and secure population.
   (d) The species is abundant and not threatened with extinction.
17. Which of the following factors is believed to be a significant driver of mass extinctions?
   (a) Volcanic eruptions
   (b) Climate change
   (c) Asteroid impacts
   (d) All of the above
18. The most well-known mass extinction event in history, which wiped out the dinosaurs, is known as:
   (a) The Permian-Triassic Extinction
   (b) The Cretaceous-Paleogene Extinction
   (c) The Devonian Extinction
   (d) The Ordovician-Silurian Extinction

19. What is the primary cause of the Sixth Mass Extinction?
    (a) Natural climate change
    (b) Volcanic eruptions
    (c) Human activities
    (d) Asteroid impacts
20. Which of the following is *not* a driver of the Sixth Mass Extinction?
    (a) Habitat destruction
    (b) Overexploitation of resources
    (c) Introduction of invasive species
    (d) Natural selection

Answers: 1-a, 2-c, 3-b, 4-c, 5-a, 6-c, 7-c, 8-b, 9-c, 10-d, 11-c, 12-c, 13-b, 14-c, 15-c, 16-b, 17-d, 18-b, 19-c, 20-d

14.8.2 Short-Answer Questions

1. What are some major threats to biodiversity worldwide?
2. How does habitat loss contribute to the decline of biodiversity?
3. What role does climate change play in threatening biodiversity?
4. How does pollution impact biodiversity and ecosystem health?
5. What are the dangers of invasive species to native biodiversity?
6. How does overexploitation of natural resources affect biodiversity?
7. What are the consequences of deforestation on global biodiversity?
8. How does habitat fragmentation pose a threat to biodiversity?
9. What are the potential impacts of agricultural practices on biodiversity?
10. How does urbanization and urban sprawl impact biodiversity?

14.8.3 Long-Answer Questions

1. Describe the major factors threatening nature's biodiversity.
2. What do you understand about IUCN Red List? What is the main purpose of the Red List and how are the species categorized according to this List?
3. Describe five mass extinction events giving emphasis on the main species affected and possible reasons.
4. Write an essay on the Sixth Mass Extinction.
5. Write short notes on:
    (a) Man-wildlife conflicts
    (b) Susceptibility to extinction
    (c) The Permian-Triassic Extinction

References

IUCN (2019) Guidelines for using the IUCN red list categories and criteria: version 14. IUCN, Gland/Cambridge. 113 pp

Parmar S, Sharma VK, Singh V (2023) Microplastics in marine ecosystem: sources, risks, mitigation technologies, and challenges. Taylor and Francis (CRC Press). 228 pp, Boca Raton and London. https://doi.org/10.1201/9781003312086

Singh V (2023) Biodiversity: concepts, crises, and conservation. NIPA, New Delhi. 175 pp

WWF (2020) In: Almond REA, Groten M, Petersen T (eds) Living planet report 2020 – bending the curve of biodiversity loss. WWF, Gland

15 Biodiversity Conservation

Biodiversity, the variety of life on Earth, is a fundamental asset that sustains the planet's ecosystems and provides numerous benefits to humanity. However, in recent decades, human activities have posed significant threats to biodiversity, leading to the decline and loss of many species worldwide. To address this pressing issue, biodiversity conservation has become a critical endeavor. This chapter explores various strategies for conserving biodiversity, including in situ and ex situ conservation methods. Additionally, we delve into the concept of megadiversity countries, centers of origin of crop plants, biodiversity hotspots, and the importance of their preservation. International efforts for biodiversity conservation have also been duly discussed.

15.1 Strategies for Biodiversity Conservation

Biodiversity conservation comprises a range of approaches aimed at preserving ecosystems, species, and genetic diversity. Two primary strategies employed are in situ conservation and ex situ conservation. In situ conservation focuses on safeguarding species and ecosystems within their natural habitats. Protected areas such as national parks, wildlife reserves, and marine sanctuaries serve as key mechanisms for in situ conservation. These protected areas provide safe havens for numerous species, allowing them to thrive and maintain their ecological roles. Additionally, indigenous and local communities play a vital role in the stewardship of these areas, employing traditional knowledge and practices to conserve biodiversity.

Ex situ conservation encompasses the preservation of species outside the natural habitats they have been thriving in. This strategy is particularly useful for species facing imminent threats of extinction or those that have already become extinct in the wild. Ex situ conservation methods include captive breeding programs, botanical gardens, seed banks, and zoos. These initiatives help protect and restore populations that can be reintroduced into their natural habitats when conditions permit.

15.2 In Situ Biodiversity Conservation Strategies: Protecting Nature Where It Thrives

Biodiversity, the variety of life on Earth, is crucial for the health and sustainability of our planet. To safeguard this invaluable natural heritage, conservation efforts have focused on protecting ecosystems and the species they support. In situ biodiversity conservation strategies aim to conserve nature in its original habitats, allowing ecosystems to thrive and species to flourish. One of the most effective ways to achieve this is through

V. Singh, *Textbook of Environment and Ecology*, https://doi.org/10.1007/978-981-99-8846-4_15

the establishment of protected and conservation areas. Let us explore the various types and categories of these areas, each playing a vital role in preserving Earth's biodiversity.

15.2.1 National Parks

National parks are large tracts of land dedicated to the conservation of natural landscapes and their ecological processes. These areas typically encompass diverse ecosystems, including forests, grasslands, wetlands, and marine habitats. National parks often serve as flagship areas for biodiversity conservation, protecting iconic species and providing opportunities for scientific research, education, and ecotourism.

15.2.2 Wildlife Sanctuaries and Reserves

Wildlife sanctuaries and reserves are specifically designated areas aimed at safeguarding wildlife and their habitats. These areas provide essential habitats for a wide range of species, including endangered and threatened ones. Wildlife sanctuaries often have specific regulations to minimize human disturbances and protect breeding grounds, feeding areas, and migration routes critical for animal populations.

15.2.3 Biosphere Reserves

Biosphere reserves are special areas aimed at reconciling the conservation of biodiversity with sustainable development. They comprise three interrelated zones (Fig. 15.1): a core area for strict protection, a buffer zone for sustainable resource use, and a transition zone for cooperation between conservation and development activities. Biosphere reserves promote scientific research, environmental education, and community participation in biodiversity conservation.

15.2.4 Marine Protected Areas (MPAs)

Marine protected areas are designated sections of oceans, seas, and coastlines that aim to conserve marine biodiversity and ecosystems. MPAs provide refuge for a variety of marine species, protect critical habitats such as coral reefs and seagrass beds, and contribute to the recovery of depleted fish stocks. They also help maintain ecological processes, including nutrient cycling and species migration.

15.2.5 Community Conserved Areas (CCAs)

Community conserved areas are landscapes or seascapes where local communities play a central role in biodiversity conservation. These areas are managed and protected by indigenous peoples and local communities, who have traditional knowledge and practices that promote sustainable resource use and conservation. CCAs foster community empowerment, strengthen cultural traditions, and enhance livelihoods while preserving biodiversity (Singh 2009). Sacred groves or sacred forests are of special significance among the Hindus in India (see Fig. 15.2). So many festivals and rituals in India are devoted to different species (Ormsby and Bhagwat 2010). Sacred forests, sacred trees, sacred seeds, sacred cow, sacred lakes, sacred rivers, etc. are the numerous notions the Indians associate with nature's biodiversity giving emphasis on their conservation.

15.2.6 Nature Reserves

Nature reserves encompass a range of protected areas established to conserve specific habitats, species, or ecological features. They may include wetland reserves, forest reserves, bird reserves, or botanical reserves, among others. Nature reserves often serve as havens for endemic and rare species, protecting their unique habitats from habitat destruction and fragmentation.

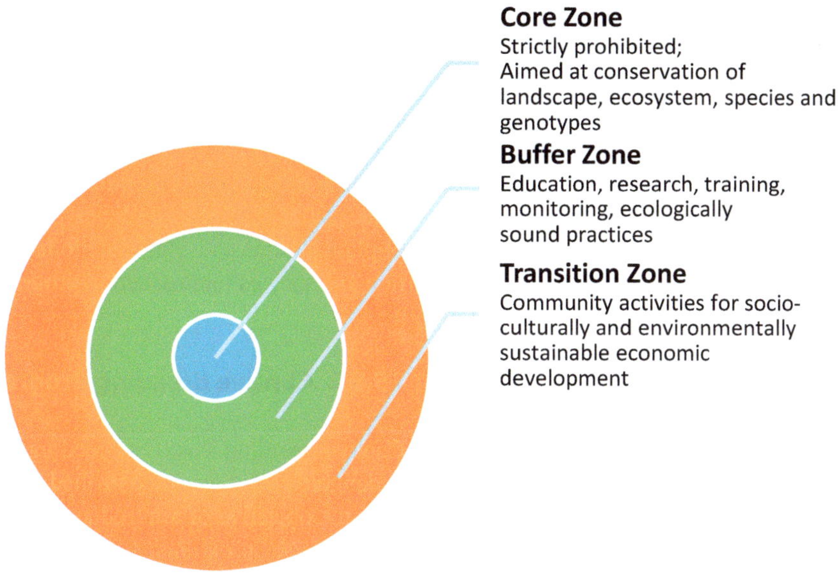

Fig. 15.1 The three zones establishing distinct purposes within a biosphere reserve

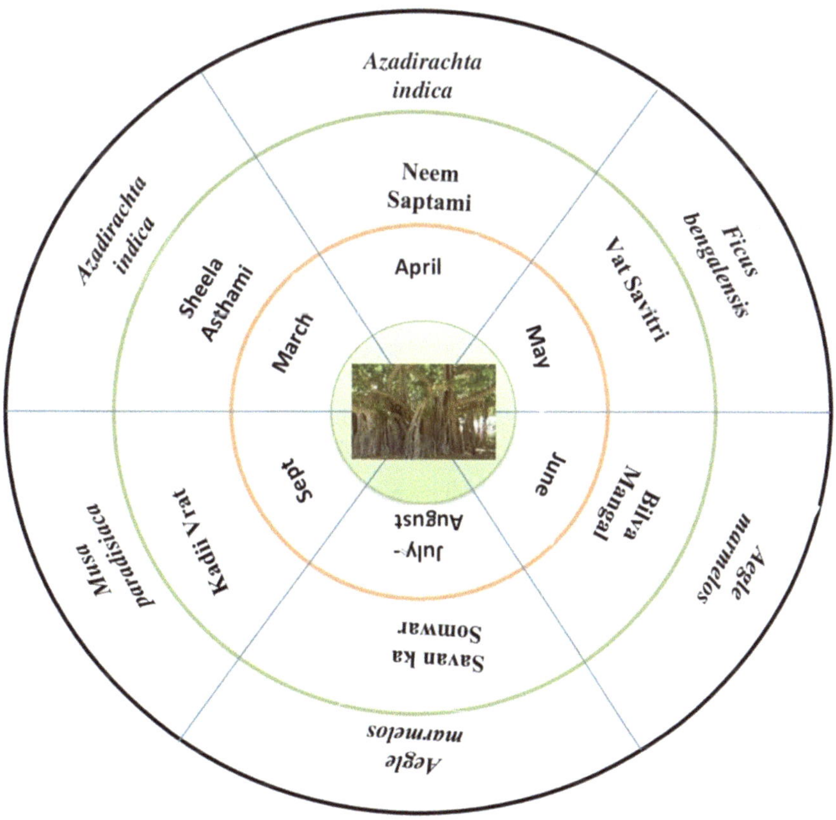

Fig. 15.2 Religious perspectives on biodiversity conservation: Hindu festivals in India linked to various plant species. (Source: Singh 2023)

15.2.7 Habitat Restoration Areas

Habitat restoration areas focus on rehabilitating degraded ecosystems to their natural state, allowing native species to recolonize and flourish. These areas involve activities such as reforestation, wetland restoration, and the removal of invasive species. By restoring habitats and ecological processes, these areas promote the recovery of biodiversity and ecosystem services.

15.3 Ex Situ Conservation Strategies

Biodiversity, the intricate web of life on Earth, is facing unprecedented threats due to habitat loss, climate change, pollution, and other human activities. To counteract the ongoing loss of species and ecosystems, conservation efforts have taken various forms, including ex situ conservation. Ex situ conservation involves the removal and protection of endangered species and their genetic resources from their natural habitats. Let us explore the significance of ex situ conservation and highlight several methods adopted in this strategy to preserve biodiversity.

15.3.1 Zoos and Wildlife Reserves

Zoological parks and wildlife reserves play a vital role in ex situ conservation. They provide a safe and controlled environment for endangered species, often serving as breeding and research centers. These facilities aim to maintain viable populations and genetic diversity of species facing extinction. Zoos engage in captive breeding programs, reintroduction efforts, and education initiatives to raise awareness about biodiversity conservation.

15.3.2 Botanical Gardens and Seed Banks

Botanical gardens conserve plant species through the cultivation and management of living collections. These gardens maintain diverse plant populations, including rare and endangered species, and serve as repositories for plant genetic resources. They often collaborate with seed banks, which store seeds under controlled conditions to ensure their long-term viability. Seed banks act as "insurance policies" against the loss of plant species and contribute to habitat restoration and species reintroduction programs.

15.3.3 Captive Breeding Programs

Captive breeding involves breeding and rearing endangered species in controlled environments with the objective of reintroducing them into the wild. These programs focus on maintaining genetic diversity and preventing the extinction of species facing severe population decline. By carefully managing captive populations, conservationists aim to produce individuals that can survive and thrive once released into suitable habitats.

15.3.4 Cryopreservation and Genetic Banks

Cryopreservation is the preservation of cells, tissues, or genetic material at extremely low temperatures. Genetic banks store frozen samples of sperm, eggs, embryos, or tissue samples to preserve the genetic diversity of endangered species. This method ensures the availability of genetic material for future research, reintroduction programs, or the potential revival of extinct species using emerging technologies.

15.3.5 Conservation Sanctuaries and Rescue Centers

Conservation sanctuaries and rescue centers serve as temporary or permanent shelters for endangered species that are confiscated from illegal trade, injured, or orphaned. These facilities provide care, rehabilitation, and appropriate living conditions while working toward the eventual release of rehabilitated individuals back into the wild. Sanctuaries also contribute to public education and awareness regarding the importance of wildlife conservation.

15.3.6 Ex Situ Habitat Restoration

In cases where habitat degradation or destruction hampers the survival of species in the wild, ex situ habitat restoration becomes necessary. This method involves recreating or rehabilitating degraded habitats within protected areas or managed landscapes. By restoring ecological conditions and reintroducing suitable plant and animal species, ex situ habitat restoration aims to establish self-sustaining populations and promote ecosystem recovery.

15.4 Megadiversity Countries

Megadiversity countries are nations recognized for their exceptionally high levels of biodiversity. The megadiversity countries boast of an extraordinary wealth of biodiversity, housing a significant portion of the Earth's species within their borders. These countries are characterized by diverse ecosystems, ranging from lush rainforests to expansive coral reefs, and serve as vital repositories of unique flora and fauna.

A total of 17 megadiversity countries have been officially recognized for their exceptional levels of biodiversity. These countries are Australia, Brazil, China, Colombia, Democratic Republic of the Congo, Ecuador, India, Indonesia, Madagascar, Malaysia, Mexico, Papua New Guinea, Peru, the Philippines, South Africa, the United States (specifically the state of California), and Venezuela.

Recognizing their megadiversity status highlights the importance of conservation efforts in these regions to safeguard their unique flora and fauna. Collectively, these megadiversity countries serve as global hotspots for conservation, highlighting the need for concerted efforts to protect and preserve these invaluable natural treasures.

15.5 Centers of Origin of Crop Plants

Centers of origin of crop plants are geographic regions where specific cultivated plants originated and evolved over time. These regions are vital reservoirs of genetic diversity and play a crucial role in global food security. Nikolai Ivanovich Vavilov, a prominent Russian botanist and geneticist, made significant contributions to the understanding of crop plant origins and the concept of centers of origin. Through extensive expeditions and research, Vavilov identified several primary centers of origin, also known as Vavilovian centers, where numerous crop plants originated and diversified. His pioneering work laid the foundation for the study of crop diversity and provided invaluable insights into the origins and evolution of cultivated plants. In this part, we will explore the Vavilovian centers and their significance in understanding agricultural diversity.

Mediterranean Centre The Mediterranean Centre, encompassing regions of Europe, North Africa, and West Asia, was one of the primary centers identified by Vavilov. This region is characterized by a mild climate and diverse landscapes, facilitating the domestication and diversification of numerous crop plants. Crops such as wheat, barley, oats, lentils, olives, grapes, and figs are believed to have originated in this center. The Mediterranean Centre has played a crucial role in shaping agricultural practices and influencing diets across the globe.

Southwestern Asia Centre The Southwestern Asia Centre, comprising the Middle East and parts of Central Asia, is another significant Vavilovian center. This region is known for the domestication of crops such as wheat, barley, peas, lentils, chickpeas, and flax. The Tigris and Euphrates river valleys, often referred to as the cradle of civilization, witnessed the emergence of agriculture and the development of early urban societies. The agricultural practices and crop diversity originating from this center have had a profound impact on human history and the spread of agriculture to other parts of the world.

Indus-Ganges Centre The Indus-Ganges Centre, encompassing the Indian subcontinent, is renowned for its rich agricultural heritage. Rice, one of the world's most important staple crops, originated in this center. Additionally, crops such as millets, lentils, chickpeas, and mangoes have been cultivated here for thousands of years. The fertile plains of the Indus and Ganges rivers provided the ideal conditions for the domestication and diversification of these crops, contributing to the remarkable agricultural diversity of the region.

Ethiopian Centre The Ethiopian Centre, covering the Horn of Africa and parts of East Africa, is recognized as a center of crop diversity. Crops such as sorghum, teff, coffee, enset (false banana), and finger millet are believed to have originated in this center. The diverse agroecological zones in this region, ranging from highlands to lowlands, have supported the cultivation of a wide range of crops, enabling local communities to adapt to various climatic conditions and contribute to the genetic diversity of these crops.

Central American and Mexican Centre The Central American and Mexican Centre is a vital Vavilovian center associated with the origins of several major crops in the Americas. Maize (corn), beans, squash, chili peppers, and tomatoes are among the crops domesticated in this center. The indigenous civilizations of Mesoamerica, such as the Maya and Aztec, heavily relied on these crops for sustenance and cultural practices. The Central American and Mexican Centre has significantly influenced global agriculture, with maize being one of the most widely cultivated and economically important crops worldwide.

Vavilov's identification of these centers of origin has provided a framework for understanding the geographical origins of cultivated plants and the diversity of crop resources. The Vavilovian centers continue to inspire research in crop genetics, conservation, and breeding, allowing scientists and farmers to access and utilize the vast genetic diversity preserved within these regions, preserving and protecting the genetic resources.

15.6 Biodiversity Hotspots

Biodiversity hotspots are biologically rich areas that face very critical levels of habitat loss and species endangerment. These regions are characterized by high levels of endemic species, meaning they are found nowhere else on Earth. Biodiversity hotspots, such as the Amazon rainforest, the Coral Triangle, and the Western Ghats of India, are of immense conservation importance. Efforts to protect and restore these hotspots are critical in preventing further species loss and preserving unique ecosystems.

Biodiversity hotspots are biogeographic regions of exceptional ecological significance, harboring an incredible array of plant and animal species. These areas, characterized by high levels of endemism and threatened habitats, serve as crucial reservoirs of biodiversity. Conservation efforts targeted at these hotspots play a vital role in safeguarding the planet's natural heritage and maintaining its delicate ecological balance. Let us explore the concept of biodiversity hotspots and provide an overview of the world's recognized hotspots.

15.6.1 Understanding Biodiversity Hotspots

Norman Myers, a British ecologist, introduced the term "biodiversity hotspot" in 1988. He extensively explored and deeply studied these global hotspots (Myers 1988, 1990, 2003). Hotspots are defined as regions that exhibit high levels of species richness and endemism and face significant habitat loss. These areas are often geographically concentrated and cover only a small fraction of the Earth's surface. Despite their relatively small size, biodiversity hotspots house a disproportionately large number of plant and animal species.

There are two main criteria that determined the status of a biodiversity hotspot according to Myers et al. (2000). These are the following:

(i) High degree of endemism: a biodiversity hotspot must contain at least 1500 species of vascular plants (> 0.05% of the world's total) as endemics.
(ii) Habitat loss: a hotspot should have lost at least 70% of its original natural vegetation.

As of now, there are 36 recognized biodiversity hotspots across the globe. Each hotspot has its unique ecological characteristics and faces distinct conservation challenges. The following is a comprehensive list of the world's biodiversity hotspots:

1. Western Ghats and Sri Lanka
2. Indo-Burma
3. Sundaland
4. Philippines
5. Wallacea
6. New Caledonia
7. Eastern Melanesia
8. Polynesia-Micronesia
9. Southwest Australia
10. California Floristic Province
11. Madrean Pine-Oak Woodlands
12. Mesoamerica
13. Caribbean Islands
14. Atlantic Forest
15. Cerrado
16. Tumbes-Choco-Magdalena
17. Tropical Andes
18. Chilean Winter Rainfall-Valdivian Forests
19. Juan Fernandez Islands
20. Guinean Forests of West Africa
21. Cape Floristic Region
22. Succulent Karoo
23. Maputaland-Pondoland-Albany
24. Eastern Afromontane
25. Coastal Forests of Eastern Africa
26. Horn of Africa
27. Madagascar and the Indian Ocean Islands
28. Mountains of Southwest China
29. Irano-Anatolian
30. Mediterranean Basin
31. Caucasus
32. Southwest Australia
33. Eastern Australia
34. New Zealand
35. New Caledonia
36. Wallacea

15.6.2 Conservation Importance and Challenges

Biodiversity hotspots are of critical significance for the conservation of global biological diversity. Their protection is crucial as they provide numerous ecosystem services, such as water regulation, carbon sequestration, and soil fertility. Moreover, hotspots often overlap with areas of high human population density, leading to intense pressures from habitat destruction, climate change, invasive species, and overexploitation. These challenges make conservation efforts within hotspots particularly urgent.

15.6.3 Conservation Strategies

Efforts to conserve biodiversity hotspots require a multifaceted approach involving governments, nongovernmental organizations, local communities, and international cooperation. Some key strategies include establishing protected areas, implementing sustainable land use practices, promoting community-based conservation initiatives, and raising awareness about the value of biodiversity and the need for its protection.

Biodiversity hotspots, in essence, are precious ecological treasures that contain a wealth of unique and irreplaceable species. These hotspots face numerous threats, making their conservation imperative. By prioritizing the preservation of biodiversity hotspots, we can safeguard the planet's ecological heritage and ensure the survival of countless species for generations to come. Through collaborative efforts and responsible stewardship, we can rise to the challenge and protect these exceptional ecosystems, ultimately preserving the remarkable diversity of life on Earth.

15.7 International Efforts for Biodiversity Conservation

International efforts for the conservation of biodiversity have gained significant momentum in recent years. Governments, organizations, and scientists from around the world have recognized the urgent need to protect and preserve the Earth's rich biodiversity for the well-being of present and future generations. Several international protocols, summits, and decisions have been instrumental in advancing global conservation efforts. Here are some crucial examples pertaining to biodiversity conservation:

15.7.1 Convention on Biological Diversity (CBD)

The CBD, adopted in 1992 during the Earth Summit in Rio de Janeiro, is a crucial international agreement that aims to conserve biodiversity, sustainably use its components, and ensure the fair and equitable sharing of benefits derived from genetic resources. It provides a framework for countries to develop national strategies and action plans for biodiversity conservation.

15.7.2 The Cartagena Protocol on Biosafety

Adopted in 2000 and operating under the CBD, the Cartagena Protocol focuses on the safe handling, transfer, and use of genetically modified organisms (GMOs). It aims to protect biodiversity by establishing clear procedures for assessing the potential risks of GMOs and regulating their transboundary movements.

15.7.3 The Nagoya Protocol

Adopted in 2010, the Nagoya Protocol on Access to Genetic Resources and the Fair and Equitable Sharing of Benefits Arising from their Utilization is also part of the CBD framework. It promotes the fair and equitable sharing of benefits derived from the utilization of genetic resources, such as plants and animals, while ensuring that access to these resources is done in a sustainable and environmentally sound manner.

15.7.4 United Nations Framework Convention on Climate Change (UNFCCC)

The UNFCCC, adopted in 1992, recognizes the interrelation between climate change and biodiversity loss. It seeks to stabilize greenhouse gas concentrations in the atmosphere and prevent dangerous human interference with the climate system. Efforts to address climate change are critical for preserving biodiversity, as many species are vulnerable to the impacts of a changing climate.

15.7.5 Conference of the Parties (COP) to the CBD

The COP is the governing body of the CBD and meets regularly to review progress, set priorities, and make decisions on biodiversity conservation. COP decisions have led to the establishment of protected areas, the integration of biodiversity considerations into various sectors, the adoption of national biodiversity strategies, and the mobilization of financial resources for conservation.

15.7.6 The Aichi Biodiversity Targets

In 2010, during the tenth meeting of the COP to the CBD in Aichi, Japan, a set of 20 ambitious targets, known as the Aichi Biodiversity Targets, were adopted. These targets provide a comprehensive framework to guide global biodiversity conservation efforts until 2020. They cover various aspects, including reducing habitat loss, preventing species extinctions, promoting sustainable resource use, and mainstreaming biodiversity across sectors.

15.7.7 Intergovernmental Science-Policy Platform on Biodiversity and Ecosystem Services (IPBES)

Established in 2012, IPBES is an independent intergovernmental body that assesses the state of biodiversity, ecosystems, and their contributions to human well-being. IPBES reports and assessments provide valuable scientific knowledge to policymakers, helping them make informed decisions for biodiversity conservation and sustainable development.

15.7.8 United Nations Decade on Ecosystem Restoration (2021–2030)

In 2021, the United Nations declared the Decade on Ecosystem Restoration, recognizing the urgent need to halt and reverse the degradation of ecosystems worldwide. By restoring ecosystems, such as forests, wetlands, and coral reefs, the Decade aims to enhance biodiversity, address climate change, and improve livelihoods.

These international protocols, summits, and decisions demonstrate the global commitment to biodiversity conservation. However, ongoing efforts are necessary to implement the agreements effectively, mobilize resources, and address emerging challenges, such as habitat loss, pollution, invasive species, and climate change, to ensure the long-term survival of Earth's diverse ecosystems and species.

Conserving biodiversity is not only a moral and ethical obligation but also essential for our own well-being. Biodiversity provides ecosystem services, such as pollination, soil fertility, and climate regulation, which are fundamental to human survival. Therefore, it is imperative that governments, organizations, and individuals work together to implement effective conservation measures, promote sustainable practices, and raise awareness about the importance of biodiversity. By doing so, we can ensure a resilient and biodiverse planet for future generations.

15.8 Summary

Biodiversity conservation is a critical effort to protect the variety of life on Earth, which is under threat due to human activities. Conservation strategies include in situ conservation, which focuses on protecting species and ecosystems in their natural habitats, and ex situ conservation, which involves preserving species outside their natural habitats. In situ conservation is achieved through protected areas such as national parks and wildlife reserves, with the involvement of indigenous and local communities. Ex situ conservation methods include captive breeding programs, botanical gardens, seed banks, and zoos. These strategies aim to preserve biodiversity and restore populations for eventual reintroduction into the wild. Overall, conservation efforts are vital to safeguarding nature and its inherent value.

A total of 17 recognized megadiversity countries are nations with exceptionally high levels of biodiversity, housing a significant portion of the Earth's species within their borders. These megadiversity countries serve as global hotspots for conservation, highlighting the need for concerted efforts to protect and preserve these invaluable natural treasures.

Centers of origin of crop plants are biogeographic regions where specific cultivated plants originated and evolved over time and are vital reservoirs of genetic diversity, playing a crucial role in global food security. Fewer than a dozen and also known as the Vavilovian centers, these centers have significance in understanding agricultural diversity and its management for human welfare.

Biodiversity hotspots, in total 36, are biogeographic regions of exceptional ecological significance, harboring an incredible array of plant and animal species. Characterized by high levels of endemism and threatened habitats, these areas serve as crucial reservoirs of biodiversity.

International protocols, summits, and decisions – notably the Convention on Biological Diversity (CBD), the Cartagena Protocol on Biosafety, the Nagoya Protocol on Access to Genetic Resources and the Fair and Equitable

Sharing of Benefits Arising from their Utilization, the United Nations Framework Convention on Climate Change (UNFCCC), Conference of the Parties (COP) to the CBD, the Aichi Biodiversity Targets, the Intergovernmental Science-Policy Platform on Biodiversity and Ecosystem Services (IPBES), and the United Nations Decade on Ecosystem Restoration (2021–2030) – demonstrate the global commitment to biodiversity conservation. The ongoing efforts are necessary to implement the agreements effectively, mobilize resources, and address emerging challenges, such as habitat loss, pollution, invasive species, and climate change, to ensure the long-term survival of Earth's diverse ecosystems and species.

15.9 Exercises

15.9.1 Multiple-Choice Questions

1. Which strategy of biodiversity conservation focuses on preserving species and ecosystems within their natural habitats?
   (a) In situ conservation
   (b) Ex situ conservation
   (c) Megadiversity conservation
   (d) Center of origin conservation
2. Which of the following is an example of in situ conservation?
   (a) Botanical gardens
   (b) Seed banks
   (c) National parks
   (d) Zoos
3. What is the primary purpose of marine protected areas (MPAs)?
   (a) Conservation of marine biodiversity and ecosystems
   (b) Captive breeding of marine species
   (c) Restoration of marine habitats
   (d) Genetic preservation of marine species
4. Community conserved areas (CCAs) are managed and protected by which group?
   (a) Government organizations
   (b) International conservation agencies
   (c) Indigenous peoples and local communities
   (d) Scientific research institutions
5. What is the primary goal of ex situ conservation?
   (a) Preservation of species in their natural habitats
   (b) Rehabilitation of degraded ecosystems
   (c) Protection of genetic material outside natural habitats
   (d) Conservation of biodiversity hotspots
6. Which ex situ conservation method involves the storage of genetic material from diverse species?
   (a) Zoos and wildlife reserves
   (b) Botanical gardens and seed banks
   (c) Captive breeding programs
   (d) Cryopreservation and genetic banks
7. What is the purpose of captive breeding programs?
   (a) Rehabilitation of injured wildlife
   (b) Restoration of degraded habitats
   (c) Conservation of marine biodiversity
   (d) Preservation of genetic diversity and preventing species extinction
8. Which ex situ conservation method involves the preservation of cells and genetic material at extremely low temperatures?
   (a) Zoos and wildlife reserves
   (b) Cryopreservation and genetic banks
   (c) Captive breeding programs
   (d) Conservation sanctuaries and rescue centers
9. What is the role of conservation sanctuaries and rescue centers in ex situ conservation?
   (a) Rehabilitation of injured wildlife
   (b) Captive breeding programs
   (c) Protection of genetic material
   (d) Temporary or permanent shelters for endangered species
10. How many megadiversity countries have been officially recognized?
   (a) 5
   (b) 10
   (c) 17
   (d) 20
11. Which country is not recognized as a megadiversity country?
   (a) Australia
   (b) Brazil
   (c) China
   (d) Russia

12. What is the significance of megadiversity countries in biodiversity conservation?
    (a) They house a significant portion of Earth's species.
    (b) They have the highest number of protected areas.
    (c) They are centers of origin for crop plants.
    (d) They have the highest number of zoos and botanical gardens.
13. Which Russian botanist and geneticist made significant contributions to the understanding of crop plant origins and the concept of centers of origin?
    (a) Ivan Pavlov
    (b) Nikolai Ivanovich Vavilov
    (c) Gregor Mendel
    (d) Dmitri Mendeleev
14. Which center of origin is characterized by a mild climate and diverse landscapes, facilitating the domestication and diversification of numerous crop plants?
    (a) Southwestern Asia Centre
    (b) Ethiopian Centre
    (c) Mediterranean Centre
    (d) Central American and Mexican Centre
15. Which center of origin is renowned for the origin of rice, one of the world's most important staple crops?
    (a) Indus-Ganges Centre
    (b) Southwestern Asia Centre
    (c) Mediterranean Centre
    (d) Ethiopian Centre
16. Which center of origin is associated with the domestication of crops such as maize (corn), beans, squash, chili peppers, and tomatoes?
    (a) Indus-Ganges Centre
    (b) Southwestern Asia Centre
    (c) Mediterranean Centre
    (d) Central American and Mexican Centre
17. What are biodiversity hotspots characterized by?
    (a) Low levels of endemism and habitat loss
    (b) High levels of genetic diversity and population density
    (c) High levels of endemic species and habitat loss
    (d) Low levels of species richness and genetic diversity
18. According to Norman Myers, how many species of vascular plants should a biodiversity hotspot contain as endemics?
    (a) At least 500 species
    (b) At least 1000 species
    (c) At least 1500 species
    (d) At least 2000 species
19. Which international agreement aims to conserve biodiversity, sustainably use its components, and ensure the fair and equitable sharing of benefits derived from genetic resources?
    (a) Convention on Biological Diversity (CBD)
    (b) Nagoya Protocol on Access to Genetic Resources
    (c) United Nations Framework Convention on Climate Change (UNFCCC)
    (d) Cartagena Protocol on Biosafety
20. Which intergovernmental body assesses the state of biodiversity, ecosystems, and their contributions to human well-being?
    (a) Intergovernmental Panel on Climate Change (IPCC)
    (b) Intergovernmental Science-Policy Platform on Biodiversity and Ecosystem Services (IPBES)
    (c) World Health Organization (WHO)
    (d) United Nations Environment Programme (UNEP)

Answers: 1-a, 2-c, 3-a, 4-c, 5-c, 6-b, 7-d, 8-b, 9-d, 10-c, 11-d, 12-a, 13-b, 14-c, 15-a, 16-d, 17-c, 18-c, 19-a, 20-b

15.9.2 Short-Answer Questions

1. What is biodiversity conservation and why is it important?
2. Differentiate between in situ and ex situ biodiversity conservation.
3. How do protected areas such as national parks contribute to biodiversity conservation?

4. What is ex situ conservation and how does it complement in situ conservation efforts?
5. What are some examples of ex situ conservation methods?
6. What is the significance of botanical gardens in biodiversity conservation?
7. How do seed banks contribute to preserving genetic diversity?
8. Name five examples of the world's biodiversity hotspots.
9. What do the Vavilovian centers indicate?
10. Name four international protocols/agreements of significance for biodiversity conservation.

15.9.3 Long-Answer Questions

1. What are the strategies and approaches used in in situ biodiversity conservation?
2. What is the concept of megadiverse countries and why are they important in biodiversity conservation?
3. What are biodiversity hotspots and why are they considered critical for conservation efforts?
4. What are the major international agreements and initiatives aimed at biodiversity conservation?
5. Write short notes on:
   (a) Biosphere reserves
   (b) Cryopreservation
   (c) Centers of origin

References

Myers N (1988) Threatened biotas: "hot spots" in tropical forests. Environmentalist 8(3):187–208

Myers N (1990) The biodiversity challenge: expanded hot-spots analysis. Environmentalist 10(4):243–256

Myers N (2003) Biodiversity hotspots revisited. Bioscience 53:916–917

Myers N, Mittermeier RA, Mittermeier CG, Fonseca GAB (2000) Biodiversity hot spots for conservation priorities. Nature 403:853–858

Ormsby AA, Bhagwat SA (2010) Sacred forests of India: a strong tradition of community-based natural resource management. Environ Conserv 37(3):320–326. https://doi.org/10.1017/S0376892910000561

Singh V (2009) Environmental services emanating from the Himalayan mountains: valuation against the backdrop of eco-philosophy and chasing the goal of global happiness. In: Gautam PL, Singh V, Melkania U (eds) Ecosystem diversity and carbon sequestration: climate change challenges and a way out for ushering in a sustainable future. Daya Publishing House, New Delhi, pp 342–355

Singh V (2023) Biodiversity: concepts, crises, and conservation. NIPA, New Delhi. 175 pp

Section IV

Environmental Disruptions

16 Air Pollution

Air pollution is a global environmental issue that poses a significant threat to human health, ecosystems, and the overall quality of life on our planet. It occurs when harmful substances are released into the atmosphere, resulting in the contamination of the air we breathe. This chapter will provide an in-depth exploration of air pollution, including its definition, causes, effects, and the various measures employed for its prevention and control.

16.1 Definition of Air Pollution

Air pollution is the presence of harmful substances in the air, either naturally occurring or human-generated, in quantities that exceed the natural capacity of the atmosphere to disperse and dilute them. These substances, known as air pollutants, can be in the form of gases, liquid droplets, solid particles, or biological agents and are released from a variety of sources.

In addition to human health, air pollution poses significant threat to the global climate. Addressing air pollution requires concerted efforts from governments, industries, and individuals to reduce emissions, promote sustainable practices, and mitigate its detrimental effects on our planet and well-being.

Phenomenal work compiled by Chameides and Perdue (1994), Finlayson-Pitts and Pitts Jr (2000), Wark et al. (2012), Vallero (2014), Lelieveld et al. (2015), Seinfeld and Pandis (2016), Zhang (2019), Dockery and Pope (2020), Pandey and Yadav (2021), and Singh (2024) speaks volumes about various dimensions and issues of air pollution.

16.2 Air Pollutants

Air pollutants can be classified based on their origin, chemical composition, and physical properties. These pollutants can be classified into primary pollutants, secondary pollutants, hazardous air pollutants, and indoor air pollutants. Understanding the classification of air pollutants is essential for identifying their sources and implementing effective strategies to reduce their emissions.

16.2.1 Primary Pollutants

Primary pollutants are directly emitted into the atmosphere in their harmful form. They include the following:

(a) **Particulate matter (PM):** Particulate matter refers to tiny solid or liquid particles suspended in the air. They are categorized based on their size. PM_{10} (particles with a diameter

V. Singh, *Textbook of Environment and Ecology*, https://doi.org/10.1007/978-981-99-8846-4_16

of 10 micrometers or less) and $PM_{2.5}$ (particles with a diameter of 2.5 micrometers or less) are of particular concern due to their ability to penetrate deep into the respiratory system.

(b) **Sulfur dioxide (SO_2):** SO_2 is primarily emitted from the burning of fossil fuels containing sulfur compounds, such as coal and oil. It contributes to the formation of acid rain and respiratory issues.

(c) **Nitrogen oxides (NO_x):** NO_x compounds are formed during high-temperature combustion processes, mainly in vehicles and power plants. They contribute to the formation of ground-level ozone, smog, and acid rain.

(d) **Carbon monoxide (CO):** CO is produced by incomplete combustion of fossil fuels and biomass. It is a poisonous gas that can lead to various health problems, especially in enclosed spaces.

(e) **Volatile organic compounds (VOCs):** VOCs are emitted from various sources, including industrial processes, solvents, and vehicle emissions. They contribute to the formation of ground-level ozone and the formation of secondary organic aerosols.

16.2.2 Secondary Pollutants

Secondary pollutants are not directly emitted but are formed through chemical reactions in the atmosphere. They include the following:

(a) **Ground-level ozone (O_3):** Ozone is formed by the reaction of sunlight with NO_x and VOCs in the presence of heat. It is a key component of smog and has harmful effects on respiratory health.

(b) **Sulfuric acid (H_2SO_4) and nitric acid (HNO_3):** These acids are formed through the oxidation of SO_2 and NO_x, respectively. They contribute to acid rain, which harms ecosystems, vegetation, and infrastructure.

(c) **Secondary organic aerosols (SOAs):** SOAs are formed through the oxidation of VOCs. They have adverse effects on air quality, visibility, and climate.

16.2.3 Hazardous Air Pollutants (HAPs)

HAPs, also known as toxic air pollutants, are substances known to cause or suspected of causing serious health effects. They include heavy metals (e.g., lead, mercury), benzene, formaldehyde, and dioxins. HAPs are primarily emitted from industrial processes, vehicle exhaust, and combustion activities.

16.2.4 Indoor Air Pollutants

Indoor air pollutants originate from various sources within buildings. Common indoor pollutants include tobacco smoke, volatile organic compounds from paints and cleaning products, biological contaminants (e.g., mold, bacteria, pollen), and radon gas.

16.3 Sources of Air Pollution

Understanding the different sources of air pollution is essential for devising effective strategies to combat its detrimental effects. Air pollutants can originate from two primary categories: point sources and nonpoint sources. Let us shed light on these sources, highlighting their characteristics and impacts on air quality.

16.3.1 Point Sources of Air Pollution

Point sources refer to stationary and localized pollution-emitting activities that release contaminants directly into the atmosphere. These sources are characterized by identifiable points of emissions, making them relatively easier to monitor and regulate. Some prominent examples of point sources include the following:

Industrial Emissions Manufacturing plants, power plants, refineries, and other industrial facilities contribute to air pollution through the release of various pollutants such as sulfur dioxide (SO_2), nitrogen oxides (NO_x), particulate matter (PM), volatile organic compounds (VOCs), and heavy metals.

Fossil Fuel Combustion Power generation, residential heating, and transportation heavily rely on fossil fuels. The combustion of coal, oil, and natural gas in power plants, vehicles, and heating systems releases significant amounts of pollutants, including carbon dioxide (CO_2), carbon monoxide (CO), and sulfur and nitrogen compounds.

Waste Disposal Incinerators, landfills, and sewage treatment plants emit pollutants such as methane (a potent greenhouse gas), particulate matter, and hazardous gases during waste disposal processes. Improper waste management practices exacerbate air pollution concerns.

Agricultural Activities Agricultural practices such as livestock rearing, crop burning, and the use of fertilizers and pesticides contribute to air pollution. Emissions from livestock operations, in particular, release ammonia (NH3) and other pollutants into the atmosphere.

16.3.2 Nonpoint Sources of Air Pollution

In contrast to point sources, nonpoint sources are diffuse and broadly distributed, making their identification and regulation more challenging. These sources emit pollutants indirectly, and their emissions are often associated with specific land use activities. Nonpoint sources include the following:

Vehicle Emissions While vehicles can be classified as point sources in specific instances (e.g., stationary idling), their overall impact on air pollution is often considered nonpoint. Vehicle emissions, including exhaust fumes, contribute to air pollution through the release of CO_2, NO_x, PM, and other pollutants.

Residential Sources Residential activities such as cooking, heating, and the use of household chemicals and cleaning products can generate pollutants like PM, VOCs, and carbon-containing compounds. Wood-burning stoves and fireplaces are additional sources of particulate matter and pollutants.

Agricultural Runoff Excessive use of fertilizers and pesticides in agriculture can result in runoff, contaminating water bodies and contributing to air pollution indirectly. The volatilization of chemicals from fields and subsequent atmospheric deposition can have detrimental effects on air quality.

Natural Sources Natural phenomena such as volcanic eruptions, dust storms, wildfires, and biogenic emissions from forests and vegetation release significant amounts of pollutants into the atmosphere. While these sources are natural and unavoidable, their intensity and impact can be exacerbated by human activities.

16.4 Photochemical Smog

Photochemical smog is a type of air pollution that occurs when sunlight reacts with certain pollutants in the atmosphere, primarily nitrogen oxides (NO_x) and volatile organic compounds (VOCs), resulting in the formation of a complex mixture of harmful chemicals. It is characterized by a hazy appearance in the atmosphere and is commonly observed in urban areas with high levels of vehicular traffic and industrial emissions.

The formation of photochemical smog involves a series of chemical reactions, which can be summarized as follows:

1. Emission of pollutants: The initial step involves the release of nitrogen oxides (NO_x) and volatile organic compounds (VOCs) into the atmosphere. These pollutants are emitted from various sources, including vehicle exhaust, industrial emissions, and certain chemical solvents.

2. Initiation: Sunlight, particularly ultraviolet (UV) radiation, provides the necessary energy to initiate the smog formation process. UV radiation from the sun strikes nitrogen dioxide (NO2) molecules, leading to their photodissociation into nitrogen monoxide (NO) and atomic oxygen (O).

$$NO_2 + \text{sunlight}(UV) \rightarrow NO + O$$

3. Chain reaction: The atomic oxygen (O) produced in the previous step reacts with molecular oxygen (O_2) to form ozone (O_3) through a chain reaction. This step is facilitated by the presence of nitrogen monoxide (NO), which regenerates the atomic oxygen (O) necessary for the continued formation of ozone.

$$O + O_2 \rightarrow O_3$$
$$NO + O_3 \rightarrow NO_2 + O_2$$

4. VOC reactions: Volatile organic compounds (VOCs) present in the atmosphere, such as hydrocarbons, react with ozone (O_3) in the presence of sunlight to form various products, including aldehydes, ketones, and peroxyacyl nitrates (PANs). These reactions are known as photochemical reactions.

$$VOC + O_3 + \text{sunlight}(UV) \rightarrow \text{Aldehydes / Ketones} + \text{other products } VOC + O_3 + \text{sunlight}(UV) \rightarrow \text{PANs} + \text{other products}$$

5. Regeneration: The nitric oxide (NO) produced in the chain reaction (step 3) reacts with ozone (O3) to regenerate nitrogen dioxide (NO2), thereby allowing the chain reaction to continue.

$$NO + O_3 \rightarrow NO_2 + O_2$$

The overall result of these reactions is the production of ozone, aldehydes, ketones, PANs, and other secondary pollutants, which contribute to the formation of photochemical smog. The presence of these pollutants in the atmosphere can have detrimental effects on human health and the environment. Ozone, for instance, is a respiratory irritant and can exacerbate respiratory conditions. Aldehydes and PANs can also cause eye and respiratory tract irritation.

Efforts to control photochemical smog involve reducing the emissions of nitrogen oxides (NO_x) and volatile organic compounds (VOCs) from various sources, implementing stricter emission standards, promoting cleaner technologies, and adopting measures to reduce vehicular congestion and traffic-related emissions.

16.5 Role of Inversion Layers

Inversion layers play a significant role in exacerbating air pollution in regions where they occur. An inversion layer refers to a meteorological phenomenon in which a layer of warm air traps cooler air near the Earth's surface. This situation is the opposite of the normal atmospheric condition where temperature decreases with increasing altitude.

16.5.1 Formation of Inversion Layers

Inversion layers form due to various factors, such as calm weather conditions, clear skies, and the absence of strong wind or weather systems. During the night, the Earth's surface cools down rapidly, and the cool air close to the ground becomes denser. The warm air above acts as a lid, preventing the vertical mixing of air. As a result, pollutants emitted from various sources, such as vehicles, industrial facilities, and power plants, get trapped within the lower atmospheric layer.

16.5.2 Effects of Inversion Layers on Air Quality

The effects of inversion layers on air quality can be detrimental. The trapped pollutants accumulate and become concentrated, leading to high levels of smog, haze, and poor air quality. This can have severe consequences for human health, including respiratory problems, increased risk of cardiovascular diseases, and aggravated asthma symptoms. Inversion layers also contribute to the formation of secondary pollutants, such as ground-level ozone, which is harmful to both human health and the environment.

16.5.3 Strategies for Dealing with Air Pollution in Inversion-Prone Regions

To address air pollution in regions prone to inversion layers, several strategies can be employed. One approach is the implementation of emission control measures that aim to reduce the release of pollutants from industrial processes, vehicles, and other pollution sources. This can involve the use of cleaner technologies, the enforcement of emission regulations, and the promotion of sustainable transportation systems.

Another strategy is the promotion of alternative energy sources, such as renewable energy, which helps reduce the reliance on fossil fuels and associated emissions. Urban planning measures can also be adopted, including the proper placement of industrial zones away from residential areas, the creation of green spaces, and the promotion of energy-efficient buildings.

Additionally, improving public awareness about the causes and impacts of air pollution can lead to behavioral changes, such as reducing personal vehicle usage, practicing sustainable lifestyles, and supporting clean air initiatives. Monitoring air quality through the use of advanced technology and implementing early warning systems can help identify pollution hotspots and trigger prompt actions.

In conclusion, inversion layers exacerbate air pollution by trapping pollutants near the Earth's surface. Their formation is influenced by meteorological conditions and can have detrimental effects on air quality and human health. Strategies for addressing air pollution in regions prone to inversion layers include emission control measures, promotion of alternative energy sources, urban planning, public awareness, and monitoring systems. Implementing these strategies collectively can help mitigate the impact of inversion layers on air pollution.

16.6 Effects of Air Pollution

Air pollution is a pressing global issue that has far-reaching consequences for various aspects of our planet. The release of harmful pollutants into the atmosphere has detrimental effects on vegetation, ecosystems, the environment, climate, human and animal health, and even historical monuments. Understanding and mitigating these effects are crucial for the well-being of both present and future generations.

16.6.1 Vegetation and Ecosystems

Air pollution significantly impacts vegetation and ecosystems. One of the most common pollutants, ground-level ozone, can impair plant growth and reduce crop yields. High levels of ozone damage plant cells, decrease photosynthesis rates, and inhibit the absorption of nutrients. Similarly, nitrogen dioxide (NO_2) and sulfur dioxide (SO_2) emissions contribute to the acidification of soil and water, disrupting the balance of nutrients and harming plant life. These pollutants also lead to the eutrophication of water bodies, causing algal blooms and depleting oxygen levels, which negatively affect aquatic ecosystems.

16.6.1.1 Effect on Forest Ecosystems

Air pollution poses a significant threat to forest ecosystems, which are vital for biodiversity, climate regulation, and the overall health of our planet. The release of pollutants into the atmosphere can have detrimental effects on forests, impacting both the vegetation and the organisms that rely on these ecosystems.

Damage to Tree Health Air pollutants, such as sulfur dioxide (SO_2), nitrogen oxides (NO_x), and ozone (O_3), can harm the health of trees. High levels of sulfur dioxide can cause visible damage to foliage, leading to leaf discoloration, stunted growth, and even premature leaf drop. Nitrogen oxides and ozone can also damage leaf tissues and affect the photosynthesis process, reducing a tree's ability to produce energy and grow. For instance, in areas with high ozone concentrations, forest trees like pines and firs have shown decreased growth rates and increased susceptibility to pests and diseases.

Forest Decline and Mortality Prolonged exposure to air pollution can lead to forest decline and mortality. Acid rain, caused by the deposition of sulfur dioxide and nitrogen oxides, damages soil quality and affects nutrient availability, making it difficult for trees to thrive. Additionally, chronic exposure to ozone weakens trees' defense mechanisms, rendering them more susceptible to insect infestations and diseases. An alarming example of forest decline due to air pollution is the Black Forest in Germany, where high levels of nitrogen deposition have caused significant damage to trees and led to widespread dieback.

Disruption of Ecosystem Services Forests provide essential ecosystem services, such as carbon sequestration, water regulation, and habitat provision. Air pollution can disrupt these services and have cascading effects on entire ecosystems. For instance, elevated levels of nitrogen deposition can alter soil nutrient balances, favoring the growth of nitrophilic plant species over others. This shift in vegetation composition can impact forest biodiversity and disrupt the delicate balance of ecological interactions. In areas heavily impacted by air pollution, such as industrial regions in China, forest ecosystems struggle to provide adequate habitat for a diverse range of species.

Forest Fires Air pollution can exacerbate the risk and severity of forest fires. Increased concentrations of particulate matter and aerosols in the atmosphere can reduce visibility, hampering firefighting efforts. Moreover, these pollutants can act as condensation nuclei, leading to the formation of cloud layers that trap heat near the forest floor, creating drier and more combustible conditions. The combination of air pollution and climate change has contributed to the intensification of forest fires, as seen in recent devastating wildfires in California, where smoke from the fires has further worsened air quality and impacted human health.

16.6.1.2 Effects on Agroecosystems

Air pollution poses significant challenges to agroecosystems, which encompass the intricate relationship between agriculture and the environment. The release of pollutants into the atmosphere can have adverse effects on crop productivity, soil health, and overall agricultural sustainability. Understanding the impacts of air pollution on agroecosystems is crucial for ensuring food security and the well-being of both rural communities and global populations.

Air pollution is a pressing concern that extends beyond cityscapes and industrial areas. It also has far-reaching effects on agroecosystems, the delicate balance of agriculture and the environment. The emission of pollutants into the atmosphere, such as nitrogen oxides (NO_x), sulfur dioxide (SO_2), volatile organic compounds (VOCs), and particulate matter, can have detrimental consequences for agricultural productivity and sustainability.

One of the major challenges posed by air pollution is the impact on crop health and productivity. High levels of ozone (O_3) can damage plant cells, impair photosynthesis, and decrease crop yields. Ozone pollution affects a wide range of agricultural crops, including staple crops like wheat, rice, and soybeans. Studies have shown that chronic exposure to ozone reduces crop productivity and compromises food security in regions with high pollution levels.

Airborne pollutants can also negatively affect soil health, which is critical for agricultural sustainability. Nitrogen compounds, primarily from agricultural and industrial activities, can deposit onto soil surfaces and contribute to soil acidification. Acidic soil conditions can hinder nutrient availability and uptake by plants, impacting crop growth and yield. Moreover, nitrogen deposition can disrupt soil microbial communities, essential for nutrient cycling and maintaining soil fertility.

Air pollution also influences the interactions between crops and pests. Elevated levels of nitrogen in the atmosphere can favor the growth of nitrophilic weeds, increasing competition with crops and reducing yields. Additionally, certain pests and diseases thrive under polluted conditions, leading to increased infestations and crop damage. For instance, aphids, a common agricultural pest, are attracted to crops that are nitrogen-rich due to air pollution, making them more prone to infestation.

Furthermore, air pollution has indirect effects on agroecosystems through climate change. Pollutants like carbon dioxide (CO_2), methane (CH_4), and nitrous oxide (N_2O) contribute to the warming of the Earth's atmosphere and subsequent climate disruptions. These changes in temperature and precipitation patterns impact agricultural systems, affecting crop growth, water availability, and pest dynamics.

16.6.1.3 Effects on Aquatic Ecosystems

Air pollution has significant impacts on aquatic ecosystems and the organisms that inhabit them. Pollutants released into the air can eventually find their way into water bodies through various pathways, leading to detrimental effects on aquatic life.

Acidification Airborne pollutants like sulfur dioxide (SO_2) and nitrogen oxides (NO_x) can dissolve in rainwater to form acid rain. When acid rain falls onto water bodies, it increases their acidity, leading to a process called acidification. Acidic water negatively affects aquatic organisms, especially those sensitive to changes in pH levels. For example, acid rain has been linked to the decline of certain fish populations, such as brook trout, which require specific pH conditions for their survival and reproduction.

Eutrophication Airborne pollutants, including nitrogen compounds from agricultural and industrial activities, can settle into water bodies. Excessive nitrogen input can lead to eutrophication, a process in which an excess of nutrients causes rapid algae growth. This excessive algae growth depletes oxygen levels in the water, creating "dead zones" where aquatic organisms struggle to survive. An example of this can be seen in the Gulf of Mexico, where excess nutrients from agricultural runoff in the Mississippi River contribute to the formation of a massive dead zone each year.

Toxic Contamination Certain air pollutants, such as heavy metals and persistent organic pollutants (POPs), can settle on the surface of water bodies or be deposited through precipitation. These contaminants can accumulate in aquatic organisms through a process called bioaccumulation. For instance, mercury emissions from industrial activities can settle in water bodies, where it can be transformed into methylmercury, a highly toxic form that accumulates in fish and can pose serious health risks to both aquatic organisms and humans consuming contaminated fish.

Oxygen Depletion Air pollution indirectly affects oxygen levels in water bodies. Particulate matter and aerosols, such as those emitted by industrial processes, can contribute to the formation of haze or smog. When these particles settle on the surface of water bodies, they can reduce the amount of sunlight penetrating the water, hindering photosynthesis by aquatic plants. As a result, oxygen production decreases, which impacts the survival of aerobic organisms that rely on dissolved oxygen for respiration.

Disruption of Aquatic Food Chains Air pollution-induced changes in aquatic ecosystems can disrupt food chains and ecological balance. For example, the excessive growth of algae due to nutrient pollution can lead to algal blooms. When these blooms eventually die off, decomposition by bacteria consumes oxygen, leading to hypoxic or anoxic conditions. This oxygen depletion can result in fish kills and other negative impacts on higher trophic levels, affecting the entire food chain.

16.6.2 Environment and Climate

Air pollution plays a crucial role in climate change. Greenhouse gases like carbon dioxide (CO_2), methane (CH_4), and nitrous oxide (N_2O) contribute to the warming of the Earth's atmosphere, leading to global temperature rise and subsequent climate disruptions. Pollutants such as black carbon and aerosols, emitted through industrial processes and burning fossil fuels, absorb and scatter sunlight, altering the Earth's radiation balance. This, in turn, affects cloud formation, precipitation patterns, and regional climates, leading to extreme weather events like heatwaves, droughts, and heavy rainfall.

16.6.3 Human and Animal Health

The impacts of air pollution on human health are extensive and well documented. Exposure to polluted air increases the risk of respiratory problems, including asthma, bronchitis, and lung cancer. Fine particulate matter ($PM_{2.5}$) and toxic gases like nitrogen oxides (NO_x) and volatile organic compounds (VOCs) can penetrate deep into the lungs, causing inflammation and long-term damage. Air pollution also contributes to cardiovascular diseases, neurological disorders, and premature death. Additionally, animals are also vulnerable to the harmful effects of air pollution, experiencing respiratory issues, reduced fertility, and damage to their natural habitats.

16.6.4 Historical Monuments

Air pollution poses a significant threat to the preservation of historical monuments and cultural heritage. Gaseous pollutants, particularly sulfur dioxide and nitrogen oxides, react with moisture in the atmosphere to form acidic compounds. These compounds corrode the surfaces of buildings, statues, and ancient artifacts, eroding their integrity and accelerating the deterioration process. The discoloration and damage caused by air pollution are particularly evident in urban areas with high pollution levels, where historical structures face constant exposure to pollutants emitted by industries, vehicles, and power plants.

16.6.5 Acid Rain

Acid rain, a grave environmental phenomenon, continues to pose a threat to ecosystems worldwide. It is a result of atmospheric pollution caused by human activities, primarily the release of harmful gases into the air. Acid rain, known for its detrimental effects on the environment, has raised concerns among scientists, policymakers, and the general public alike.

Acid rain refers to rainfall or any form of precipitation that contains high levels of acidic components, such as sulfuric acid and nitric acid. These acids are formed when industrial emissions, such as sulfur dioxide (SO_2) and nitrogen oxides (NO_x), react with atmospheric moisture, resulting in the formation of sulfuric acid (H_2SO_4) and nitric acid (HNO_3). These emissions primarily originate from industrial processes, power plants, and vehicular emissions.

16.6.5.1 Causes of Acid Rain

Fossil Fuel Combustion The burning of fossil fuels, such as coal, oil, and gas, releases sulfur dioxide and nitrogen oxides into the atmosphere. Power plants, industries, and transportation are the main contributors to these emissions.

Industrial Emissions Many industrial processes release SO_2 and NO_x directly into the atmosphere, further exacerbating acid rain. Chemical manufacturing, metal smelting, and mining operations contribute significantly to these emissions.

Vehicle Emissions The combustion of gasoline and diesel fuels in automobiles leads to the release of nitrogen oxides and sulfur dioxide. The increasing number of vehicles on the road contributes to the overall pollution levels.

16.6.5.2 Effects of Acid Rain

Environmental Impact Acid rain has severe consequences for ecosystems. It damages forests, soils, and bodies of water, disrupting the natural balance. Acidified lakes and rivers can harm aquatic life, leading to the depletion of fish populations. It also affects plant growth, as acidic soil inhibits nutrient absorption.

Forest Decline Acid rain damages forests by leaching essential nutrients from the soil and weakening trees' defenses against diseases and pests. It also stunts tree growth and leads to the decline of entire forest ecosystems.

Infrastructure Damage Acid rain corrodes buildings, bridges, and statues made of materials like limestone and marble, which are particularly vulnerable to acidic reactions. This results in the deterioration of architectural heritage and increased maintenance costs.

Human Health Concerns While direct exposure to acid rain does not pose significant health risks, the pollutants responsible for acid rain, such as sulfur dioxide and nitrogen oxides, can contribute to the formation of fine particulate matter ($PM_{2.5}$). Prolonged exposure to $PM_{2.5}$ can lead to respiratory problems and other health issues.

16.6.6 Stratospheric Ozone Layer Depletion

The stratospheric ozone layer plays a critical role in protecting life on Earth by filtering out harmful ultraviolet (UV) radiation from the sun. However, due to human activities and the release of certain air pollutants, this vital shield is being depleted.

The stratospheric ozone layer is located approximately 10–50 kilometers above the Earth's surface. It consists of a high concentration of ozone (O_3) molecules, which absorb and block a significant portion of the sun's UV radiation. The release of certain air pollutants, particularly chlorofluorocarbons (CFCs), halons, carbon tetrachloride, and other ozone-depleting substances (ODS), contributes to the thinning of this protective layer.

16.6.6.1 Causes of Stratospheric Ozone Layer Depletion

Ozone-Depleting Substances (ODS) Industrial processes, aerosol propellants, refrigeration systems, and fire suppression equipment release ODS into the atmosphere. The most common ODS are chlorofluorocarbons (CFCs), which were widely used in air conditioning, refrigeration, and aerosol products before their harmful effects were recognized.

Halons Halons, used primarily in fire suppression systems, contribute to ozone depletion. While their use has been phased out, their long atmospheric lifetime continues to affect the ozone layer.

Industrial Emissions Certain industrial processes release nitrous oxide (N_2O) and nitrogen oxides (NO_x), which indirectly contribute to ozone depletion by accelerating the destruction of ozone molecules.

16.6.6.2 Consequences of Stratospheric Ozone Layer Depletion

Increased UV Radiation Thinning of the ozone layer allows higher levels of UV radiation to reach the Earth's surface. Overexposure to UV radiation can lead to skin cancer, cataracts, weakened immune systems, and other adverse health effects in humans. It also affects marine life, crops, and ecosystems, leading to reduced biodiversity and ecological imbalances.

Climate Change The release of ozone-depleting substances and the resulting thinning of the ozone layer can contribute to climate change. Some ODS, such as CFCs, are potent greenhouse gases, exacerbating the greenhouse effect and global warming.

Global efforts to combat stratospheric ozone layer depletion include the Montreal Protocol and technology and policy interventions. Adopted in 1987, the Montreal Protocol is an international treaty aimed at phasing out the production and use of ozone-depleting substances. It has been successful in reducing the production and consumption of ODS and has played a crucial role in preventing further ozone layer depletion. Governments and industries have taken steps to replace ozone-depleting substances with safer alternatives. This includes the development and adoption of ozone-friendly technologies, such as hydrochlorofluorocarbons (HCFCs) and hydrofluorocarbons (HFCs), which have lower ozone depletion potentials.

16.7 Prevention and Control of Air Pollution

16.7.1 Regulatory Measures

Governments play a crucial role in implementing and enforcing policies to control air pollution. These measures include setting emission standards, promoting cleaner technologies, and implementing pollution control measures in industries and vehicles.

16.7.2 Transition to Clean Energy

Shifting from fossil fuels to renewable energy sources like solar, wind, and hydroelectric power can significantly reduce air pollution. Encouraging energy efficiency and promoting sustainable transportation systems, such as electric vehicles and public transport, also contribute to pollution reduction.

To safeguard our planet and future generations, it is imperative that we take proactive steps to prevent and control air pollution. In this article, we will explore some innovative approaches and striking examples of efforts aimed at mitigating this invisible foe.

16.7.3 Renewable Energy Sources

Shifting from fossil fuels to renewable energy sources like solar, wind, and hydroelectric power can significantly reduce air pollution. Encouraging energy efficiency and promoting sustainable transportation systems, such as electric vehicles and public transport, also contribute to pollution reduction.

Transitioning to renewable energy sources is a paramount step in curbing air pollution. China, the world's largest emitter of greenhouse gases, has made remarkable progress in this area. By investing heavily in wind and solar energy, China has reduced its reliance on coal-fired power plants, resulting in significant reductions in air pollution. For instance, the city of Lanzhou went from having some of the worst air quality in China to becoming a role model for clean energy, thanks to the development of a massive solar power station.

16.7.4 Green Transportation Initiatives

Transportation is a major contributor to air pollution, particularly in densely populated urban areas. Copenhagen, Denmark, has become a shining example of combating this issue. The city has implemented an extensive network of bicycle lanes and prioritized pedestrian-friendly infrastructure, leading to a remarkable decrease in car usage. As a result, Copenhagen boasts some of the cleanest air among major cities, with citizens enjoying reduced exposure to harmful pollutants.

16.7.5 Innovative Urban Planning

Cities are at the forefront of the battle against air pollution, and their design plays a pivotal role in shaping environmental outcomes. The city of Medellín in Colombia has undergone a transformative journey by integrating nature into its urban fabric. By constructing green corridors and vertical gardens, Medellín has not only enhanced air quality but also revitalized neglected neighborhoods. These nature-based solutions serve as a powerful testament to the positive impact urban planning can have on air pollution control.

16.7.6 Tackling Indoor Air Pollution

While outdoor air pollution often takes center stage, indoor air pollution is equally concerning. In India, the "Ujjwala" scheme has brought clean cooking solutions to millions of households. By providing LPG connections to rural families, the initiative aims to replace traditional biomass fuels like wood and cow dung, which are major sources of indoor pollution. This effort not only improves air quality within homes but also reduces the health risks associated with prolonged exposure to harmful smoke.

16.7.7 Global Collaboration for Change

Addressing air pollution necessitates global collaboration and concerted efforts. The European Union has made significant strides in this regard by implementing stringent emission standards. As a result, European cities like Stockholm, Sweden, have witnessed remarkable improvements in air quality. This success story serves as a reminder that united action and cross-border cooperation can yield tangible results in the fight against air pollution.

16.7.8 Waste Management Strategies

Effective waste management practices, including recycling, composting, and proper disposal methods, help minimize the release of pollutants from waste.

16.7.9 Afforestation and Green Spaces

Planting trees and creating green spaces in urban areas can act as natural air filters, absorbing pollutants and improving air quality.

16.7.10 Public Awareness and Education

Educating the public about the causes and impacts of air pollution is essential. Encouraging individuals to adopt environmentally friendly practices, such as reducing energy consumption and using public transportation, can contribute to pollution reduction.

Air pollution remains a persistent threat to human health and the environment, but it is a challenge that can be overcome through collective action and innovative solutions. The examples highlighted in this article demonstrate the immense potential we have to prevent and control air pollution. By embracing renewable energy, promoting sustainable transportation, implementing nature-based urban planning, addressing indoor air pollution, and fostering global collaboration, we can pave the way for a cleaner and healthier future for all. Let us seize the opportunity to safeguard our planet and ensure a breathable atmosphere for generations to come.

Air pollution is a complex issue that requires collective efforts from governments, industries, and individuals to mitigate its adverse effects. By understanding the causes, effects, and preventive measures discussed in this chapter, we can work toward creating a cleaner and healthier environment for present and future generations. Only through a concerted global commitment to combat air pollution can we ensure a sustainable and thriving planet.

16.8 Summary

Air pollution is a global issue with significant implications for human health, ecosystems, and the overall quality of life. The chapter provides a comprehensive overview of air pollution, including its definition, causes, effects, and prevention and control measures. Air pollution is the presence of harmful substances in the air that exceed the natural capacity of the atmosphere to disperse and dilute them. Pollutants can be classified as primary, secondary, hazardous, or indoor air pollutants based on their origin and properties. Understanding the sources of air pollution is crucial, including point sources, such as industrial emissions and fossil fuel combustion, as well as nonpoint sources like vehicle emissions and agricultural activities. Photochemical smog is a type of air pollution that forms when sunlight reacts with pollutants like nitrogen oxides and volatile organic compounds, leading to the production of harmful chemicals. Effects of air pollution on ecosystems, vegetation, agroecosystems, environment, climate, human health, and historical monuments are pretty conspicuous. Measures to control air pollution involve reducing emissions, implementing stricter standards, promoting cleaner technologies, and addressing specific sources or pollution.

16.9 Exercises

16.9.1 Multiple-Choice Questions

1. What is the definition of air pollution?
   (a) The presence of harmful substances in the air that exceed natural capacity
   (b) The release of gases into the atmosphere
   (c) The contamination of water bodies by pollutants
   (d) The emission of chemicals from industries
2. Which of the following is a primary pollutant?
   (a) Ground-level ozone
   (b) Sulfuric acid
   (c) Nitric acid
   (d) Particulate matter
3. Which pollutant is primarily emitted from the burning of fossil fuels containing sulfur compounds?
   (a) Particulate matter
   (b) Sulfur dioxide
   (c) Carbon monoxide
   (d) Volatile organic compounds
4. How are secondary pollutants formed?
   (a) They are emitted directly into the atmosphere
   (b) They are released from industrial processes
   (c) They are formed through chemical reactions in the atmosphere
   (d) They are produced by volcanic eruptions
5. Which of the following is a hazardous air pollutant?
   (a) Carbon monoxide
   (b) Sulfur dioxide
   (c) Lead
   (d) Nitrogen oxides
6. Where do indoor air pollutants originate from?
   (a) Vehicle emissions
   (b) Industrial processes
   (c) Outdoor air pollution
   (d) Various sources within buildings
7. What are point sources of air pollution characterized by?
   (a) Diffuse and broadly distributed emissions
   (b) Stationary and localized pollution-emitting activities
   (c) Natural phenomena like volcanic eruptions
   (d) Agricultural activities
8. What are the primary sources of vehicle emissions?
   (a) Industrial emissions
   (b) Agricultural activities
   (c) Fossil fuel combustion
   (d) Natural sources
9. Which type of air pollution occurs when sunlight reacts with nitrogen oxides and volatile organic compounds?
   (a) Acid rain
   (b) Indoor air pollution
   (c) Photochemical smog
   (d) Particulate matter
10. What is the primary role of sunlight in the formation of photochemical smog?

(a) To provide energy for chemical reactions
(b) To cause damage to vegetation
(c) To regulate temperature in the atmosphere
(d) To emit harmful ultraviolet radiation

11. How does air pollution affect vegetation and crop yields?
(a) Increases nutrient absorption in plants
(b) Enhances photosynthesis rates
(c) Decreases plant growth and crop yields
(d) Improves soil fertility
12. Which air pollutant is known to damage tree health and cause forest decline?
(a) Carbon monoxide
(b) Sulfur dioxide
(c) Ground-level ozone
(d) Volatile organic compounds
13. What effect does air pollution have on forest ecosystems?
(a) Enhances biodiversity
(b) Increases tree growth rates
(c) Disrupts nutrient balances in soil
(d) Promotes forest regeneration
14. How does air pollution contribute to climate change?
(a) It reduces greenhouse gas emissions
(b) It alters atmospheric composition
(c) It increases cloud formation
(d) It promotes global cooling
15. What are the primary health effects of air pollution on humans?
(a) Increased lung capacity
(b) Enhanced immune system
(c) Improved cognitive function
(d) Respiratory and cardiovascular diseases
16. How can emissions from industrial sources be reduced?
(a) Implementing stricter emission standards
(b) Increasing industrial production
(c) Encouraging the use of fossil fuels
(d) Relocating industries to urban areas
17. Which approach helps control air pollution from vehicles?
(a) Promoting electric vehicles and public transportation
(b) Encouraging private vehicle ownership
(c) Reducing fuel efficiency standards
(d) Increasing road infrastructure
18. What is the purpose of promoting cleaner technologies in addressing air pollution?
(a) To increase pollution levels
(b) To improve energy efficiency
(c) To support fossil fuel consumption
(d) To reduce environmental regulations
19. Which sector is responsible for a significant share of air pollution emissions?
(a) Agriculture
(b) Renewable energy
(c) Waste management
(d) Education
20. What is a key measure to prevent and control air pollution?
(a) Encouraging deforestation
(b) Supporting unsustainable practices
(c) Implementing pollution control technologies
(d) Increasing industrial emissions

Answers: 1-a, 2-d, 3-b, 4-c, 5-c, 6-d, 7-b, 8-c, 9-c, 10-a, 11-c, 12-c, 13-c, 14-b, 15-d, 16-a, 17-a, 18-b, 19-a, 20-c

16.9.2 Short-Answer Questions

1. What are the main sources of indoor air pollution?
2. Name two primary pollutants commonly found in urban air.
3. How are photochemical smog and acid rain formed?
4. List three health effects of air pollution on humans.
5. What are the primary factors contributing to outdoor air pollution in urban areas?
6. Explain the difference between primary and secondary air pollutants.
7. What are the major sources of particulate matter in the atmosphere?
8. Describe the process of ozone depletion and its implications.
9. How does air pollution impact aquatic ecosystems and marine life?
10. Discuss two strategies or approaches to mitigate air pollution.

16.9.3 Long-Answer Questions

1. Describe the mechanisms by which air pollutants can affect respiratory health, including the types of pollutants involved and their specific impacts on the respiratory system.
2. Discuss the concept of inversion layers and their role in exacerbating air pollution. Explain how they form, their effects on air quality, and potential strategies for dealing with air pollution in regions prone to inversion layers.
3. Air pollution is known to have detrimental effects on the environment. Choose one specific environmental impact of air pollution, such as the deterioration of ecosystems, and provide a detailed explanation of the processes involved, the key pollutants responsible, and the long-term consequences.
4. Evaluate the effectiveness of different air pollution control measures and strategies, such as emission regulations, technological advancements, and public awareness campaigns. Discuss their strengths and weaknesses, and propose any additional measures that could enhance their efficiency.
5. Write short notes on the following:
   (a) Photochemical smog
   (b) Air pollution effects on forest ecosystems
   (c) Transition to clean energy

References

Chameides W, Perdue EM (eds) (1994) Environmental chemistry of the elements. CRC Press (Taylor and Francis), Boca Raton

Dockery DW, Pope CA (eds) (2020) Outdoor air pollution. Oxford University Press, Oxford

Finlayson-Pitts BJ, Pitts JN Jr (2000) Chemistry of the upper and lower atmosphere: theory, experiments, and applications. Academic Press, New York

Lelieveld J et al (eds) (2015) Particulate matter: atmospheric tracers, emissions, and health effects. Springer, Singapore

Pandey A, Yadav R (eds) (2021) Environmental pollution and control: soil, air, and water. Springer, Singapore

Seinfeld JH, Pandis SN (2016) Atmospheric chemistry and physics: from air pollution to climate change, 3rd edn. Wiley, New Jersey

Singh V (2024) Environmental disruptions: planet earth in the vicious cycle of pollution, global warming, and climate change. NIPA, New Delhi. 156 pp

Vallero DA (2014) Fundamentals of air pollution, 5th edn. Academic Press, New York

Wark K, Warner CF, Davis WT (2012) Air pollution: its origin and control, 4th edn. Pearson, London

Zhang R (2019) Atmospheric chemistry and physics: from air pollution to climate change, 2nd edn. Oxford University Press, Oxford

17 Water Pollution

Water is one of the most vital resources on our planet, essential for the survival and sustenance of all living organisms. However, due to human activities and various environmental factors, water pollution has become a significant global issue. This chapter aims to explore the causes, effects, and potential solutions to water pollution, highlighting the urgent need for sustainable water management practices.

17.1 What Is Water Pollution?

Water pollution refers to the contamination, degradation, or alteration of water bodies – such as lakes, rivers, oceans, groundwater, and even smaller bodies like ponds and wetlands – due to the introduction of harmful substances or excessive amounts of natural or synthetic pollutants. It occurs when various pollutants are discharged into water sources, leading to adverse effects on the environment, human health, and aquatic life.

17.2 Water Pollutants

Pollutants can come from various sources, both natural and human-induced. Natural sources include volcanic eruptions, erosion of rocks and minerals, and decomposition of organic matter. However, human activities are the primary contributors to water pollution. These activities involve industrial processes, agricultural practices, urban development, and inadequate waste management systems. Water pollutants can be classified into several categories as shown in Table 17.1.

17.3 Types of Water Pollution

Water pollution can be categorized into several types based on the sources of contamination:

(a) **Point source pollution:** Pollution that comes from a single, identifiable source, such as industrial effluents or sewage outfalls
(b) **Nonpoint source pollution:** Pollution that originates from diffuse sources, such as agricultural runoff or urban stormwater
(c) **Groundwater pollution:** Contamination of undergroundwater sources due to the infiltration of pollutants from the surface or subsurface activities
(d) **Surface water pollution:** Contamination of water bodies like rivers, lakes, and oceans by pollutants entering directly or indirectly through runoff

V. Singh, *Textbook of Environment and Ecology*, https://doi.org/10.1007/978-981-99-8846-4_17

Table 17.1 Various types of water pollutants

Water pollutants	The characteristics of the pollutants
Organic pollutants	These include substances derived from living organisms or their byproducts. Examples are sewage, agricultural runoff containing pesticides and fertilizers, and organic chemicals like oil and petroleum products. These pollutants can lead to oxygen depletion in water bodies, harming aquatic organisms.
Inorganic pollutants	These are nonliving substances originating from industrial discharges, mining operations, and agricultural activities. Inorganic pollutants include heavy metals like lead, mercury, and cadmium, as well as acids, salts, and other chemical compounds. They can accumulate in organisms and disrupt their physiological processes.
Nutrients	Excessive levels of nutrients, primarily nitrogen and phosphorus, can cause water pollution through eutrophication. This occurs when these nutrients enter water bodies from agricultural runoff or untreated sewage. Eutrophication leads to excessive growth of algae and aquatic plants, which depletes oxygen levels and harms other organisms.
Sediments	Soil erosion, construction activities, and deforestation can result in sediment runoff, leading to water pollution. Excessive sedimentation can cloud water, reduce light penetration, and negatively impact aquatic habitats.
Pathogens	Bacteria, viruses, and other microorganisms from human and animal waste can contaminate water sources, causing waterborne diseases such as cholera, typhoid, and hepatitis. Poor sanitation and inadequate wastewater treatment contribute to the presence of pathogens in water bodies.

17.4 Causes of Water Pollution

17.4.1 Industrial Activities

Industries release various pollutants into water bodies, including toxic chemicals, heavy metals, and organic compounds. Improper waste disposal and industrial accidents can have severe consequences for water quality and aquatic life.

17.4.2 Agricultural Practices

Intensive agricultural practices contribute to water pollution through the excessive use of fertilizers, pesticides, and herbicides. These chemicals can enter water bodies through runoff, leading to eutrophication, algal blooms, and the depletion of oxygen, causing harm to aquatic ecosystems.

17.4.3 Domestic and Municipal Waste

Improper disposal of domestic waste and inadequate wastewater treatment systems result in the release of untreated or partially treated sewage into water bodies. This contamination introduces harmful pathogens, organic matter, and nutrients, leading to the spread of waterborne diseases and ecological imbalances.

17.4.4 Mining Activities

Mining operations, particularly those involving extraction of metals and minerals, produce substantial amounts of waste materials, known as tailings. These tailings contain hazardous substances like heavy metals and sulfides that can contaminate nearby water bodies, affecting both aquatic life and human communities.

17.4.5 Oil Spills

Accidental or deliberate oil spills from offshore drilling, transportation vessels, or oil refineries can cause significant water pollution. Oil slicks coat the water surface, impacting marine ecosystems, birds, and other wildlife. The long-term effects of oil spills can be devastating and require extensive cleanup efforts.

17.5 Groundwater Pollution

Groundwater, a vital natural resource, plays a crucial role in sustaining ecosystems and meeting the water demands of millions of people worldwide. However, this hidden treasure is increasingly under threat from groundwater pollution. Contamination of groundwater poses significant risks to both the environment and human health. Understanding the causes, consequences, and potential solutions to groundwater pollution is essential in safeguarding this invaluable resource.

Groundwater pollution can originate from a variety of sources, both natural and human-induced, such as industrial activities, agricultural practices, landfills and waste disposal, underground storage tanks, sewage and septic system, etc. The consequences of groundwater pollution are far-reaching and pose significant risks to both the environment and human well-being.

17.6 Effects of Water Pollution

17.6.1 Environmental Effects

Water pollution has far-reaching consequences for aquatic ecosystems. Pollutants disrupt the delicate balance of aquatic organisms, leading to the decline or extinction of certain species. Eutrophication, caused by excessive nutrient input, disrupts the ecological equilibrium and results in oxygen depletion, dead zones, and harmful algal blooms.

Dissolved oxygen (DO) is a critical component of water quality and plays a vital role in the health and well-being of aquatic ecosystems. It refers to the amount of oxygen gas dissolved in water, which is necessary to support the respiration and survival of aquatic organisms. DO levels are influenced by various natural and human-induced factors, and their fluctuations can have significant impacts on aquatic life.

The primary sources of dissolved oxygen in water bodies are atmospheric diffusion and photosynthesis by aquatic plants and algae. Atmospheric diffusion occurs when oxygen from the air comes into contact with the water surface, leading to oxygen dissolution. Photosynthesis, mainly performed by aquatic plants, phytoplankton, and algae, is a process that utilizes sunlight to convert carbon dioxide and water into oxygen and organic matter. During the day, photosynthesis contributes to oxygen production, increasing DO levels.

Water pollution can influence the levels of dissolved oxygen in a water body. Warmer water has a lower capacity to hold dissolved oxygen compared to colder water. Hot waters discharged from various industries, when drained into nature water bodies, lower down their DO levels.

When organic matter accumulates in water bodies, bacteria and other microorganisms break it down through a process called decomposition. Decomposition consumes oxygen, leading to a decrease in DO levels. Excessive organic matter can cause oxygen depletion and create hypoxic or anoxic conditions, which are harmful to aquatic life.

Insufficient levels of dissolved oxygen can have adverse effects on aquatic organisms. Fish, in particular, are highly sensitive to changes in DO levels. When DO drops below certain thresholds, fish may exhibit stress responses, reduced growth rates, and impaired reproduction. Severe oxygen depletion can lead to fish kills, where large numbers of fish die due to oxygen-starved conditions.

DO levels in a water body serve as valuable indicators of water quality. On the basis of this, two parameters used for determining water quality are biochemical oxygen demand (BOD) and chemical oxygen demand (COD).

17.6.1.1 Biochemical Oxygen Demand

Biochemical oxygen demand (BOD) is a measure of the amount of oxygen consumed by microorganisms in the biological degradation of organic matter present in water. BOD is a key parameter used to assess the level of organic pollution in water bodies, particularly in relation to wastewater and surface water quality.

When organic matter, such as sewage, agricultural runoff, or industrial effluents, enters a water body, microorganisms (mainly bacteria) break down the organic compounds as they consume oxygen in the process. The more organic matter present, the greater the demand for oxygen by these microorganisms. BOD quantifies the amount of dissolved oxygen required by these microorganisms over a specified period to degrade the organic matter.

The BOD value is expressed in milligrams of oxygen per liter of water (mg/L) and indicates the amount of oxygen consumed by microorganisms during the incubation period. A higher BOD value indicates a greater amount of organic pollution in the water. It is worth noting that BOD measurements provide an estimate of the oxygen demand and the potential for pollution rather than directly measuring the concentration of pollutants themselves.

By monitoring and managing BOD levels, water resource managers, environmental agencies, and wastewater treatment plants can ensure the protection and conservation of water bodies, mitigate pollution impacts, and maintain healthy aquatic ecosystems.

17.6.1.2 Chemical Oxygen Demand

Chemical oxygen demand (COD) is a measure of the amount of oxygen required to chemically oxidize the organic and inorganic compounds present in water. COD is widely used as a parameter to assess the level of pollution in water bodies, particularly in industrial and wastewater treatment processes.

Unlike BOD, which measures the oxygen consumed by biological processes, COD measures the oxygen demand resulting from both biological and chemical reactions. COD provides a more comprehensive and rapid evaluation of water quality, as it includes both biodegradable and nonbiodegradable organic substances, as well as certain inorganic compounds.

Expressed in milligrams of oxygen per liter of water (mg/L), the COD provides an estimate of the amount of oxygen required to chemically oxidize the organic matter present in the water sample. High COD values indicate a greater concentration of pollutants, both biodegradable and nonbiodegradable, in the water.

The COD test provides a rapid and reliable method for evaluating the pollution load in water bodies and wastewater. By monitoring and managing COD levels, industries, municipalities, and environmental agencies can take appropriate measures to minimize pollution, protect water resources, and ensure the sustainability of aquatic ecosystems.

17.6.2 Biological Magnification

Biological magnification, also known as biomagnification or bioaccumulation, is a concerning phenomenon that occurs in ecosystems affected by water pollution. It describes the process by which certain pollutants, particularly persistent organic pollutants (POPs) and heavy metals, become increasingly concentrated and magnified as they move up the food chain.

In water bodies, pollutants such as pesticides, industrial chemicals, and heavy metals are often discharged directly or indirectly into the environment. These pollutants can enter the aquatic food chain through various pathways. Phytoplankton, aquatic plants, and algae can absorb pollutants from the water, and small organisms like zooplankton consume them. As the pollutants move up the food chain, they become more concentrated and reach higher trophic levels.

The primary reason for the magnification of pollutants is their tendency to accumulate in the fatty tissues of organisms. While the concentration of pollutants may be low in the water, they can be absorbed by organisms through ingestion or gill respiration. Once inside an organism's body, the pollutants are not easily metabolized or excreted, leading to their accumulation over time.

As smaller organisms, such as zooplankton, consume contaminated food or water, they accumulate a certain level of pollutants in their tissues. When these organisms are consumed by larger organisms, such as small fish, the pollutants are transferred and further concentrated in the predator's body. This process continues as larger fish consume smaller fish, leading to even higher concentrations of pollutants. At the top of the food chain, large predators like sharks or humans may accumulate significant levels of these pollutants.

The consequences of biological magnification are concerning for both the environment and human health:

Environmental Impact Biological magnification can have detrimental effects on aquatic ecosystems. High concentrations of pollutants can disrupt the reproductive systems, growth, and behavior of organisms. It can lead to population declines and loss of biodiversity. Additionally, the accumulation of pollutants in sediments can have long-term impacts on the health and functioning of aquatic habitats.

Human Health Concerns Human consumption of organisms high in the food chain, such as predatory fish or shellfish, can expose individuals to elevated levels of pollutants. These pollutants can have toxic effects on human health, causing various diseases and disorders. For example, certain POPs, such as PCBs (polychlorinated biphenyls) and dioxins, have been linked to developmental issues, hormone disruption, and cancer.

Biological magnification underscores the need for comprehensive approaches to water pollution management. By addressing the sources of pollution, minimizing the release of harmful substances, and protecting ecosystems, we can work toward reducing the accumulation of pollutants in aquatic organisms and safeguarding both the environment and human health.

17.6.3 Eutrophication

Eutrophication is a process that occurs in aquatic ecosystems, such as lakes, rivers, and coastal areas, when there is an excessive input of nutrients, primarily nitrogen and phosphorus. This excessive nutrient load leads to an overabundance of plant and algal growth, resulting in changes to the ecosystem's structure and functioning.

The main sources of nutrients that contribute to eutrophication are human activities, including agricultural runoff, sewage discharges, and industrial effluents. These activities can introduce excessive amounts of nitrogen and phosphorus into water bodies, often in the form of fertilizers, detergents, or organic waste. The elevated nutrient levels stimulate the growth of algae, aquatic plants, and other vegetation.

The eutrophication process can be divided into two phases: the nutrient enrichment phase and the algal bloom phase.

1. **Nutrient enrichment phase:** Excessive nutrients enter the water body, leading to increased concentrations of nitrogen and phosphorus. These nutrients act as fertilizers for aquatic plants and algae, promoting their growth. Initially, the growth may be slow, but as nutrient levels continue to rise, the system becomes increasingly enriched.
2. **Algal bloom phase:** As the nutrient levels reach a certain threshold, algal growth becomes rapid and intense. The water body may experience a sudden proliferation of algae, forming what is known as an algal bloom. Algal blooms can cause water discoloration, turning the water green, brown, or red, depending on the dominant species. These blooms can be harmful to the ecosystem and have several negative impacts:
   (a) **Oxygen depletion:** During the bloom, algae undergo photosynthesis and release oxygen, leading to a temporary increase in dissolved oxygen levels. However, when the algal bloom eventually dies off, the dead algae sink to the bottom and are decomposed by bacteria. This decomposition process consumes oxygen, leading to a depletion of dissolved oxygen levels in the water. Oxygen depletion can harm fish and other aquatic organisms that require oxygen to survive, leading to fish kills and other ecological imbalances.
   (b) **Light limitation:** Dense algal blooms can block sunlight from reaching the deeper layers of the water, reducing light penetration. This can inhibit the growth of submerged aquatic plants,

which are important for providing oxygen, food, and habitat for various organisms.

(c) **Toxin production:** Certain species of algae can produce toxins, such as cyanobacteria (blue-green algae). These toxins, known as harmful algal toxins or cyanotoxins, can be harmful to humans, animals, and aquatic life. They can contaminate drinking water sources and pose a risk to public health.

(d) **Disruption of aquatic ecosystems:** Eutrophication can alter the balance of aquatic ecosystems. The excessive growth of algae and plants can lead to a decrease in biodiversity, as some species may outcompete others for resources. Additionally, the decomposition of dead algae consumes nutrients and releases carbon dioxide, further altering the water chemistry.

By addressing the underlying causes of eutrophication and implementing appropriate management strategies, we can mitigate its impacts and protect the health and integrity of aquatic ecosystems.

17.6.4 Human Health Impacts

Contaminated water sources pose a serious threat to human health. Waterborne diseases such as cholera, typhoid, and dysentery are prevalent in areas with poor water quality and inadequate sanitation. Prolonged exposure to pollutants in water, such as heavy metals or pesticides, can lead to chronic health issues, including organ damage, cancer, and developmental disorders.

Various human diseases caused by drinking water contaminated with different types of pathogens are listed in Table 17.2.

Table 17.2 Many human diseases caused by consuming water contaminated with a variety of pathogens

Pathogen	Disease/ailment
Bacteria	
Escherichia coli (*E. coli*)	Gastroenteritis, diarrhea, abdominal cramps, and urinary tract infections
Salmonella	Salmonellosis (nausea, vomiting, diarrhea, fever) and typhoid fever
Vibrio cholera	Cholera (severe diarrhea, dehydration, electrolyte imbalance)
Campylobacter	Campylobacteriosis (diarrhea, abdominal pain, fever)
Shigella	Shigellosis (diarrhea, stomach cramps, fever)
Legionella	Legionnaires' disease (pneumonia-like symptoms, fever, cough)
Helicobacter pylori	Gastritis, peptic ulcers, and stomach cancer
Streptococcus	Streptococcal infections (sore throat, pneumonia, skin infections)
Viruses	
Hepatitis A	Hepatitis A (jaundice, fatigue, nausea, abdominal pain)
Norovirus	Norovirus infection (vomiting, diarrhea, stomach cramps)
Rotavirus	Rotavirus gastroenteritis (severe diarrhea, vomiting)
Adenovirus	Respiratory infections, conjunctivitis (pink eye), and gastroenteritis
Parasites	
Giardia	Giardiasis (diarrhea, abdominal cramps, bloating)
Cryptosporidium	Cryptosporidiosis (diarrhea, stomach cramps, fever)
Entamoeba histolytica	Amoebiasis (bloody diarrhea, abdominal pain, liver abscess)
Cyclospora	Cyclosporiasis (diarrhea, loss of appetite, weight loss)
Schistosoma	Schistosomiasis (abdominal pain, blood in urine, liver damage)

It is important to note that the severity and manifestation of these diseases can vary depending on the specific strain or species of the pathogen and individual factors such as overall health and immune system strength. Ensuring access to clean and safe drinking water is crucial for preventing waterborne diseases caused by these pathogens.

Various diseases/ailments cause due to drinking water contaminated by heavy metals are presented in Table 17.3.

Table 17.3 Diverse illnesses resulting from the consumption of water contaminated with metals

Heavy metal	Diseased/ailment caused due to drinking water contaminated with
Lead	Lead poisoning, cognitive impairments, developmental delays, neurological disorders, anemia, kidney damage, and reproductive problems
Arsenic	Arsenicosis (skin lesions, skin cancer), peripheral vascular disease (black foot), cardiovascular diseases, diabetes, kidney damage, respiratory issues, and neurological effects
Mercury	Minamata disease (neurological disorder), developmental delays, cognitive impairments, kidney damage, respiratory issues, cardiovascular problems, and reproductive disorders
Cadmium	Kidney damage and bone diseases (e.g., itai-itai or ouch-ouch), lung damage, cardiovascular problems, reproductive disorders, cancer (lung, prostate)
Chromium	Respiratory problems, lung cancer, kidney damage, liver damage, and skin conditions
Copper	Gastrointestinal issues, liver damage, Wilson's disease (copper accumulation in organs), neurological disorders, and kidney damage
Zinc	Gastrointestinal issues, nausea, vomiting, diarrhea, stomach cramps, and copper deficiency (excessive zinc intake can interfere with copper absorption)
Nickel	Respiratory problems, lung cancer, skin allergies, kidney damage, and gastrointestinal issues

It is important to note that the severity and manifestation of these diseases and ailments can vary depending on the concentration and duration of exposure to the specific heavy metal. Preventing and minimizing heavy metal pollution in drinking water sources is crucial for safeguarding human health.

17.6.5 Minamata Disease

Minamata disease is a neurological disorder caused by long-term exposure to methylmercury, a highly toxic form of mercury. The disease received its name from the City of Minamata, Japan, where a severe outbreak occurred in the mid-twentieth century. The cause of the outbreak was industrial wastewater containing methylmercury that was discharged into Minamata Bay by a chemical factory.

The contaminated water and marine life in the bay resulted in the ingestion of methylmercury by the local population, primarily through the consumption of contaminated fish and shellfish. Methylmercury bioaccumulated in the bodies of these organisms, eventually reaching high concentrations.

Minamata disease is characterized by various symptoms, including sensory disturbances, muscle weakness, tremors, difficulty speaking and hearing, and in severe cases paralysis, coma, and even death. It also led to severe developmental issues in children born to affected mothers, known as congenital Minamata disease.

The outbreak of Minamata disease raised awareness about the devastating effects of mercury pollution and led to significant changes in environmental regulations and industrial practices worldwide. It highlighted the importance of controlling and minimizing the release of toxic substances into the environment to prevent such catastrophic health impacts. The lessons learned from Minamata disease continue to guide efforts to protect human health and the environment from mercury contamination.

17.6.6 Hazards of Groundwater Pollution

Polluted groundwater poses significant health risks and can lead to various diseases and ailments. One such disease is methemoglobinemia, commonly known as blue baby syndrome. It occurs when infants ingest water contaminated with high levels of nitrate or other pollutants. Nitrate interferes with the oxygen-carrying capacity of red blood cells, causing a bluish discoloration of the skin and depriving the body of oxygen.

Additionally, groundwater contamination with high levels of fluoride can result in skeletal fluorosis. Prolonged consumption of fluoride-contaminated water leads to the accumulation of fluoride in bones, causing joint pain, stiffness, and skeletal deformities.

Another disease associated with polluted groundwater is black foot disease, which is prevalent in certain regions with high levels of arsenic in the groundwater. Chronic exposure to arsenic can lead to skin lesions, gangrene, and even cancer.

These examples highlight the critical importance of ensuring the quality and safety of groundwater sources. Proper monitoring, treatment, and management of groundwater supplies are necessary to prevent these diseases and protect the health of communities reliant on groundwater for drinking and daily use.

17.6.7 Economic Consequences

Water pollution affects various economic sectors. Impaired water quality reduces the availability of clean water for agriculture, industry, and domestic use. This can lead to decreased crop yields, increased production costs, and compromised public health, impacting the overall economic productivity and stability of affected regions.

Water pollution can have significant economic consequences, affecting various sectors and causing both short-term and long-term impacts. Some of the key economic consequences of water pollution are the following:

Impact on Public Health Water pollution can lead to the spread of waterborne diseases, resulting in increased healthcare costs, loss of productivity, and reduced quality of life for affected individuals. Treating and managing waterborne illnesses can strain healthcare systems and result in economic burdens on governments and households.

Loss of Livelihoods Industries that rely on clean water for their operations, such as agriculture, fisheries, and tourism, can suffer from water pollution. Contaminated water can reduce crop yields, damage aquatic ecosystems and fisheries, and discourage tourists from visiting polluted areas. This can lead to job losses, reduced income for workers, and economic decline in affected regions.

Cost of Water Treatment Water pollution often necessitates expensive treatment processes to make water safe for consumption. The costs of building and operating water treatment facilities, implementing advanced treatment technologies, and maintaining infrastructure can be substantial and borne by governments, water utilities, and consumers through increased water bills.

Impact on Tourism and Recreation Water pollution can tarnish the reputation of tourist destinations known for their pristine water bodies, affecting tourism revenues. Polluted beaches, rivers, or lakes may deter visitors and result in decreased tourism-related income, including accommodation, restaurants, and recreational activities.

Decline in Property Values Water pollution can negatively impact property values, particularly for properties located near contaminated water bodies. The perception of polluted water can make properties less desirable and lead to decreased market prices, affecting homeowners and property investors.

Environmental Remediation Costs Cleaning up polluted water bodies and restoring ecosystems can be a costly and time-consuming process. Governments or responsible parties may incur expenses related to remediation efforts, including cleanup operations, environmental monitoring, and restoration of affected areas.

Regulatory Compliance and Fines Industries and businesses found responsible for water pollution may face legal consequences, including fines, penalties, and lawsuits. Compliance with water quality regulations can also impose additional costs on industries, such as implementing pollution control measures or upgrading facilities to meet environmental standards.

Impact on Agriculture Water pollution can harm agricultural productivity by contaminating irrigation water or degrading soil quality. This can lead to reduced crop yields, increased use of fertilizers and pesticides to compensate for nutrient deficiencies, and additional costs for farmers.

Addressing water pollution and its economic consequences requires investments in pollution prevention, water treatment infrastructure, and sustainable practices (Vesilind et al. 2013, Kassotaki et al. 2021, Xiong et al. 2022). Implementing stringent regulations, promoting environmentally friendly technologies, and raising awareness about the importance of clean water are crucial steps toward mitigating the economic impacts of water pollution (Parmar et al. 2023).

17.7 Solutions to Water Pollution

17.7.1 Regulatory Measures

Governments and regulatory bodies play a crucial role in addressing water pollution. Implementing and enforcing strict environmental regulations, such as effluent standards for industries, can help minimize pollution levels. Monitoring and penalizing noncompliant entities are essential for maintaining water quality standards.

17.7.2 Sustainable Agriculture Practices

Promoting sustainable agricultural practices, such as organic farming, precision irrigation techniques, and integrated pest management, can reduce chemical inputs and minimize agricultural runoff. Effective soil erosion control measures and buffer zones along water bodies also aid in preventing sedimentation and nutrient pollution.

17.7.3 Improved Wastewater Treatment

Investing in robust wastewater treatment infrastructure and promoting the use of advanced treatment technologies can significantly reduce the discharge of untreated sewage into water bodies (Zhang et al. 2022). Implementing tertiary treatment processes and reusing treated wastewater for non-potable purposes can help conserve water resources.

17.7.4 Sewage Treatment

Sewage treatment is the process of removing contaminants and pollutants from wastewater to make it safe for disposal or reuse. It involves several stages, including primary, secondary, and tertiary treatments.

17.7.4.1 Primary Treatment

Primary treatment is the initial stage of sewage treatment, which focuses on the physical separation of solid and floating materials from the wastewater. The wastewater enters a primary settling tank, also known as a sedimentation tank or clarifier. In this tank, heavy particles such as sand, gravel, and other large debris settle to the bottom, forming a sludge layer, while lighter materials like oils, grease, and scum float to the top.

The settled sludge is removed and sent for further treatment or disposal. The floating scum and grease are skimmed off the surface. The partially treated wastewater, known as effluent, flows out of the settling tank and moves on to the next stage, secondary treatment.

17.7.4.2 Secondary Treatment

Secondary treatment focuses on the biological degradation of organic matter present in the wastewater. It aims to remove dissolved and colloidal organic compounds that were not effectively eliminated during the primary treatment. This stage employs microorganisms to break down the remaining organic material.

There are two common methods of secondary treatment: activated sludge process and trickling filter process.

Activated Sludge Process In this process, the wastewater is mixed with a culture of microorganisms in an aeration tank. The microorganisms, mainly bacteria, consume the organic matter as their food source, converting it into carbon dioxide, water, and additional microbial biomass. Aeration provides the necessary oxygen for microbial growth and promotes the breakdown of organic pollutants. After aeration, the mixture flows into a secondary settling tank, where the newly formed microbial sludge settles. Some of the settled sludge is recirculated back to the aeration tank to maintain the microbial population, while the excess sludge is removed for further treatment or disposal. The treated wastewater, or secondary effluent, undergoes further treatment in the tertiary stage.

Trickling Filter Process In this process, the wastewater trickles over a bed of rocks or plastic media coated with microbial films called biofilms. The microorganisms in the biofilms degrade the organic matter as the wastewater passes over them. The treated wastewater then flows into a secondary settling tank, where the microbial sludge is settled and removed. Similar to the activated sludge process, the excess sludge is further treated or disposed of, and the remaining effluent proceeds to the tertiary treatment.

17.7.4.3 Tertiary Treatment

Tertiary treatment, also known as advanced or final treatment, focuses on removing any remaining impurities and achieving a high level of water quality suitable for various applications, such as reuse or release into sensitive environments. The specific methods used in tertiary treatment (Fig. 17.1) can vary depending on the desired level of treatment and the intended reuse of the water. Some common tertiary treatment processes include the following:

Filtration The effluent is passed through different types of filters, such as sand filters or membrane filters, to remove remaining suspended solids, fine particles, and microorganisms.

Disinfection This step involves the application of chemical disinfectants, such as chlorine or ultraviolet (UV) light, to kill or inactivate any remaining pathogens (disease-causing microorganisms) present in the water.

Nutrient Removal In certain cases, such as in environmentally sensitive areas, excess nutrients like nitrogen and phosphorus are removed from the effluent through processes like biological nutrient removal or chemical precipitation.

Advanced Oxidation Some advanced treatment methods, like ozonation or advanced oxidation processes, are employed to further degrade persistent organic compounds, pharmaceuticals, or other trace contaminants that may still be present.

After the tertiary treatment, the water is considered to be of high quality and can be reused for various purposes such as irrigation, industrial processes, or even drinking water, depending on the local regulations and requirements.

It is important to note that the level of treatment and the specific processes employed may vary depending on the local regulations, the capacity of the treatment plant, and the quality standards set for the treated wastewater.

17.7.5 Public Awareness and Education

Raising awareness among the general public about the importance of water conservation, pollution prevention, and sustainable water management is crucial. Educating communities about proper waste disposal, responsible water use, and the impacts of water pollution empowers individuals to make informed choices and take proactive measures (Parmar et al. 2023, Singh 2024).

In conclusion, water pollution poses a significant threat to ecosystems, human health, and the overall well-being of our planet. Addressing this

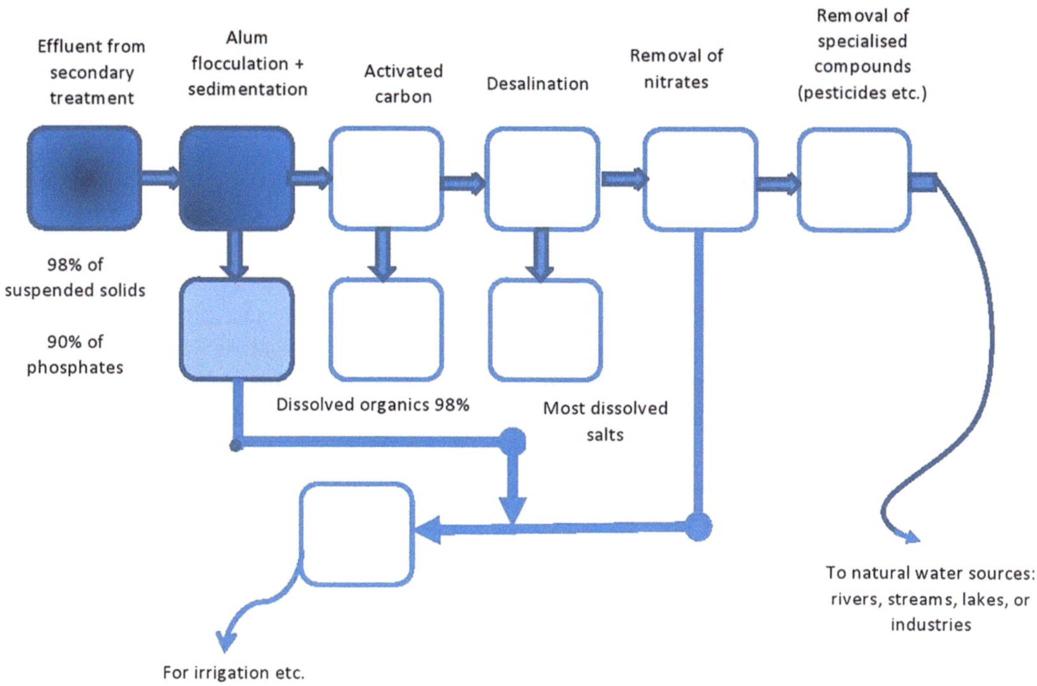

Fig. 17.1 Illustration of tertiary sewage treatment in an effluent treatment plant (Source: Singh 2024)

issue requires collective efforts from individuals, communities, industries, and governments. By implementing effective pollution control measures, promoting sustainable practices, and prioritizing the conservation of water resources, we can safeguard our water bodies and ensure a sustainable future for generations to come.

17.8 Summary

The chapter on water pollution provides an overview of the causes, effects, and potential solutions to this global issue. Water pollution refers to the contamination, degradation, or alteration of water bodies due to the introduction of harmful substances or excessive amounts of natural or synthetic pollutants. It can be categorized into types such as point-source pollution, nonpoint source pollution, groundwater pollution, and surface water pollution.

Various human activities contribute to water pollution, including industrial processes, agricultural practices, domestic waste disposal, mining activities, and oil spills. These activities release pollutants such as toxic chemicals, heavy metals, organic compounds, and oil into water bodies, leading to adverse effects on the environment, human health, and aquatic life.

The chapter also discusses the specific case of groundwater pollution, which poses significant risks to the environment and human well-being. Groundwater pollution can originate from industrial activities, agricultural practices, landfills, sewage systems, and more.

Water pollution has extensive environmental effects, including disruption of aquatic ecosystems, eutrophication, dissolved oxygen depletion, and biological magnification. Eutrophication occurs when excessive nutrients, primarily nitrogen and phosphorus, lead to an overabundance of plant and algal growth, causing changes to the ecosystem. Biological magnification refers to the increasing concentration of certain pollutants, such as persistent organic pollutants (POPs) and heavy metals, as they move up the food chain.

The effects of water pollution extend to human health, with contaminated water sources posing a threat. Waterborne diseases like cholera, typhoid, and dysentery are prevalent in areas with poor water quality and inadequate sanitation. Prolonged exposure to pollutants in water, such as heavy metals or pesticides, can lead to chronic health issues, including organ damage, cancer, and developmental disorders.

This chapter highlights the importance of monitoring and managing water quality indicators such as dissolved oxygen, biochemical oxygen demand (BOD), and chemical oxygen demand (COD) to assess pollution levels and protect water resources. It emphasizes the need for comprehensive approaches to water pollution management, addressing pollution sources, minimizing the release of harmful substances, and protecting ecosystems.

Overall, the chapter emphasizes the urgent need for sustainable water management practices to mitigate water pollution and ensure the conservation of this vital resource.

17.9 Exercises

17.9.1 Multiple-Choice Questions

1. What is water pollution?
   (a) Contamination of water bodies due to excessive rainfall
   (b) Introduction of harmful substances into water bodies
   (c) Natural process of altering the composition of water
   (d) Excessive growth of aquatic plants in water bodies
2. Which of the following is not a source of water pollution?
   (a) Industrial activities
   (b) Agricultural practices
   (c) Snowfall
   (d) Urban development
3. What is point-source pollution?
   (a) Pollution originating from multiple sources
   (b) Pollution that comes from a single identifiable source
   (c) Pollution caused by natural disasters
   (d) Pollution resulting from oil spills
4. How do industrial activities contribute to water pollution?
   (a) By releasing toxic chemicals and heavy metals
   (b) By decreasing the water temperature
   (c) By promoting the growth of algae
   (d) By reducing the oxygen levels in water bodies
5. What is the main cause of eutrophication?
   (a) Excessive input of nutrients like nitrogen and phosphorus
   (b) Industrial accidents
   (c) Oil spills
   (d) Groundwater contamination
6. Which type of pollution occurs when pollutants infiltrate undergroundwater sources?
   (a) Groundwater pollution
   (b) Nonpoint source pollution
   (c) Surface water pollution
   (d) Industrial pollution
7. What is the primary consequence of oxygen depletion in water bodies?
   (a) Increased fish reproduction
   (b) Decreased algal growth
   (c) Harm to aquatic life
   (d) Enhanced water clarity
8. What is the measure of the amount of oxygen consumed by microorganisms in the biological degradation of organic matter in water?
   (a) Biochemical oxygen demand (BOD)
   (b) Chemical oxygen demand (COD)
   (c) Dissolved oxygen (DO)
   (d) Biological magnification
9. How do pollutants become increasingly concentrated as they move up the food chain?
   (a) By undergoing photosynthesis
   (b) Through biological magnification
   (c) By increasing dissolved oxygen levels
   (d) Through algal blooms
10. What is the main source of nutrients that contribute to eutrophication?
    (a) Natural processes such as weathering
    (b) Human activities like agricultural runoff
    (c) Oil spills from offshore drilling

   (d) Volcanic activity
11. What is the neurological disorder caused by long-term exposure to methylmercury?
   (a) Cholera
   (b) Dysentery
   (c) Minamata disease
   (d) Typhoid
12. What is the term used to describe pollution originating from diffuse sources?
   (a) Groundwater pollution
   (b) Nonpoint source pollution
   (c) Point source pollution
   (d) Surface water pollution
13. What is the primary consequence of algal blooms in water bodies?
   (a) Increased dissolved oxygen levels
   (b) Enhanced biodiversity
   (c) Light penetration to deeper layers of water
   (d) Oxygen depletion and harm to aquatic life
14. Which of the following diseases is not commonly associated with contaminated water sources?
   (a) Cholera
   (b) Typhoid
   (c) Dysentery
   (d) Malaria
15. What is the primary purpose of monitoring water quality indicators?
   (a) To assess pollution levels
   (b) To promote algal growth
   (c) To increase water temperature
   (d) To protect endangered species
16. What does COD stand for in water quality assessment?
   (a) Carbon organic decomposition
   (b) Chemical oxygen demand
   (c) Carbonate oxygen degradation
   (d) Contamination odor detection
17. Which approach is essential for managing water pollution effectively?
   (a) Encouraging algal growth
   (b) Promoting point source pollution
   (c) Maximizing oil spills
   (d) Protecting ecosystems
18. What is the recommended approach to minimize water pollution from industrial activities?
   (a) Encouraging excessive nutrient input
   (b) Increasing pollutant release
   (c) Reducing biodiversity conservation efforts
   (d) Efficient waste disposal and treatment
19. What is the main goal of sustainable water management?
   (a) To deplete natural water resources
   (b) To promote water pollution
   (c) To ensure conservation of water resources
   (d) To increase algal blooms
20. Why is water pollution a global concern?
   (a) It only affects aquatic life.
   (b) It threatens ecosystems and human well-being.
   (c) It has no impact on human health.
   (d) It increases biodiversity and species conservation.

Answers: 1-b, 2-c, 3-b, 4-a, 5-a, 6-a, 7-c, 8-a, 9-b, 10-b, 11-c, 12-b, 13-d, 14-d, 15-a, 16-b, 17-d, 18-d, 19-c, 20-b

17.9.2 Short-Answer Questions

1. What is the main source of oil pollution in water bodies?
2. Name one agricultural practice that contributes to water pollution.
3. What is the primary environmental impact of thermal pollution?
4. Give an example of four heavy metals that can contaminate water bodies.
5. What is the significance of a clean water body?
6. Define the term "eutrophication."
7. What are some common indicators used to assess water quality?
8. Name one method to control nonpoint source pollution.
9. What is the cause of Minamata disease?
10. Briefly explain the concept of biomagnification in water ecosystems.

17.9.3 Long-Answer Questions

1. Explain the concept of point source pollution and nonpoint source pollution in the context

of water pollution. Provide examples of each and discuss the challenges associated with controlling and mitigating these different types of pollution.
2. Describe the process of eutrophication and its effects on water ecosystems. Discuss the main causes of eutrophication and the ecological consequences of excessive nutrient enrichment in lakes, rivers, and coastal areas.
3. Explain the concept of biomagnification in aquatic food chains. Discuss how persistent pollutants, such as heavy metals and certain pesticides, can accumulate in organisms at higher trophic levels. Describe the potential risks posed by biomagnification to human health and the strategies employed to mitigate this issue.
4. Explain the process of sewage treatment and its significance in addressing water pollution. Discuss the main steps involved in sewage treatment, including primary, secondary, and tertiary treatment processes, and their respective objectives. Additionally, highlight the importance of proper sewage treatment in protecting public health and maintaining the ecological balance of water ecosystems.
5. Write short notes on the following:
   (a) Groundwater pollution
   (b) BOD and COD
   (c) Economic implications of water pollution

References

Kassotaki E, Paranychianakis NV, Nikolaidis NP (2021) Advances in wastewater treatment and resource recovery: selected papers from the 3rd International Conference on Environmental Science and Sustainable Energy (ICESSE-2019). Springer, Cham

Parmar S, Sharma VK, Singh V (2023) Microplastics in the marine ecosystem: sources, risks, mitigation technologies, and challenges. CRC Press (Taylor and Francis). (ISBN: 9781003312086), Boca Raton/London, p 228. https://doi.org/10.1201/9781003312086

Singh V (2024) Environmental disruptions: planet earth in the vicious cycle of pollution, global warming, and climate change. NIPA, New Delhi. 156 pp

Vesilind PA, Morgan SM, Heine LG (2013) Introduction to environmental engineering. Cengage Learning, Stamford

Xiong Y, Guo H, Li X (2022) Water quality monitoring and management: techniques, methods, and applications. Springer, Singapore

Zhang Y, Huang J, Cui B (2022) Emerging organic contaminants in water and wastewater: analysis, treatment, and environmental impacts. Springer, Cham

18 Soil Pollution

Soil, a vital component of our ecosystem, is the foundation of life on Earth. It supports the growth of plants, sustains countless organisms, and provides essential nutrients for food production. However, in recent times, soil pollution has emerged as a pressing environmental concern. Soil pollution refers to the contamination of soil with harmful substances that adversely affect its fertility, productivity, and overall health. Soil pollution is a significant environmental concern that arises from the accumulation of various contaminants in the soil, leading to adverse effects on soil fertility, ecosystems, and human health (Rastogi et al. 2011, 2019, Negi et al. 2019, Singh 2020). This chapter aims to provide a comprehensive overview of soil pollution, including its causes, types of contaminants, impacts, and potential solutions. By understanding the issues associated with soil pollution, we can take appropriate measures to mitigate its effects and promote sustainable soil management practices.

18.1 Sources of Soil Pollution

There are various sources of soil pollution, including both natural and anthropogenic. The natural factors have always been present in the biosphere but are not of significant concern. Anthropogenic factors, which often trigger even natural ones, are the primary sources of soil pollution.

18.1.1 Natural Sources

While anthropogenic activities contribute significantly to soil pollution, natural sources also play a role in the contamination of soil. Natural disasters such as volcanic eruptions and earthquakes can release hazardous substances into the soil. Volcanic ash, rich in heavy metals and toxic elements, can cause significant soil degradation. Similarly, earthquakes can disrupt underground storage tanks or pipelines, leading to the release of pollutants such as oil and gas into the soil.

Furthermore, geological processes like weathering and erosion can contribute to the presence of naturally occurring contaminants in the soil. Certain areas may naturally contain high levels of metals, such as arsenic or lead, which can leach into the soil over time. Additionally, geological formations and deposits may contain radionuclides, which pose risks to soil quality and human health.

18.1.2 Anthropogenic Sources

Human activities have had a profound impact on soil quality and have become the primary source of soil pollution worldwide. Numerous anthropogenic activities contribute to soil contamination, including industrial processes, agricultural practices, improper waste management, and urbanization.

V. Singh, *Textbook of Environment and Ecology*, https://doi.org/10.1007/978-981-99-8846-4_18

18.1.3 Industrial Activities

Industrial processes release a wide range of toxic substances into the environment, often resulting in soil pollution. Chemical spills, inadequate disposal of industrial waste, and improper handling of hazardous materials can all lead to soil contamination. Heavy metals, solvents, petroleum hydrocarbons, and pesticides are some of the common pollutants originating from industrial sources.

18.1.4 Agricultural Practices

Agriculture, while essential for food production, can also contribute significantly to soil pollution. The excessive and improper use of chemical fertilizers, pesticides, and herbicides can accumulate in the soil, causing long-term damage. Inadequate irrigation practices and poor soil management can lead to soil erosion, further exacerbating the issue.

18.1.5 Mining and Construction

Mining operations heavily requiring blasting and bulldozing of mining sites for extraction of minerals lead to soil disturbance and alter soil structure and composition and even soil ecosystem functions. Construction activities generate enormous waste material and lead to soil compaction and also turn the soil ecosystems into concrete deserts.

18.1.6 Improper Waste Management

Improper waste management practices, including indiscriminate dumping of solid waste and disposal of hazardous materials, pose a significant threat to soil health. Landfills that are not properly constructed or managed can contaminate the soil with heavy metals, toxins, and organic pollutants.

18.1.7 Urbanization

Urbanization, driven by rapid population growth, often leads to soil pollution as well. Construction activities, improper disposal of construction waste, and urban runoff contribute to soil degradation. Urban areas are also susceptible to contamination from vehicle emissions, industrial runoff, and improper disposal of household chemicals.

18.2 Types of Contaminants

Heavy Metals Heavy metals like lead, mercury, cadmium, chromium, arsenic, etc. accumulate in the soil through industrial activities, mining, and agriculture that can persist in the environment for a long time and pose risks to human health.

Organic Pollutants Pesticides, herbicides, industrial solvents, petroleum hydrocarbons, etc. introduce toxic chemicals into the soil, affecting soil microorganisms and ecosystems, and can bioaccumulate in the food chain, impacting human and animal health.

Nutrient Imbalance Excessive use of chemical fertilizers leads to nutrient imbalances. Nitrogen and phosphorus runoff from agricultural fields contributes to eutrophication. Nutrient imbalance disrupts natural nutrient cycling and reduces soil fertility over time.

Radioactive Substances Uranium, radium, cesium, and other radioactive isotopes released from nuclear power plants, mining activities, and radioactive waste disposal pose serious health risks and long-term environmental contamination.

18.3 Impacts of Soil Pollution

Soil pollution is not a standalone occurrence; its implications extend widely to water, air, plants, animals, and human systems (Table 18.1). The health of soil is vital for all terrestrial life forms.

Table 18.1 Major effects of soil pollution

Soil pollutants' major effects	Consequences of soil pollution
Soil degradation	• Reduced soil fertility and nutrient cycling capacity • Loss of soil structure, leading to erosion and decreased water-holding capacity • Decline in soil biodiversity and beneficial microorganisms
Air pollution	• Entry of soil pollutants into air due to strong winds and human activities • Air pollutant deposition on foods and water sources • Respiratory and cardiovascular problems due to fine dust particles (PM_{10}, $PM_{2.5}$) • Health hazards due to heavy metals blown into air
Water contamination	• Leaching of pollutants into groundwater and surface water bodies • Contaminated water poses risks to human and aquatic life • Algal blooms, fish kills, and disruption of aquatic ecosystems
Crop contamination	• Uptake of contaminants by plants, leading to food chain contamination • Accumulation of heavy metals and organic pollutants in crops • Threat to food safety and human health
Ecological imbalance	• Soil pollution disrupts soil-plant interactions and ecological processes • Impacts on soil-dwelling organisms, including earthworms, insects, and microorganisms • Disruption of the biogeochemical cycles and natural balance in ecosystems and loss of biodiversity

When the soil is in a healthy state, terrestrial organisms thrive, but when it becomes contaminated, all terrestrial beings suffer the consequences. The repercussions of soil sickness are far-reaching, eventually leading to the death of the soil itself. Dead soils transform into unproductive deserts. Throughout history, civilizations that neglected their soil disappeared from the face of the Earth. Our socioeconomic progress is deeply intertwined with soil health. Unhealthy soil negatively impacts socioeconomic development. The adverse effects of soil pollution are interconnected (Fig. 18.1).

Soil pollution profoundly impacts biogeochemical cycles, including both sedimentary and gaseous cycles. Impoverished soil yields poor crop quality. Cultivated crops and natural vegetation growing on polluted soils display low productivity. Over time, soil pollution disrupts the functioning of the soil ecosystem.

Polluted soils hinder the growth of many sensitive plant species, leading to biodiversity loss and the potential extinction of valuable species. The most severe consequence of polluted soils is the reduction in photosynthetic efficiency within ecosystems, resulting in decreased carbon sequestration, increased carbon emission, and diminished carbon storage capacity of ecosystems and ultimately contributing to global warming and climate change. The effects of soil pollution are not isolated incidents; they trigger a chain reaction that impacts the entire biosphere.

Toxic chemicals and potentially harmful pollutants are absorbed by plants that grow on contaminated soils. These pollutants then enter the human body through food, causing various metabolic disorders. Contaminated soils contribute to health issues in humans, similar to the effects of air and water pollution discussed in other chapters.

In lectures on waste management, there is often an excessive focus on the use of urban wastewater, sewage sludge, and livestock manure in agriculture. However, utilizing untreated waste carries significant risks, as it can introduce antimicrobial substances into the soil, resulting in the presence of antimicrobial-resistant bacteria (Kuppusamy et al. 2018). Given the current state of waste generation and subsequent soil pollution, it is necessary to reconsider O'Neill's (2014) estimation that antimicrobial-resistant infections may become the leading cause of death worldwide by 2050.

18.4 Solutions to Soil Pollution

Soil pollution – largely as a consequence of human activities – poses a significant threat to our environment and the well-being of all living

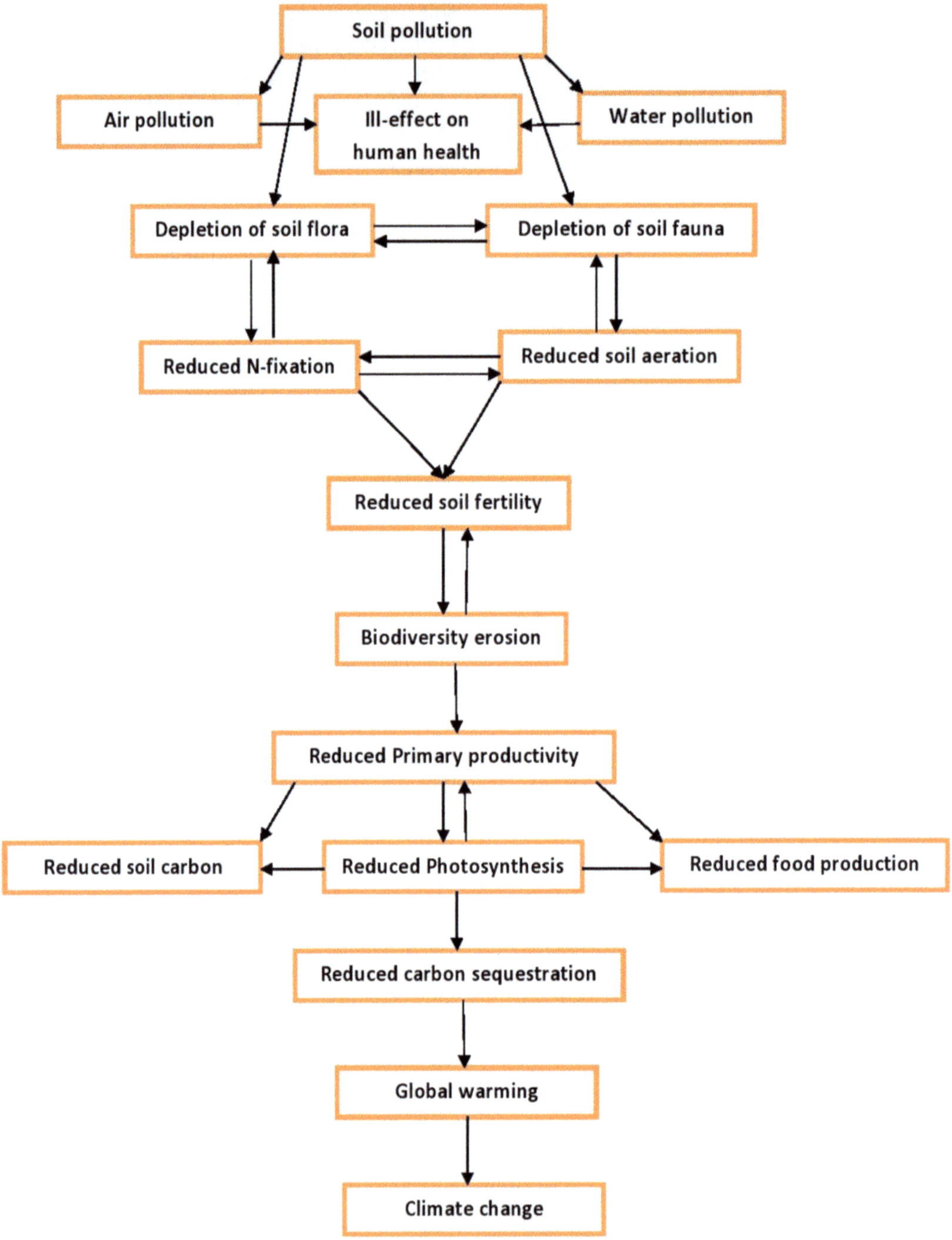

Fig. 18.1 The causes and consequences of soil pollution. (Source: Singh 2024)

organisms. The escalating soil pollution crisis demands immediate action and the implementation of sustainable solutions. Let us explore some effective strategies and initiatives to combat soil pollution and foster a healthier planet for future generations.

18.4.1 Implementing Responsible Agricultural Practices

One of the primary causes of soil pollution is the excessive use of agrochemicals such as fertilizers and pesticides in modern agriculture. Adopting sustainable farming techniques can help minimize soil pollution. These practices include integrated pest management, organic farming, crop rotation, and agroforestry. By reducing chemical inputs, preserving soil fertility, and improving water management, responsible agriculture not only mitigates soil pollution but also promotes biodiversity and enhances food security.

18.4.2 Encouraging Soil Remediation Technologies

Soil remediation technologies offer a ray of hope for contaminated sites. Techniques like bioremediation, phytoremediation, and electrokinetic remediation employ natural processes or specific plants to degrade or remove pollutants from the soil. Bioremediation harnesses the power of microorganisms to break down pollutants, while phytoremediation uses plants to extract and store contaminants. These eco-friendly solutions provide cost-effective and sustainable methods to restore polluted soil back to health.

18.4.3 Enhancing Waste Management Practices

Improper waste disposal is a significant contributor to soil pollution. Effective waste management systems, including recycling, composting, and waste-to-energy processes, can significantly reduce the accumulation of hazardous materials in landfills. By encouraging waste reduction at the source and implementing proper disposal and treatment methods, we can prevent the leaching of toxic substances into the soil, safeguarding its quality and fertility.

18.4.4 Promoting Soil Monitoring and Regulation

Monitoring soil quality and enforcing regulations are vital steps in preventing and controlling soil pollution. Governments and environmental agencies should establish comprehensive soil monitoring programs to identify contamination sources and assess the health of soil ecosystems. Strict regulations should be implemented to control industrial emissions, limit the use of harmful chemicals, and enforce proper waste management practices. Public awareness campaigns and educational programs can also play a crucial role in promoting responsible soil management practices.

18.4.5 Formulating Sustainable Land Use Planning

Effective land use planning is essential for sustainable soil management. It involves considering soil quality, biodiversity, and ecosystem services when making decisions about urban development, infrastructure projects, and agricultural expansion. Protecting arable land, creating green spaces, and implementing zoning regulations can prevent soil degradation, erosion, and contamination. By integrating soil health into land use policies, we can ensure the long-term sustainability of our ecosystems.

In conclusion, soil pollution is a complex issue with wide-ranging implications for human health, ecosystems, and food security. It demands immediate attention and concerted efforts from individuals, communities, industries, and governments to adopt sustainable practices, remediation techniques, and regulatory measures. By addressing the causes and impacts of soil pollution, we can restore soil health, preserve biodiversity, and ensure a sustainable future for generations to come.

18.5 Summary

Soil pollution, the contamination of soil with harmful substances, has become a critical environmental issue affecting soil fertility, ecosystems, and human health. This chapter provides a comprehensive overview of soil pollution, including its sources, types of contaminants, impacts, and potential solutions. Natural sources such as volcanic eruptions and geological processes contribute to soil pollution, but anthropogenic activities are the primary culprits. Industrial activities, agricultural practices, improper waste management, and urbanization all play a significant role in soil contamination.

The types of contaminants found in polluted soil include heavy metals, organic pollutants, nutrient imbalances, and radioactive substances. Soil pollution has far-reaching impacts, affecting water, air, plants, animals, and human systems. It disrupts biogeochemical cycles, reduces crop productivity, leads to biodiversity loss, and contributes to climate change. Contaminated soil also poses health risks as pollutants are absorbed by plants and enter the human body through the food chain.

To address soil pollution, several solutions can be implemented. Responsible agricultural practices such as organic farming, integrated pest management, and agroforestry reduce the use of agrochemicals and promote soil health. Soil remediation technologies like bioremediation and phytoremediation use natural processes to remove pollutants from the soil. Enhancing waste management practices through recycling, composting, and proper disposal prevents hazardous materials from accumulating in landfills and leaching into the soil. Soil monitoring, regulation, and awareness campaigns are essential for preventing and controlling soil pollution. Finally, sustainable land use planning that considers soil quality and biodiversity helps protect soil from degradation, erosion, and contamination.

By implementing these solutions, we can mitigate the effects of soil pollution and promote sustainable soil management practices. It is crucial to take immediate action to protect our soil, a vital resource for the health of our ecosystems and the well-being of current and future generations.

18.6 Exercises

18.6.1 Multiple-Choice Questions

1. What is soil pollution?
   (a) Contamination of water bodies by soil sediments
   (b) Contamination of soil with harmful substances
   (c) Disruption of soil erosion processes
   (d) Introduction of artificial soil additives
2. Which of the following is not a natural source of soil pollution?
   (a) Volcanic eruptions
   (b) Earthquakes
   (c) Weathering and erosion
   (d) Industrial activities
3. Which type of pollution is primarily caused by human activities?
   (a) Natural pollution
   (b) Anthropogenic pollution
   (c) Volcanic pollution
   (d) Geological pollution
4. What is the primary source of soil pollution in agriculture?
   (a) Excessive use of agrochemicals
   (b) Natural disasters
   (c) Erosion and weathering
   (d) Urbanization
5. Which soil remediation technique uses microorganisms to break down pollutants?
   (a) Bioremediation
   (b) Phytoremediation
   (c) Electrokinetic remediation
   (d) Chemical remediation
6. What is the main goal of waste management practices in preventing soil pollution?
   (a) Recycling waste materials
   (b) Proper disposal of hazardous materials
   (c) Composting organic waste
   (d) All of the above
7. What is the purpose of soil monitoring programs?
   (a) Identify contamination sources
   (b) Assess soil health
   (c) Enforce regulations
   (d) All of the above
8. Which sector contributes the most to soil pollution?
   (a) Industrial sector
   (b) Agricultural sector
   (c) Residential sector
   (d) Mining sector
9. Which of the following is not a type of contaminant found in polluted soil?
   (a) Heavy metals
   (b) Radioactive substances
   (c) Fossil fuels

   (d) Organic pollutants
10. How does soil pollution impact crop productivity
   (a) Enhances crop quality
   (b) Increases nutrient absorption
   (c) Reduces crop yield
   (d) Accelerates growth rate
11. What is the primary purpose of sustainable land use planning in relation to soil pollution?
   (a) Protect arable land
   (b) Create green spaces
   (c) Prevent soil erosion
   (d) All of the above?
12. Which of the following is not an example of responsible agricultural practice to mitigate soil pollution?
   (a) Organic farming
   (b) Crop rotation
   (c) Excessive use of chemical fertilizers
   (d) Integrated pest management
13. Which soil remediation technique uses plants to extract and store contaminants?
   (a) Bioremediation
   (b) Phytoremediation
   (c) Electrokinetic remediation
   (d) Chemical remediation
14. What is a common source of chemical pollutants leading to soil erosion?
   (a) Rainwater
   (b) Organic farming
   (c) Soil erosion
   (d) Pesticide application
15. What is the purpose of soil monitoring programs?
   (a) Identify contamination sources
   (b) Assess soil health
   (c) Enforce regulations
   (d) All of the above
16. What would happen if the soil is sick, that is, polluted?
   (a) Food crops growing on it would hardly be affected.
   (b) Soil flora and fauna will not be affected.
   (c) Plants, animals, and human beings dependent on it would not stay healthy.
   (d) Its productivity will not be affected.
17. What is affected by soil pollution?
   (a) Carbon sequestration
   (b) Global climate
   (c) Terrestrial ecosystem functioning
   (d) All of the above
18. Which of the following statements about soil pollution is not correct?
   (a) Soil pollution is potentially linked with climate change.
   (b) Soil pollution leads to poor crop production but it does not affect soil structure and composition.
   (c) Green Revolution, or modern agriculture, is often criticized for exacerbating soil pollution due to indiscriminate agrochemical applications.
   (d) Soil is an ecosystem.
19. Which cropping system is not healthy for the soil?
   (a) Monocultures
   (b) Agro-biodiversity-based cultures
   (c) Agroforestry
   (d) Silviculture
20. Which of the following natural phenomena make a soil ecosystem healthy and productive?
   (a) Photosynthesis
   (b) Chemosynthesis
   (c) Nitrogen fixation
   (d) All of the above

Answers: 1-b, 2-d, 3-b, 4-a, 5-a, 6-d, 7-d, 8-a, 9-c, 10-c, 11-d, 12-c, 13-b, 14-d, 15-d, 16-c, 17-d, 18-b, 19-a, 20-d

18.6.2 Short-Answer Questions

1. Define soil pollution.
2. What are the main natural sources of soil pollution?
3. Name three anthropogenic activities that contribute to soil pollution.
4. Give some examples of a soil remediation technique.
5. How can proper waste management practices help prevent soil pollution?
6. Why are agrochemicals treated as soil pollutants?
7. Describe one negative impact of soil pollution on crop productivity.

8. How does soil pollution affect biodiversity?
9. How does soil pollution affect human health?
10. What are some key strategies for sustainable land use planning to combat soil pollution?

18.6.3 Long-Answer Questions

1. Discuss the different types of contaminants found in polluted soil and their sources, and explain the potential risks they pose to soil health, ecosystems, and human well-being.
2. Explore the various techniques and approaches used in soil remediation, such as bioremediation, phytoremediation, and electrokinetic remediation. Provide examples of real-world applications and discuss their effectiveness in restoring polluted soil.
3. In the context of agricultural practices, analyze the impact of excessive use of agrochemicals on soil pollution. Discuss the long-term consequences of chemical fertilizers, pesticides, and herbicides on soil fertility, biodiversity, and human health.
4. Examine the role of waste management systems in preventing soil pollution. Discuss the importance of recycling, composting, and proper disposal methods in reducing the accumulation of hazardous materials in landfills and the subsequent contamination of soil.
5. Write short notes on the following:
   (a) Types of soil pollutants
   (b) Soil monitoring
   (c) Sustainable land use planning as a soil pollution solution

References

Kuppusamy S, Kakarla D, Venkatswarlu K, Megharaj M, Yoon YE, Lee YB (2018) Veterinary antibiotics (VAs) contamination as a global agro-ecological issue: a critical view. Agric Ecosyst Environ 257:47–49. https://doi.org/10.1016/j.agee.2018.01.026

Negi S, Singh V, Melkania U, Rai JPN (2019) Estimation of heavy metal pollution index for groundwater around integrated industrial Estate of Tarai Region. Environ Ecol 37(IB):429–435

O'Neill J (2014) Antimicrobial resistance: tackling a crisis for the health and wealth of nations. Review on Antimicrobial Resistance, London

Rastogi A, Seth P, Singh V (2011) Pulp and paper industry and environmental disaster: death of a stream and heavy metal accumulation in wheat crop in Himalayan foothills. Lambert Academic Publishing, Saarbruecken. 138 pp

Rastogi A, Singh V, Arunachalam A (2019) Allelopathic effects of aqueous leaf extract of Jatropha curcas L. on food crops in the Himalayan foothills. J Emerg Technol Innov Res 6(6):866–875

Singh V (2020) Environmental plant physiology: botanical strategies for a climate smart planet. CRC Press (Taylor and Francis), London. 320 pp

Singh V (2024) Environmental disruptions: planet earth in the vicious cycle of pollution, global warming, and climate change. NIPA, New Delhi. 156 pp. isbn: 9788119002665

Noise Pollution

19

In the modern era, where cities pulsate with life and technology permeates every aspect of our existence, one often overlooked menace silently permeates our surroundings: noise pollution. The harmony of nature and the soothing melodies of life are drowned out by a cacophony of sound, impacting not only our environment but also our physical, mental, and social well-being. This chapter sheds light on the concept of noise pollution and its various dimensions, sources, effects, and potential solutions to mitigate its detrimental impacts.

19.1 Defining Noise Pollution

Noise pollution, or sound pollution, refers to the excessive and unwanted sounds that disrupt the natural balance of our acoustic environment. While sound is an integral part of our lives, noise becomes pollution when it exceeds safe levels and becomes intrusive, causing annoyance, discomfort, or even health issues.

19.2 Dimensions of Noise Pollution

Noise pollution manifests itself in various dimensions, encompassing different aspects of our lives. These dimensions include environmental, indoor, transport, and occupational noise pollution. Explained in Fig. 19.1, these noise pollution dimensions affect environment, humans, and all other species in various ways, in various areas, and under certain circumstances. Several researchers have significantly contributed to the study of noise pollution, with notable works including those by Bronzaft et al. (2017). Kang and Schulte-Fortkamp (2021), Cook (2022), Murphy et al. (2023), and Singh (2024), each shedding light on various aspects of this environmental and human concern.

19.3 Measurement and Levels of Noise Pollution

Noise pollution is typically measured using a device called a sound level meter (SLM). Sound level meters measure sound pressure levels in decibels (dB), which quantifies the intensity of sound. These devices capture and analyze sound waves to provide objective measurements of noise levels in the environment. With ambient sound intensity equal to the reference intensity, the noise level is considered equal to 0 dB.

The range of the noise level can extend from zero to 140 dB (Table 19.1). Sound up to 30 dB is categorized as faint, 40–50 dB moderate, 60–80 dB very loud, 90 dB extremely loud, and 120–140 dB as painful. The sound crossing 120 dB causes physical discomfort. Normal conversation sound ranges between 35 dB and

V. Singh, *Textbook of Environment and Ecology*, https://doi.org/10.1007/978-981-99-8846-4_19

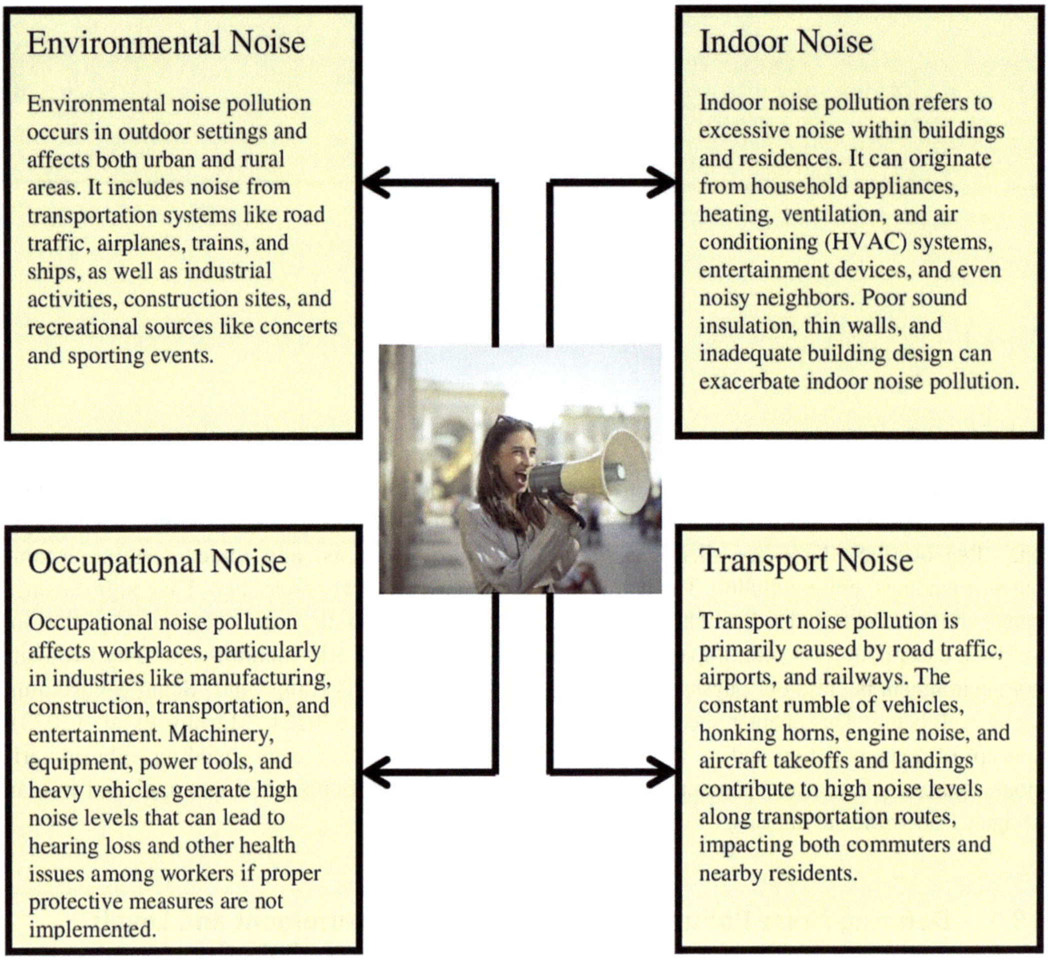

Fig. 19.1 Various dimensions of noise pollution (Source: Singh 2024)

Table 19.1 Sound intensity, characteristics, and the sources generating sound

Sound range (dB)	Sound characteristic	Sources
30	Faint	Whispering, quiet library
40–50	Moderate	Quiet room
60–80	Very loud	Vacuum cleaner, busy street
90	Extremely loud	Lawnmower, truck traffic
120–140	Painful	Jet plane take off, amplified rock music

60 dB. Prolonged exposure to 80 dB or higher may cause hearing impairment. Exposure to sound intensity of 120 dB for a long period would be painful. Constant exposure to an environment of noise pollution in some factories may make the workers lose their ability to hear soft sounds.

19.4 Effects of Noise Pollution

The effects of noise pollution extend beyond mere annoyance and disrupt our physical, psychological, and social well-being. Here are some notable impacts:

19.4.1 Physical Health Effects

Prolonged exposure to high levels of noise can lead to hearing loss, increased blood pressure, cardiovascular issues, sleep disturbances, and impaired cognitive function. Individuals living or working near noisy environments, such as airports or factories, are particularly vulnerable to these health risks. Prolonged exposure to excessive noise levels can have detrimental effects on our physical health:

(i) **Hearing loss:** One of the most apparent consequences of noise pollution is hearing loss. Continuous exposure to loud sounds can damage the delicate structures in our ears, leading to permanent hearing impairment or even complete deafness.

(ii) **Cardiovascular issues:** Studies have shown a clear link between chronic noise exposure and cardiovascular problems. The stress response triggered by loud noises increases blood pressure, heart rate, and the risk of heart diseases such as hypertension and coronary artery disease.

(iii) **Sleep disturbances:** Noise pollution disrupts our sleep patterns, leading to sleep deprivation and its associated health issues. Even intermittent noise can prevent us from reaching deep, restorative sleep, resulting in fatigue, irritability, and impaired cognitive function.

(iv) **Impaired cognitive function:** Noise pollution can have a significant impact on cognitive function and overall well-being. Here are several ways in which noise pollution can impair cognitive function:

- **Distraction and reduced concentration:** Excessive noise can be a major distraction, disrupting our ability to concentrate and focus on tasks. When exposed to constant or intermittent noise, such as traffic noise or construction sounds, it becomes challenging to maintain attention and engage in complex cognitive activities.
- **Increased stress levels:** Noise pollution triggers a stress response in the body, leading to the release of stress hormones like cortisol. Prolonged exposure to high levels of noise can result in chronic stress, which has been linked to impaired cognitive function. Stress can interfere with memory, attention, and decision-making abilities.
- **Sleep disruption:** Disrupted sleep especially during nighttime caused by noise pollution has a detrimental impact on cognitive abilities, including memory, learning, problem-solving, and creativity. It impairs the brain's ability to consolidate information, leading to decreased cognitive performance during wakefulness.
- **Interference with communication:** Decreased communication effectiveness due to noise pollution can impair cognitive functions related to language processing, comprehension, and social interaction.
- **Mental fatigue:** Constant exposure to noise can lead to mental fatigue, even if individuals are not consciously aware of it. The brain constantly processes and filters noise, which can drain cognitive resources and reduce mental energy. This can result in reduced cognitive performance and an increased risk of errors.
- **Impaired memory and learning:** Noise pollution can negatively affect memory and learning processes. Studies have shown that exposure to high levels of noise can impair both short-term and long-term memory. It can interfere with information encoding, retrieval, and consolidation, making it more difficult to remember and learn new information.

19.4.2 Psychological and Emotional Effects

Noise pollution has profound effects on our mental health. Constant exposure to loud noise can cause chronic stress, anxiety, irritability, and decreased concentration and productivity. It can also exacerbate existing mental health conditions, such as depression and mood disorders, and can even cause schizophrenia. Noise pollution disrupts our sense of calm and triggers a constant state of alertness, leading to elevated stress hormone levels. The chronic exposure to noise disrupts our emotional equilibrium and contributes to the deterioration of mental well-being.

19.4.3 Communication and Social Effects

Excessive noise hampers effective communication, leading to misunderstandings and frustration. It can disrupt social interactions, decrease quality of life, and contribute to social isolation. Additionally, it negatively impacts learning environments, hindering concentration and academic performance, especially in schools located in noisy areas.

19.4.4 Environmental Effects

Noise pollution disrupts the natural habitats of animals, interfering with their communication, reproduction, and overall behavior. It can lead to changes in migration patterns, alter predator-prey relationships, and disturb the balance of ecosystems.

19.5 Mitigating Noise Pollution

Reducing noise pollution requires a combination of individual and collective efforts, as well as technological advancements and urban planning considerations. Here are some potential solutions:

19.5.1 Regulatory Measures

Governments and local authorities play a vital role in establishing and enforcing noise regulations. These can include zoning laws to separate residential areas from industrial zones, imposing restrictions on noise levels emitted by vehicles and machinery, and implementing curfews for noisy activities.

19.5.2 Noise Barriers and Insulation

Constructing physical barriers, such as noise walls and acoustic fences, can help reduce the propagation of sound. Additionally, incorporating soundproofing materials in buildings can minimize indoor noise levels and provide a quieter living or working environment.

19.5.3 Improved Urban Planning

Urban planners can design cities with noise reduction in mind, considering factors such as green spaces, buffer zones, and the location of transportation routes. By carefully designing infrastructure and urban layouts, noise pollution can be mitigated.

19.5.4 Public Awareness and Education

Raising awareness about the harmful effects of noise pollution is crucial. Educational campaigns can promote responsible use of sound-emitting devices, encourage the use of quieter alternatives, and foster a culture of respecting silence and tranquility.

19.5.5 Technological Innovations

Advancements in technology offer promising solutions to combat noise pollution. For instance, the development of quieter engines for vehicles, noise-cancelling technologies, and improved sound insulation materials can significantly reduce noise emissions.

19.5.6 Sound Level Guidelines

When it comes to safe limits of noise pollution, different countries and regions may have specific guidelines and regulations in place. The World Health Organization (WHO) provides some general recommendations for community noise levels (Table 19.2).

Typical noise pollution levels for residential, commercial, industrial, and hospital areas during daytime and nighttime are approximate and can vary depending on local regulations and specific circumstances.

Table 19.2 Acceptable ambient noise levels by zone

Area	Daytime[a]	Nighttime[a]
Residential	55 dB or lower	45 dB or lower
Commercial	65 dB or lower	55 dB or lower
Industrial	75 dB or lower	70 dB or lower
Silent	50 dB or lower	40 dB or lower
Hospital	45 dB or lower	35 dB or lower

[a]When it comes to noise pollution, the specific daytime and nighttime hours may fluctuate in accordance with local regulations and guidelines

It is important to note that these are general guidelines, and specific regulations can vary depending on the location and the type of area (e.g., residential, commercial, industrial). Additionally, exposure to excessive noise levels for prolonged periods can have negative health effects, even if they are below the defined limits.

Regulatory authorities in different countries often establish their own noise limits and guidelines based on factors such as land use, time of day, and the type of area. It is advisable to consult local regulations or authorities for precise information regarding safe limits and applicable standards in a particular area.

19.6 Summary

The chapter on noise pollution highlights the detrimental impact of excessive and unwanted sound, the noise pollution, on environment and human well-being. It explores the different dimensions of noise pollution, especially environmental, indoor, occupational, and transport noise pollution. The measurement and levels of noise pollution are discussed. The effects of noise pollution on physical health, such as hearing loss and cardiovascular issues, as well as cognitive function, sleep disturbances, and mental well-being are examined. Communication and social effects, as well as environmental impacts, are also addressed. Potential solutions to mitigate noise pollution, including regulatory measures, noise barriers and insulation, improved urban planning, public awareness and education, and technological innovations. By prioritizing efforts to reduce noise pollution, a quieter and more harmonious future can be achieved, benefiting our overall well-being.

19.7 Exercises

19.7.1 Multiple-Choice Questions

1. What is noise pollution?
   (a) Excessive use of sound for communication
   (b) Unwanted and disruptive sounds
   (c) The absence of sound in an environment
   (d) Pleasant and soothing melodies
2. Which of the following is an example of environmental noise pollution?
   (a) Loud music at a concert
   (b) Noisy neighbors in an apartment building
   (c) Heavy machinery in a factory
   (d) Birds chirping in a park
3. What device is typically used to measure noise pollution?
   (a) Sound level meter
   (b) Noise-canceling headphones
   (c) Decibel analyzer
   (d) Acoustic amplifier
4. Which of the following is a potential health effect of prolonged exposure to noise pollution?
   (a) Improved cognitive function
   (b) Reduced blood pressure
   (c) Hearing loss
   (d) Restful sleep
5. Which organization provides general recommendations for community noise levels?
   (a) World Health Organization (WHO)
   (b) Environmental Protection Agency (EPA)
   (c) International Organization for Standardization (ISO)
   (d) United Nations (UN)
6. What is the recommended noise level for residential areas during the daytime?
   (a) 45 dB or lower
   (b) 55 dB or lower
   (c) 65 dB or lower
   (d) 75 dB or lower
7. Which aspect of cognitive function can be impaired by noise pollution?
   (a) Memory and learning
   (b) Attention and concentration

(c) Problem-solving and creativity
(d) All of the above

8. What psychological effect can noise pollution have?
   (a) Increased productivity
   (b) Decreased concentration
   (c) Improved communication
   (d) Enhanced relaxation
9. How can noise pollution disrupt social interactions?
   (a) By improving communication effectiveness
   (b) By fostering a sense of calm
   (c) By causing misunderstandings and frustration
   (d) By increasing social engagement
10. How can noise pollution impact animal habitats?
    (a) By disrupting their communication and behavior
    (b) By promoting their communication
    (c) By enhancing migration patterns
    (d) By improving predator-prey relationships
11. What are some potential solutions to mitigate noise pollution?
    (a) Regulatory measures and noise barriers
    (b) Increased use of loudspeakers
    (c) Expansion of industrial zones
    (d) Promotion of noisy activities
12. What role do governments and local authorities play in reducing noise pollution?
    (a) They encourage excessive noise levels.
    (b) They enforce noise regulation.
    (c) They promote noisy activities.
    (d) They ignore the issue of noise pollution.
13. What can be done to reduce indoor noise pollution?
    (a) Incorporating soundproofing materials
    (b) Increasing the use of loud appliances
    (c) Encouraging noisy neighbors
    (d) Removing walls and partitions
14. How can urban planning contribute to noise reduction?
    (a) By creating more transportation routes
    (b) By avoiding green spaces
    (c) By designing cities with noise reduction in mind
    (d) By increasing industrial activities in residential areas
15. What is an effective way to raise awareness about noise pollution?
    (a) Educational campaigns
    (b) Promoting the use of loudspeakers
    (c) Encouraging noisy activities
    (d) Ignoring the issue
16. Which of the following is an example of a technological innovation to combat noise pollution?
    (a) Quieter engines for vehicles
    (b) Louder construction equipment
    (c) Noisy entertainment devices
    (d) Inefficient sound insulation materials
17. What is the overall impact of noise pollution on well-being?
    (a) Positive and uplifting
    (b) Neutral and indifferent
    (c) Negative and detrimental
    (d) Irrelevant and inconsequential
18. What can noise pollution cause in terms of physical health?
    (a) Improved sleep quality
    (b) Hearing loss
    (c) Decreased blood pressure
    (d) Increased cognitive function
19. How can noise pollution affect cognitive function?
    (a) By improving memory and learning
    (b) By enhancing concentration and focus
    (c) By promoting creativity and problem-solving
    (d) By impairing attention and memory
20. What is the ultimate goal of reducing noise pollution?
    (a) Creating a chaotic and disruptive environment
    (b) Restoring harmony and tranquility
    (c) Promoting excessive sound levels
    (d) Ignoring the negative impacts of noise

Answers: 1-b, 2-c, 3-a, 4-c, 5-a, 6-b, 7-d, 8-b, 9-c, 10-a, 11-a, 12-b, 13-a, 14-c, 15-a, 16-a, 17-c, 18-b, 19-d, 20-b

19.7.2 Short-Answer Questions

1. What is the definition of noise pollution?
2. Name four dimensions of noise pollution.
3. How is noise pollution typically measured?
4. What are some potential health effects of prolonged exposure to noise pollution?
5. What organization provides general recommendations for community noise levels?
6. What are the recommended noise levels for residential areas during daytime and nighttime?
7. How can noise pollution impact cognitive function?
8. What are some psychological and emotional effects of noise pollution?
9. How can noise pollution disrupt social interactions?
10. How does noise pollution affect animal habitats?

19.7.3 Long-Answer Questions

1. Explain the potential health risks associated with prolonged exposure to high levels of noise pollution. How does noise pollution contribute to hearing loss, cardiovascular issues, sleep disturbances, and impaired cognitive function? Provide supporting evidence and discuss preventive measures.
2. Noise pollution can have significant effects on mental health and well-being. Discuss the psychological and emotional impacts of noise pollution, including chronic stress, anxiety, decreased concentration, and productivity. How does noise pollution exacerbate existing mental health conditions and what strategies can be employed to mitigate these effects?
3. Noise pollution not only affects individuals' physical and mental health but also disrupts communication and social interactions. Elaborate on how excessive noise hampers effective communication, contributes to social isolation, and affects learning environments, particularly in schools located in noisy areas. Discuss the implications and potential solutions for these social effects of noise pollution.
4. Noise pollution has adverse effects on both human and animal habitats. Discuss how noise pollution disrupts the natural behavior and communication of animals, leading to changes in migration patterns, altered predator-prey relationships, and disturbance of ecosystems. How can we minimize the environmental impacts of noise pollution on wildlife and restore ecological balance?
5. Write short notes on the following:
   (a) Dimensions of noise pollution
   (b) Levels of noise pollution
   (c) Impaired cognitive function

References

Bronzaft AL, McAndrew FT, Sykes DM (2017) Environmental noise pollution: noise mapping, public health, and policy. In: Springer handbook of noise and vibration control. Springer, pp 685–715

Cook RF (2022) Noise pollution: a modern plague. Putnam's Sons, G.P

Kang J, Schulte-Fortkamp B (2021) Urban sound environment. CRC Press, Boca Raton

Murphy E, Kang J, Hansen CH (eds) (2023) Environmental noise pollution: noise mapping, measurement, and management. Taylor & Francis, New York

Singh, Vir. 2024. Environmental disruptions: planet earth in the vicious cycle of pollution, global warming, and climate change. New Delhi: NIPA. 156 pp

Global Warming and Climate Change

20

Global warming and climate change have emerged as pressing environmental concerns with far-reaching implications for the planet and its inhabitants. The increasing concentration of greenhouse gases in the atmosphere, primarily as a result of human activities, has led to a rise in global temperatures, altering weather patterns and affecting ecosystems worldwide. This chapter delves into the causes and consequences of global warming and climate change.

20.1 Understanding Global Warming

Global warming refers to the long-term increase in Earth's average surface temperature. The primary driver of this phenomenon is the greenhouse effect, whereby certain gases in the atmosphere trap heat from the sun, preventing it from escaping back into space. The main greenhouse gases responsible for global warming are carbon dioxide (CO_2), methane (CH_4), nitrous oxide (N_2O), and fluorinated gases. Human activities, such as burning fossil fuels, deforestation, and industrial processes, have significantly increased the levels of these gases in the atmosphere, amplifying the greenhouse effect.

20.1.1 Impacts of Climate Change

Climate change manifests through various environmental and socioeconomic impacts. Rising temperatures have led to the melting of glaciers and polar ice caps, contributing to sea-level rise. This, in turn, poses a threat to coastal communities, increases the frequency and intensity of coastal flooding, and exacerbates erosion. Changes in precipitation patterns result in more frequent and severe droughts, floods, and storms, impacting agriculture, water resources, and infrastructure. Additionally, climate change disrupts ecosystems, leading to shifts in species' geographical ranges, loss of biodiversity, and increased vulnerability to invasive species.

20.1.2 Feedback Mechanisms and Tipping Points

Global warming can trigger feedback mechanisms and tipping points that exacerbate the impacts of climate change. For instance, as temperatures rise, thawing permafrost releases large amounts of methane, a potent greenhouse gas, further enhancing global warming. The loss of reflective surfaces, such as ice and snow, reduces

V. Singh, *Textbook of Environment and Ecology*, https://doi.org/10.1007/978-981-99-8846-4_20

the Earth's albedo, causing more sunlight to be absorbed and accelerating temperature increases. These positive feedback loops can amplify the warming trend and create a cascade of effects with potentially catastrophic consequences.

20.2 Greenhouse Gases

Greenhouse gases (GHGs) play a crucial role in regulating Earth's surface temperature to maintain a suitable range for the well-being and functioning of living organisms. The accumulation of GHGs in the atmosphere beyond a certain threshold leads to global warming. The composition of these gases in the atmosphere directly influences the average global temperature. GHGs possess the ability to absorb and emit radiation in the thermal infrared range.

The primary GHGs in the globe's atmosphere are carbon dioxide (CO_2), methane (CH_4), nitrous oxide (N_2O), ozone (O_3), and water vapor (H_2O). Chlorofluorocarbons (CFCs) and hydrofluorocarbons (HFCs) are categorized as secondary GHGs.

The non-GHGs, namely, nitrogen (N_2), oxygen (O_2), and argon (Ar), constitute most of the atmosphere, accounting for 78%, 21%, and 0.9%, respectively, making up approximately 99.9% of the atmospheric composition. GHGs account for less than 0.1% of the atmosphere, with the following abundance order: $H_2O > CO_2 > CH_4 > N_2O >$ CFCs > HFCs.

Other GHGs like ozone (O_3), carbon monoxide (CO), and nitrogen oxides (NO_x) are considered short-lived and are typically not included among the main GHGs. Aerosols, such as black carbon and mineral dust, which exhibit variability in different locations and times, are also not commonly considered as GHGs.

While water vapor has the potential to contribute to the direct greenhouse effect by 36–72%, it is usually not a major concern when discussing the impact of other GHGs. This is due to the fluctuating concentration of atmospheric water vapor (10–50,000 ppm) across regions, within the same region, and varying with seasons, daily fluctuations, and even within the same day. Moreover, the water cycle is vital for the ecological integrity of the biosphere, and increased water vapor concentration in the atmosphere generally has positive thermodynamic implications. Table 20.1 presents the sources of GHG emissions and concentrations of the main GHGs.

Apart from the major GHGs mentioned earlier, the IPCC list (IPCC 2001a, b) includes other gases such as perfluorocarbons, sulfur hexafluoride (SF_6), tetrafluoromethane (CF_4), hexafluoroethane (C_2F_6), and nitrogen trifluoride (NF_3). However, due to their extremely low proportions in the atmosphere, they are often excluded from the main list of greenhouse gases.

Within the atmosphere, 99.9% is composed of non-greenhouse gases. The dominant gases, nitrogen (N_2) and oxygen (O_2), consist of two similar atoms in their molecules, resulting in no net charge distribution during their vibration. Argon, being a monoatomic gas, does not possess vibrational modes and is also unaffected by infrared radiation. Some other gases with two atoms of different elements, such as carbon monoxide (CO) and hydrogen chloride (HCl), do absorb infrared radiation. However, these gases have short lifespans due to their reactivity or solubility and therefore make minimal contributions to the greenhouse effect.

20.3 Greenhouse Effect

A significant portion of solar radiation entering Earth's gaseous envelope reaches the planet's surface. Some of this radiation is absorbed by the Earth's surface, while the remaining energy is radiated back into space. Shortwave radiation reaches the surface, whereas longwave radiation (infrared) is emitted back. During this process, greenhouse gases absorb a portion of the returning shortwave radiation, contributing to the warming of Earth's atmosphere. The atmosphere then radiates some of the absorbed energy back to the surface, creating a downward flux known as the greenhouse flux, which helps maintain the warmth of the Earth. This phenomenon is known as the greenhouse effect, and Table 20.1 describes the gases with atoms of different elements that have a significant impact on global warming (see also Fig. 20.1).

Table 20.1 The main greenhouse gases, their origins, and properties

Greenhouse gas	Source of emission	Concentration in 1750	Concentration in 2011*/2019**	% increase since 1750	Atmospheric life time (years)
Carbon dioxide (CO_2)	• Chiefly produced by the burning of fossil fuels • Released by plants and animals during respiration	280 ppm	415 ppm**	48	5–200
Methane (CH_4)	• Caused by anaerobic decomposition in the presence of methanogens, a group of bacteria • Released from garbage dumps, swamps, flooded rice fields, and enteric fermentation	700 ppb	1801 ppb*	157	12
Nitrous oxide (N_2O)	• Industry, biomass burning, and agriculture are the main sources • Breakdown of nitrogenous fertilizers in the soil, burning of N-rich fuels, livestock wastes, and nylon production	270 ppb	324 ppb*	20	114
Chlorofluorocarbons (CFC-11) + Hydrofluorocarbons (HCFC-23)	• Synthesized gaseous compounds of carbon and halogens • Leaking of air conditioners and refrigeration units, propellants in aerosol spray cans, production of plastic foams, and evaporation of industrial solvents • Extensively utilized as refrigerants, aerosol propellants, fire extinguishers, and insulators	0	262 ppt*	262	45–260

Source: Singh (2024)
ppm parts per million (10^6) = μmol/mol, *ppb* parts per billion (10^9) = nmol/mol, *ppt* parts per trillion (10^{12}) = pmol/mol

The greenhouse effect, facilitated by GHGs, is crucial for sustaining life in the biosphere. The average global temperature of 15 °C is a result of this effect. Without GHGs in the atmosphere, the planet's average temperature would be −20 °C, making life as we know it nearly impossible.

As the greenhouse effect and the warming of the Earth's atmosphere and surface depend on the presence of GHGs, their concentration relative to non-GHGs determines the extent of global warming, which is commonly referred to as the greenhouse effect.

20.4 Causes of Global Warming

Since the onset of the Industrial Age in 1750, human activities have significantly increased the presence of GHGs, particularly carbon dioxide, in the atmosphere. These activities

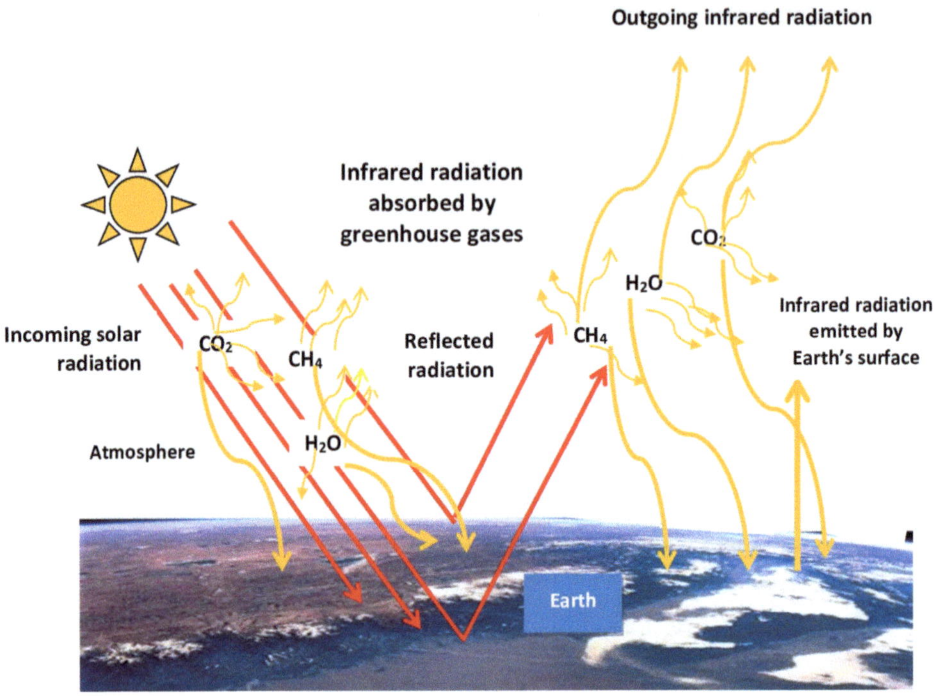

Fig. 20.1 Illustration of global warming caused by greenhouse gases. (Source: Singh 2024)

include the extensive burning of fossil fuels, industrial processes, and land use changes such as deforestation, agriculture, urbanization, and soil degradation. Consequently, from 1750 to 2019, there has been a 48% increase in CO_2 levels. Methane (CH_4), nitrous oxide (N_2O), and chlorofluorocarbons (CFCs) have experienced increases of 157%, 20%, and 262%, respectively, during the same period (Table 20.1).

To assess the severity of their effects, different GHGs are standardized to CO_2 equivalents based on their global warming potential (GWP) over a specific period, with a CO_2 value of 1 unit. Table 20.2 presents the 100-year GWP for various GHGs as reported by the IPCC.

Figure 20.2 illustrates the sector-wise distribution of GHG emissions in percentage. Figure 20.3 illustrates emissions from electricity and heat production emanating from sectors of final energy use, indicating indirect emissions. Taking these indirect emissions into account, carbon emissions from industry and buildings increase to 32% and 18.4%, respectively. The GWP values of different GHGs vary, resulting in different impacts on global warming. For example, N_2O has a much higher GWP value than CO_2, but its overall impact on the atmosphere is comparatively less due to its lower

Table 20.2 Global warming potential of some GHGs over a period of 100 years

GHG	Formula	100-year GWP
Carbon dioxide	CO_2	1
Methane	CH_4	28
Fossil methane[a]	CH_4	30
Nitrous oxide	N_2O	265
CFC-11	CCl_3F	4660
CFC-12	CCl_2F_2	10,200
CFC-13	$CClF_3$	13,900
HCFC-21	$CHCl_2F$	148
HCFC-22	$CHClF_2$	1760
HFC-23	CHF_3	12,400
PFC-14	CF_4	7390
PFC-116	C_2F_6	12,200
PFC-218	C_3F_8	8830
Nitrogen trifluoride	NF_3	16,100
Sulfur hexafluoride	SF_6	23,500

Source: Compiled from the Fifth Assessment Report of IPCC (Myhre et al. 2013)

[a]Metric values of methane of fossil origin include the oxidation to CO_2

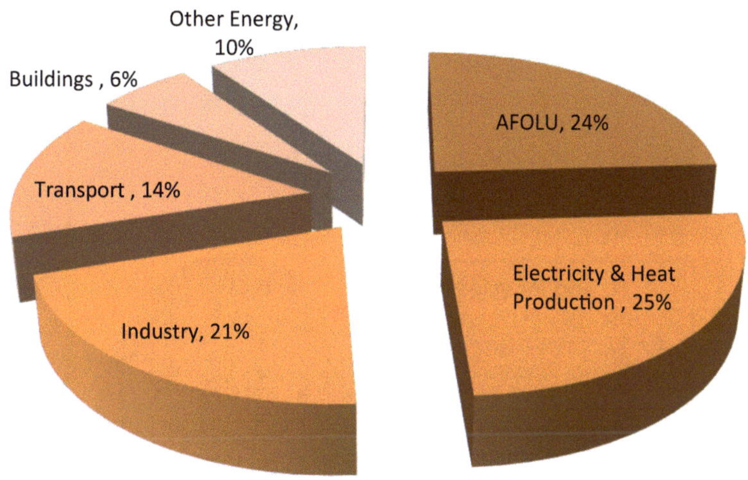

Fig. 20.2 Anthropogenic GHG emissions by economic sectors. (*AFOLU* Agriculture, forestry and other land use)

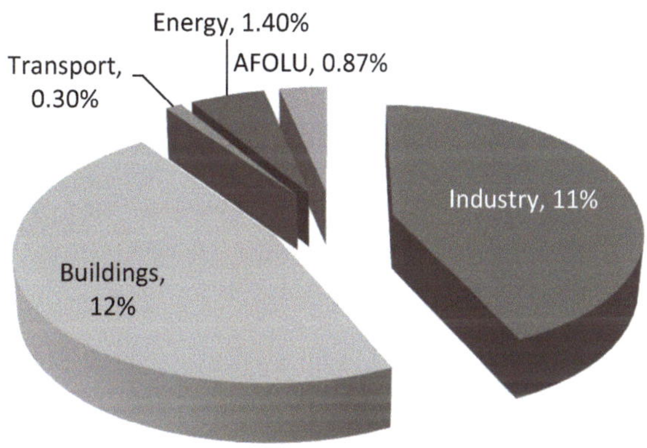

Fig. 20.3 Electricity and heat production emissions due to sectors of final energy use

atmospheric concentration. The impact of each GHG is expressed in terms of the equivalent amount of CO_2 that would have the same warming effect.

CO_2 remains the major anthropogenic GHG, accounting for 76% of emissions, followed by methane (16%), nitrous oxide (6%), and fluorinated gases (2%) (IPCC 2014).

Furthermore, various other dimensions of environmental disruptions exacerbate the enhanced greenhouse effect and worsen the state of our environment. These dimensions include land use change, biodiversity erosion, nitrogen and phosphorus cycles, global freshwater use, ocean acidification, chemical changes, atmospheric aerosol loading, and stratospheric ozone layer depletion. Interrelated and triggering effects of these dimensions in terrestrial systems significantly impact global warming and climate change (Fig. 20.4). These changes result from human interventions, and each change directly or indirectly contributes to a significant decline in ecosystems' photosynthetic capacities.

20.5 Global Warming and Climate Change

Global warming and climate change are distinct phenomena but closely interconnected and often used interchangeably. While global warming refers to the increasing global temperatures, climate change encompasses broader and more complex types of changes, including shifts in temperature patterns, precipitation, humidity, wind patterns, weather events, and seasonal variations over a prolonged period. Climate changes have occurred throughout geological epochs spanning millions of years. However, the changes witnessed today, driven by human activities, are unprecedented.

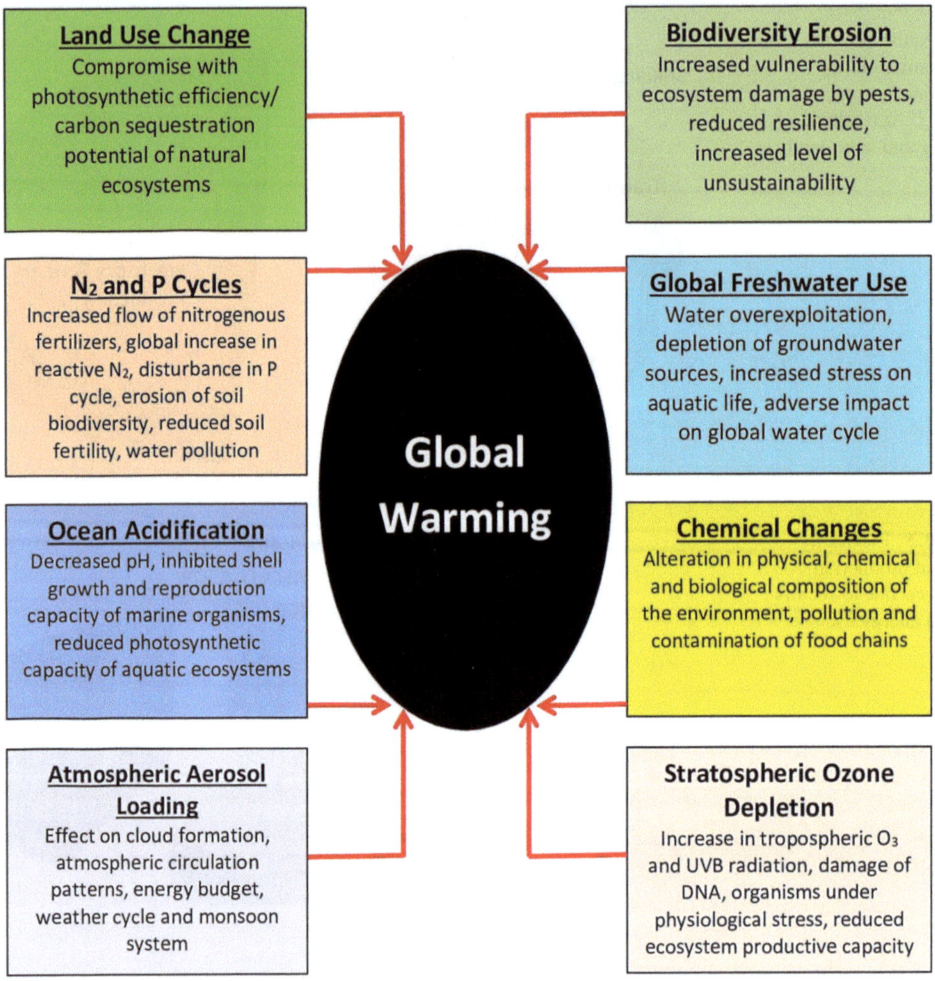

Fig. 20.4 Global warming-enhancing factors (Source: Singh 2024)

We are currently living in an era of climate change, which is primarily driven by global warming. Global warming is not a static process but induces thermodynamic changes affecting all aspects of the biosphere. The changes in the composition and chemistry of the Earth's atmosphere, caused predominantly by human activities, have far-reaching consequences. The average climate of the planet consists of diverse regional climates and microclimates. Organisms across all five kingdoms of life – Monera, Protista, Fungi, Plantae, and Animalia – thrive in different climate types. The variety of climates is crucial for maintaining biodiversity in terms of ecosystems, species, and genotypes.

The anthropogenic factors discussed thus far, although significant, are not exclusive contributors to climate change. Climate change results from the interactive influences of multiple factors. The amount of light received on the Earth's surface is a key determinant of climate, and various factors influencing or influenced by light shape the Earth's climate system. Figure 20.5 illustrates ten factors, both cosmic and terrestrial, that exert substantial influence on the planet's climate order.

20.6 Impacts of Climate Change

The impact of climate change is being felt and will continue to affect the entire world. However, the extent of its effects may vary across regions and areas within the same region. Climate change

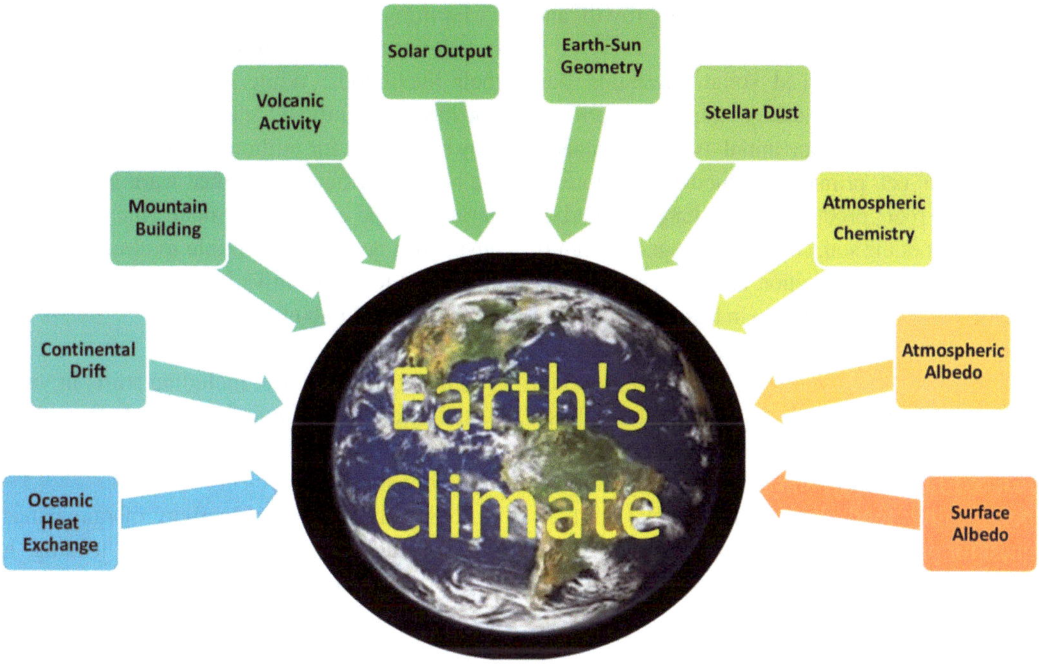

Fig. 20.5 The ten elements, both terrestrial and cosmic, affecting the Earth's climate system

has far-reaching consequences for the physical environment, living organisms, various species, ecosystems (both natural and human-made), and human societies. These impacts are already visible, with extreme weather events, shrinking Arctic Sea ice, glacier retreat, and rising sea levels being the most notable ones. Climate change is leaving its mark on all ecosystems worldwide, and socioeconomic systems and human life in general are increasingly vulnerable to its processes. Let us explore the different areas experiencing the impacts of climate change in more detail.

20.6.1 Impacts on Agriculture

Agriculture, which includes crop cultivation, forestry, and livestock production, is one of the major socioeconomic systems that are vulnerable to rising atmospheric CO_2 levels and changing climate patterns.

Climate change is primarily driven by the increasing concentrations of CO_2 in the atmosphere, and CO_2 serves as a source of nourishment for plants. Higher levels of CO_2 result in increased rates of photosynthesis in green plants, known as the CO_2 fertilization effect. This effect leads to enhanced biomass production, improved water use efficiency, increased tolerance to low light conditions, and higher optimum temperature for photosynthesis.

The impact of the CO_2 fertilization effect on plant growth depends on factors such as the availability of adequate water, minerals, temperature, and plant varieties. When all these factors are present, the influence of increased CO_2 levels is more significant on C_3 plants like wheat, rice, barley, cotton, potato, and pigeon pea. Since most of the human food is derived from C_3 plants, it might seem that higher atmospheric CO_2 levels would greatly benefit agriculture. However, the relationship between CO_2 and crop production is not linear. The negative environmental effects of increased CO_2 concentrations can impede the CO_2 fertilization effect. Concentrations of CO_2 above 400 ppm can cause stomatal closure, leading to reduced transpiration rates and hindered gas exchange between the atmosphere and plant leaves. This environmental state, coupled with elevated temperatures due to global warming, can lower photosynthetic rates and negatively impact plant growth and metabolic activities.

Research has revealed some interesting outcomes of CO_2-induced changes in plant growth patterns. These include an initial but unsustainable increase in net primary productivity of forests, increased root production resulting in carbon enrichment of the soil, changes in plant community structure, minimal alteration in the microbial community, no significant change in leaf area index, and a sharp decline in nitrogen levels in forest soils.

While an initial increase in CO_2 levels can benefit plants by boosting carbohydrate synthesis and biomass production, it can also lead to reduced rates of protein synthesis, resulting in poor-quality food. Insects and pests, in their quest for protein, may extract more biomass from plants. This emerging scenario could lead to a dramatic increase in crop damage caused by insect infestations.

Long-term increases in CO_2 levels pose several risks to overall plant performance (Singh 2024):

- Erratic weather patterns with excessive rainfall within a short period, followed by devastating floods and recurring droughts, would significantly harm agriculture.
- Increased evaporation rates due to higher temperatures would subject plants to water stress, affecting their physiology and productivity.
- A reduced number of snowfall days would negatively impact horticultural crops in temperate regions and high mountainous regions in the tropics where snowfall occurs.
- Decomposition rates of organic matter and fertilizers in the soil would increase.
- Soil health deterioration would hinder plant growth and agricultural yields.
- Insect and pest populations would proliferate, leading to detrimental effects on crop production.
- Non-native plants, better adapted to the changing environmental conditions, would invade and displace native plants and food crops.
- The increased frequency of cyclones, hurricanes, and extreme weather events would further exacerbate challenges in agriculture.

Farming communities worldwide, particularly in less developed countries, rely on livestock for their livelihoods. While domestic animals may not have a significant global economic impact, they are crucial for millions of farmers. Livestock are both contributors to global warming and vulnerable to the effects of climate change. Animal husbandry is more susceptible to climate change compared to crop cultivation in various ways, as indicated by Singh et al. (2017):

- Changes in land use, including shifts in vegetation composition and animal diets, and the inability of small-scale farmers to cope with fodder shortages
- Reduced primary productivity of rangelands, grasslands, forages, and crops, leading to shortages and reduced nutritive value of fodder
- Alterations in plant species composition, such as an increased proportion or density of invasive species in forests, rangelands, and pastures with no fodder value
- Decline in the quality of plant material due to increased lignification caused by higher temperatures, resulting in reduced digestibility of fodder
- Scarcity of drinking water, which adversely affects animal health and production potential
- Increased incidence of vector-borne diseases like malaria and tick-borne diseases
- Higher chances of helminth infections
- Increased outbreaks of diseases
- Rise in heat-related mortality and morbidity among livestock

20.6.2 Climate Change and Human Diseases

Climate change is intensifying existing human diseases and paving the way for the emergence of new, unprecedented diseases. As global temperatures rise, mosquitoes, for instance, are migrating to new habitats where they were not previously found. This shift in mosquito habitats will bring diseases like malaria, encephalitis, meningitis, dengue fever, chikungunya, and West Nile virus to new areas. The extent of the expansion of mosquito habitats depends on the degree of climate change. Historically, malaria was not a threat to people in temperate and highland regions of tropical countries. However, as the globe warms, malaria will become a menace in these previously unaffected areas. Similarly, other mosquito-borne diseases will also spread to new regions.

Many common infectious diseases are highly sensitive to climate variations, including those transmitted by insects (vector-borne communicable diseases), such as malaria, dengue, hantavirus, and cholera, as well as water- and foodborne diseases like giardiasis and cholera. Rising temperatures and increased flooding can lead to severe outbreaks of these diseases. Vector-borne and communicable diseases will expand into new areas as climate change continues to worsen.

The dynamics of human infectious diseases, including those transmitted by vectors, water, food, and air, are influenced by the complex relationship between humans, pathogens, and the transmission environment. Various factors related to climate change, such as temperature, humidity, precipitation, wind, and pollutants, create a more favorable environment for disease transmission. Once diseases affect individuals, they can spread throughout society. The only way for society to address these diseases is through adaptation.

20.6.3 Long-Term Implications of Climate Change

The long-term implications of ongoing climate change are extremely serious. The reduced capacity of terrestrial and aquatic ecosystems to absorb CO_2 will lead to a continuous accumulation of CO_2 in the atmosphere. Stabilizing the climate system requires significant reductions in CO_2 emissions. However, by the time fossil fuel-based energy resources are fully replaced by clean energy sources in alignment with global climate agreements, which seems unlikely in the near future, it may be too late to prevent the global temperature from surpassing the target of 1.5–2 degrees Celsius by 2100. As a result, the symptoms of rising temperatures and intensified global warming effects will become even more evident and severe. Negative consequences of climate change on natural ecosystems, agriculture, forestry, livestock, and all other socioeconomic sectors will be formidable.

Some notable indicators of climate change on a global scale include the significant reduction of snow and ice cover in polar and nonpolar regions, as well as the increased thawing of permafrost. Heat waves, heavy summer rains, prolonged droughts in arid areas, and reduced water flow in rivers and streams during drought periods will become increasingly unbearable. Excessive snow melt in glaciers and polar regions will contribute to rising sea levels, leading to the inundation of coastal areas, including major cities, by seawater. The displacement of populations to distant regions will become inevitable, causing further distress in various parts of the world.

The incidence of devastating wildfires will escalate, putting remaining natural ecosystems under severe stress and dramatically reducing primary productivity. Agricultural productivity will decline, necessitating changes in cropping patterns. While parts of Europe, Siberia, and the Western regions may witness more land available for agriculture, the long-term advantages will be limited.

Climate change will have a harsh impact on the entire globe, but certain regions will experience more drastic consequences. Earth's poles, the Arctic region, the Himalayas, and the Alps will be particularly affected by significant changes in global climate patterns.

The Antarctic, Arctic, and Greenland ice sheets hold significant potential to mitigate global warming by exerting a cooling effect on the planet. However, these cryosphere environments are highly vulnerable to climate change. The West Antarctic ice sheet (WAIS) may collapse within a millennium, posing one of the most severe consequences of climate change. According to a study by Winkelmann et al. (2015), the Arctic ice sheet could completely melt over the next millennium, contributing to a sea level rise of 58 m, with 30 m realized within the first 1000 years. The Greenland ice sheet, containing a vast amount of freshwater in ice form, has the potential to raise sea levels by 7 m. Greenland's ice melt may be complete in over 1000 years.

Another likely consequence of global warming is ocean anoxia, where the capacity of water bodies to dissolve oxygen declines due to warming. This will have extremely deleterious effects on aquatic life.

Significant reserves of methane clathrate or methane hydrate, which contain large amounts of methane within their crystals, are present in ocean floor sediments and permafrost regions. The release of greenhouse gases (GHGs) from methane clathrate due to persistent global warming is a cause for concern. Archer and Buffett (2005) estimate that upon releasing approximately 2000 Gt of anthropogenic carbon into the atmosphere, 2000–4000 Gt of carbon could be released from methane clathrate reserves in the oceans and permafrost regions.

In Latin America, tropical forests will gradually give way to savannahs in eastern Amazonia. North America will experience reduced snow cover in western mountain ranges. Europe will face increased risks of inland floods and coastal floods, as well as rising sea levels and more frequent storms, leading to devastating effects. In Asia, the Himalayas, which hold the largest mass of freshwater in the form of snow and ice outside of the polar regions, will no longer receive significant snowfall in lower areas, while the higher Himalayas will see a thinning snowpack. This will result in a drastic reduction in freshwater availability in Central and Southeast Asia.

Over time, the long-term consequences of climate change will accelerate the ongoing Sixth Mass Extinction.

20.7 Summary

The book chapter explores the topic of global warming and climate change, discussing their causes, impacts, and long-term implications. It begins by explaining the greenhouse effect and how human activities have increased the concentration of greenhouse gases in the atmosphere, leading to global warming. The chapter then delves into the various impacts of climate change, including rising temperatures, sea-level rise, changes in precipitation patterns, and disruptions to ecosystems. It also highlights feedback mechanisms and tipping points that can amplify the effects of global warming.

The chapter provides an overview of the different greenhouse gases and their role in regulating Earth's temperature. It emphasizes that while water vapor has the potential to contribute to the greenhouse effect, it is not a major concern compared to other greenhouse gases due to its fluctuating concentration. The chapter also discusses the sources and concentrations of the main greenhouse gases in the atmosphere.

Furthermore, the chapter examines the causes of global warming, primarily human activities such as burning fossil fuels and deforestation. It explains how these activities have increased the levels of greenhouse gases in the atmosphere over time.

The impacts of climate change are explored in detail, focusing on agriculture, livestock production, and human diseases. The chapter discusses how changing climate patterns and increased CO_2 levels can have both positive and negative effects on plant growth and crop production. It also highlights the vulnerability of animal husbandry to climate change and the spread of vector-borne diseases due to shifting habitats.

Lastly, the chapter addresses the long-term implications of climate change, including the continued accumulation of CO_2 in the atmosphere, reduction of snow and ice cover, rising sea levels, increased frequency of extreme weather events, and the potential collapse of ice sheets. It discusses regional impacts in different parts of the world, emphasizing the vulnerability of polar regions, the Himalayas, and coastal areas.

Overall, the chapter provides a comprehensive overview of global warming and climate change, emphasizing the urgent need to address these issues to mitigate their far-reaching consequences on the environment, ecosystems, and human societies.

20.8 Exercises

20.8.1 Multiple-Choice Questions

1. What is the primary driver of global warming?
   (a) Deforestation
   (b) Greenhouse effect
   (c) Industrial processes
   (d) Melting of polar ice caps
2. Which of the following greenhouse gases has the highest abundance in the atmosphere?
   (a) Methane (CH_4)
   (b) Carbon dioxide (CO_2)
   (c) Nitrous oxide (N_2O)
   (d) Water vapor (H_2O)
3. What is the main impact of climate change on coastal communities?
   (a) Increased rainfall
   (b) Sea-level rise
   (c) Reduced temperatures
   (d) Soil degradation
4. What are feedback mechanisms in relation to global warming?
   (a) Processes that amplify the impacts of climate change
   (b) Mechanisms that reduce the concentration of greenhouse gases
   (c) Processes that stabilize global temperatures
   (d) Mechanisms that counteract the greenhouse effect
5. Which of the following is not considered a greenhouse gas?
   (a) Nitrogen (N_2)
   (b) Carbon dioxide (CO_2)
   (c) Methane (CH_4)
   (d) Ozone (O_3)
6. What is the primary contributor to increased levels of greenhouse gases in the atmosphere since the Industrial Age?
   (a) Burning fossil fuels
   (b) Deforestation
   (c) Agricultural activities
   (d) Volcanic eruptions
7. How do rising temperatures impact agriculture?
   (a) Increase in crop yields
   (b) Reduced water use efficiency
   (c) Decreased rates of photosynthesis
   (d) Enhanced tolerance to low light conditions
8. Which factor contributes to the spread of vector-borne diseases due to climate change?
   (a) Decreased temperatures
   (b) Migration of insects to new habitats
   (c) Improved healthcare system
   (d) Reduced precipitation
9. What are the long-term implications of climate change on the polar regions?
   (a) Increased snowfall
   (b) Reduced sea-level rise
   (c) Thawing of permafrost
   (d) Expansion of ice sheets
10. What is the significance of the greenhouse effect?
    (a) It maintains the average global temperature suitable for life
    (b) It reduces global temperatures
    (c) It prevents the absorption of solar radiation
    (d) It accelerates the melting of polar ice caps
11. Which greenhouse gas has the highest global warming potential?
    (a) Carbon dioxide (CO_2)
    (b) Methane (CH_4)
    (c) Nitrous oxide (N_2O)
    (d) Water vapor (H_2O)
12. What is the primary impact of climate change on livestock production?
    (a) Reduced primary productivity of rangelands
    (b) Increased water availability
    (c) Decreased incidence of vector-borne diseases
    (d) Enhanced forage nutritive value
13. What is the major consequence of increasing CO_2 levels above 400 ppm?
    (a) Increased transpiration rates in plants
    (b) Enhanced gas exchange between atmosphere and leaves
    (c) Stomatal closure and hindered photosynthetic rates
    (d) Improved water use efficiency in crops

14. How do changes in precipitation patterns affect agriculture?
    (a) Increased crop yields
    (b) Improved water resources
    (c) More frequent and severe drought
    (d) Enhanced soil fertility
15. Which region is particularly vulnerable to the collapse of ice sheets due to climate change?
    (a) Amazon rainforest
    (b) Himalayas
    (c) Sahara Desert
    (d) Great Barrier Reef
16. How do feedback mechanisms exacerbate the impacts of global warming?
    (a) By reducing greenhouse gas emissions
    (b) By stabilizing global temperatures
    (c) By amplifying the greenhouse effect
    (d) By decreasing atmospheric CO_2 levels
17. What is the primary cause of rising sea levels?
    (a) Increased rainfall
    (b) Melting of polar ice caps
    (c) Enhanced coastal erosion
    (d) Soil degradation
18. What is the relationship between global warming and climate change?
    (a) They are interchangeable terms.
    (b) Global warming is a cause of climate change.
    (c) Climate change is a cause of global warming.
    (d) They are unrelated phenomena.
19. How does global warming affect the frequency and intensity of extreme weather events?
    (a) It reduces their occurrence.
    (b) It has no impact on extreme weather events.
    (c) It increases their frequency and intensity
    (d) It only affects certain regions.
20. What are the consequences of increased CO_2 levels on plant growth and crop production?
    (a) Enhanced photosynthetic rates and improved crop quality
    (b) Reduced biomass production and poor-quality food
    (c) Increased transpiration rates and improved gas exchange
    (d) Decreased rates of protein synthesis and increased resistance to pests

Answers: 1-b, 2-d, 3-b, 4-a, 5-a, 6-a, 7-b, 8-b, 9-c, 10-a, 11-b, 12-a, 13-c, 14-c, 15-b, 16-c, 17-b, 18-b, 19-c, 20-b

20.8.2 Short-Answer Questions

1. What is global warming?
2. What are the primary greenhouse gases responsible for global warming?
3. What are some impacts of climate change?
4. How do feedback mechanisms and tipping points worsen the impacts of climate change?
5. Name the main greenhouse gases in the Earth's atmosphere.
6. Which gases are categorized as secondary greenhouse gases?
7. What are the main non-greenhouse gases in the atmosphere?
8. How do greenhouse gases contribute to the greenhouse effect?
9. What are the causes of global warming?
10. How does climate change differ from global warming?

20.8.3 Long-Answer Questions

1. Discuss the role of human activities in contributing to global warming and climate change, highlighting specific examples such as burning fossil fuels and deforestation. Explore the implications of these activities on greenhouse gas emissions and the amplification of the greenhouse effect.
2. Explain the concept of feedback mechanisms and tipping points in the context of global warming and climate change. Provide examples of positive feedback loops and their potential to exacerbate the impacts of climate change, including the release of methane from thawing permafrost and the loss of reflective surfaces. Discuss the potential consequences

of these feedback mechanisms and their implications for the future.

3. Describe the composition and significance of greenhouse gases in the Earth's atmosphere. Discuss the primary greenhouse gases and their sources, as well as secondary greenhouse gases. Analyze their abundance and their respective contributions to the greenhouse effect, highlighting the role of water vapor and the considerations for its impact on global warming.
4. Assess the long-term implications of climate change on various sectors and ecosystems, including agriculture, human health, and the cryosphere. Discuss the potential consequences of rising temperatures, sea-level rise, and changes in precipitation patterns on these sectors. Analyze the vulnerabilities and challenges posed by climate change and the potential need for adaptation strategies to mitigate its impacts.
5. Write short notes on the following:
   (a) Greenhouse gases
   (b) Climate change impact on agriculture
   (c) Climate change and human health

References

Archer S, Buffett B (2005) Time-dependent response of the global ocean clathrate reservoir to climatic and anthropogenic forcing. Geochem Geophys Geosyst 6(3):Q03002

IPCC (Intergovernmental Panel on Climate Change). 2001a. In: Houghton JT, Ding Y, Griggs DJ, Noguer M, van der Linden PJ, Dai X, Maskell K, Johnson CA (Eds) Climate change 2001: the scientific basis. Contributions of working group I to the third assessment report of the intergovernmental panel on climate change. Cambridge University Press, Cambridge and New York. 881 pp

IPCC (Intergovernmental Panel on Climate Change) (2001b) Climate change 2001: the scientific basis. Cambridge University Press, London and New York

IPCC (Intergovernmental Panel on Climate Change) (2014) Summary for policymakers. In: Edenhofer O, Pichs-Madruga R, Sakona Y, Farahani E, Kadner S, Seyboth K, Adler A, Baum I, Brunner S, Eickemeier P, Kriemann B, Savolainen J, Schlomer S, von Stechow C, Zwickel T, Minx JC (eds) Climate change 2014: mitigation of climate change. Contribution of working group III to the fifth assessment report of the intergovernmental panel on climate change. Cambridge University Press, Cambridge and New York

Myhre G, Shindell D, Breon F-M, Collins W, Fuglestvedt J, Huang J, Koch D, Lamarque J-F, Lee D, Mendoza B, Nakajima T, Robock A, Stepens G, Takemura T, Jhang H (2013) Anthropogenic and natural radiative forcing. In: Stocker TF, Quin D, Plattner G-K, Tignor M, Allen SK, Boschung J, Nauels A, Xia Y, Bex V, Midgley PM (eds) Climate change 2013: the physical science basis. Contributions of working group I to the fifth assessment report of the intergovernmental panel of climate change. Cambridge University Press, Cambridge and New York

Singh V (2024) Environmental disruptions: planet earth in the vicious cycle of pollution, global warming, and climate change. NIPA, New Delhi. 156 pp

Singh V, Rastogi A, Nautiyal N, Negi V (2017) Livestock and climate change: the key actors and sufferers of global warming. Indian J Anim Sci 87(1):11–20

Winkelmann R, Levermann A, Ridgwell A, Caldeira K (2015) Combustion of available fossil fuel resources sufficient to eliminate the Antarctic ice sheet. Sci Adv 1(8):e1500589

Section V

Environmental Management

21 Solid Waste Management

Human lifestyles are generating mountains of solid wastes: useless, unwanted, and undesirable materials. Solid wastes are a great nuisance posing a threat to the environment and public health. The industrialization has provided a huge number of products for human use. At the same time, it has also created a vicious problem of solid waste pollution. Getting rid of solid wastes or garbage is a critical issue facing all human settlements – educational, industrial, commercial, and residential areas. Improper disposal of solid wastes can cause unhygienic conditions and environmental pollution which may lead to an outbreak of diseases.

Solid waste management, itself a discipline, refers to the control of generation, storage, collection, transfer, processing, and disposal of solid waste. This process also facilitates recycling of the items that do not end up in garbage or trash. Britannica defines solid waste management as "collecting, treating, and disposing of solid material that is discarded because it has served its purpose or is no longer useful."

The tasks of solid waste management are not simple ones. They pose complex technical challenges and involve administrative and social and economic problems that have also to be solved. Solid waste management involves a set of technologies applicable for harnessing energy out of the garbage. Many countries and many cities in a country have developed state-of-the-art-type solid waste management systems that help them generate electricity fulfilling many of their essential needs.

There are six functional components in the whole process of solid waste management: (i) identification of waste, (ii) on-site handling and storage, (iii) waste collection (collection from different sources), (iv) waste transfer, (v) waste processing (sorting of reusable and recyclable wastes), and (vi) disposal (e.g., at landfill site).

21.1 Classification of the Wastes

Various sorts of materials are discarded by people, turning them into waste. These wastes can range from being harmless to being extremely toxic. It depends on the sources the solid wastes are generated from. Thus, there can be five categories of waste as described in Table 21.1.

According to WHO, biomedical waste can be divided into four groups, namely:

(i) Infectious, that is, wastes from surgeries and any material containing pathogens
(ii) Pathological, that is, tissues, organs, drugs, etc.
(iii) Radioactive, that is, contaminated with a radioactive substance
(iv) Others, that is, wastes from the hospital, housekeeping, kitchen, etc.

V. Singh, *Textbook of Environment and Ecology*, https://doi.org/10.1007/978-981-99-8846-4_21

Table 21.1 Various types of solid wastes and their main characteristics

Solid waste type	Characteristic	Examples
Biodegradable	Can be degraded by microorganisms; give characteristic smell; make stinking atmosphere	Food wastes, vegetable and fruit peels, tea leaves, egg shells, leaf litter, crop residues, dung, etc.
Nonbiodegradable	Cannot be degraded by microorganisms; do not rot; do not stink	Glass, metal scraps, plastic, polythene, asbestos, ceramics, aluminum cans, etc.
Toxic	Poisonous in nature; harmful to health; can kill the plants, animals, and microorganisms	Pesticides, herbicides, acids, alkalis, radioactive substances, etc.
Nontoxic	Nonpoisonous in nature	Glass, metal, ceramics, asbestos, ceramics, paper, etc.
Biomedical	Released from hospitals, medical institutes, clinics, and pharmaceutical companies	Syringes, cotton, plastic and glass bottles, anatomical and pathological wastes, etc.

21.2 Various Sources of Solid Wastes

The generation of solid wastes is a continuous process and an indistinguishable part of contemporary human civilizations. There is no place inhabited by human beings where solid wastes are not generated. They come from our homes, restaurants, hotels, picnic spots, honeymoon huts, hospitals, dispensaries, industries, educational institutes, commercial establishments, aerodromes, and railway stations, from bus stands, and every place infested with human activities. The generation of solid wastes, in essence, is an inherent characteristic of modern lifestyles. A list of the major sources of solid waste and the types of waste generated is provided in Table 21.2.

21.3 Causes of Solid Waste Generation

In the history of human civilization, waste has never been a problem. Whatever human being extracted from the natural resources, it ended up in its recycling. This problem emerged in the aftermath of the industrial age. There are many reasons for solid waste generation in huge and often unmanageable amounts.

21.3.1 Popollution

Environmental pollution is increasing in proportion to an increase in population. The burgeoning population is generating mountains of waste products. The word "popollution" is a fusion of popula-

Table 21.2 Primary origins and categories of generated solid wastes

Major source	Solid waste produces
Residences (households, family establishments)	Garbage of food wastes, papers, plastics, leather, glass, cardboard, tire, electronic items, used up domestic gadgets, old mattresses, tires, etc.
Institutions (universities, colleges, schools, government centers, military barracks, etc.)	Food wastes, plastics, wood, rubber, paper, metals, hazardous products, etc.
Municipal services (cities and towns)	Street cleaning wastes, landscaping wastes, sludge, and wastes from recreational areas
Industries (heavy and light manufacturing industries, factories, mills, fabrication plants, recycling plants, canning plants, etc.)	Food wastes, housekeeping wastes, packaging wastes, plastics, hazardous wastes, woody material, heavy metals, construction and demolition material, ashes, etc.
Commercial establishments (markets, hotels, restaurants, go-downs, stores, and office buildings)	Glass, woods, wools, cloth pieces, plastics, food wastes, papers, packaging material, plastics, bottles, medicines, hazardous wastes, etc.
Construction and demolition sites (building construction, road repair, building renovation, building demolition sites)	Concrete, bricks, steel materials, rubber, wood, glass, plastics, copper wires, dirt, etc.
Treatment plants (processing plants, mineral extraction plants, chemical plants, power plants, refineries, etc.)	Metals, plastics, ashes, coal powder, plastics, oils, husk, peels, hazardous wastes, etc.
Medical establishments (hospitals, dispensaries, medical research institutes, medical universities, pharmaceutical companies, health centers, clinics, dispensaries, medical stores, medical shops, etc.)	Medicines, syringes, gloves, bandages, plastics, chemicals, etc.
Agriculture (crop fields, gardens, vineyards, poultry farms, feedlots, dairies, etc.)	Fertilizer bags, pesticide containers, plastics, dung, manure, feathers, spoilt eggs, food wastes, husk, crop residues, etc.

tion and pollution. The larger the population size the higher the rates of waste production.

21.3.2 Economic Affluence and Modern Lifestyles

The industrialization has transformed a resource-centered economy into a market-based economy. People have a more direct interest in the count of their cash or bank balance rather than in the state of natural resources they have to depend upon for their survival and sustainable livelihoods. Now the roots of one's survival and contentment lie in the market. Affluent society has inculcated a use-and-throw lifestyle. People's economic prosperity always needs ever-increasing rates of resource extraction. The higher the level of economic affluence, the greater the degree of consumerism and consequent waste generation. Waste is a "grace" for a consumer society. There is hardly any room for frugality in consumer society. Almost every product bought from the market generates some kind of waste. The socioeconomic status of a consumer society is, thus, inevitably linked with waste generation.

21.3.3 Technologies

Technologies have transformed our lifestyles. In a bid to bring comfort to life, technologies have made it rather more cumbersome with uncontrolled production of waste and environmental pollution. Every material procured from the market comes with packaging. The packaging itself has become an independent industry that matters. This not only ensures the safety of soft items but also adds to the attractiveness of the items and also advertises them. Technologies associated with the packaging of market products have exacerbated consumerism in a big way.

21.4 Effects of Poor Waste Management

Accumulation of waste material coupled with poor solid waste management can cause various problems to the environment, humans, and other organisms. Various sorts of effects can range from spoilage of aesthetic values to physicochemical and biological ones. Hazards caused by solid wastes at a place can spell out widespread consequences and assume global proportions. The consequences might even assume the proportions of an epidemic or a pandemic.

The problem of solid waste accumulation in developing countries is especially worse and is increasingly posing a challenge to the city authorities due to a variety of reasons. Poor waste management due to constraints imposed by budget and lack of technologies invites multiple problems – ecological, environmental, socioeconomic, ethical, and aesthetic ones.

21.4.1 Spoilage of Aesthetic Looks

Uncontrolled dumping of wastes followed by improper disposal systems spoils the landscape with surrounding environments filled with bad smells. In many places, people are seen littering their surroundings with domestic garbage. Accumulation of garbage near human settlements and around houses eclipses the aesthetic values of the environment. Dumping of solid wastes on roadsides and near railway tracks is a common case of landscape spoilage in many places. Spoilage of aesthetic looks of the environment has bad impacts on the human mind as well as on public health.

21.4.2 Environmental Pollution

Improper disposal of waste causes widespread environmental pollution. Surrounding airs are laden with foul smells and gases released upon the decomposition of organic wastes. Pollutants flow into surface waters and cause widespread water pollution. Toxic substances in the waste leach into the ground and contaminate groundwater. Solid wastes can change the physicochemical and biological properties of the soils and decline their productivity. Pollutants of the wastes can affect community structure in terrestrial aquatic and soil ecosystems.

21.4.3 Contamination of Food Chains

Urban and industrial wastes contain a variety of toxic chemicals. These chemicals contaminate air, water, and soils and enter into food chains. Thus, the health of the ecosystems is affected and all the organisms are severely affected by a variety of toxic pollutants in the wastes.

21.4.4 Health Hazards

Improper dumping of solid wastes in or near residential areas and in vacant lands elsewhere is bound to create unhygienic conditions leading to the outbreak of various diseases, notably gastroenteritis, cholera, and jaundice. Solid wastes can provide a breeding ground for mosquitoes that can bring spurt in malaria. Wastes also provide a breeding refuge for rodents that can carry the bacteria *Yersinia pestis*, the causal agents of plague. The decomposition of the organic wastes produces inflammable gases like methane. Their ignition under uncontrolled conditions can lead to widespread damage to the environment. Burning of wastes produces poisonous gases, like H2S, SO2, and CO that cause various health problems as we have discussed in the Chap. 16 on Air Pollution.

21.5 Methods of Solid Waste Management

Solid wastes do not always cause environmental problems and health hazards if managed properly and wisely. Disposal of some wastes following proper management techniques can be of high ecological and socioeconomic value. Much waste can also be converted into precious products. Several methods of solid waste management are applied for different purposes as discussed under the following subheads.

21.5.1 Incineration

This is an industrial process of controlled burning of wastes at very high temperatures (about 850 °C) in the presence of oxygen. Incinerators used to carry out this process are designed in such a way that they do not give off extreme amounts of heat. After incineration, the volume of the original waste is reduced significantly to 20–30%. This process is intended to significantly reduce the volume of the waste so that the final product, the ash, needs a very limited land area for disposal, although this method provides a good facility for waste disposal but also releases many toxic gases into the atmosphere.

21.5.2 Pyrolysis

In this process, solid wastes are chemically decomposed by heat in the absence of oxygen. This is usually carried out under pressure and temperatures up to 430 °C. Pyrolysis reduces solid wastes into solid residues enriched in carbon, chemicals, and gases. This is an irreversible process and is generally used in charring wood. In the absence of oxygen, organic matter does not undergo combustion but is converted into useful products.

21.5.3 Composting

This process involves the degradation of organic wastes, especially animal excreta, by microorganisms in the presence of oxygen. In the process of composting, the organic waste is converted into compost and CO_2. Compost is of vital significance in improving and maintaining soil fertility and in increasing and sustaining crop productivity. Apart from enhancing soil organic matter (SOM) and soil organic carbon (SOC), compost serves as a key input in organic farming.

21.5.4 Vermicomposting or Vermiculture

Vermicomposting involves the use of earthworms and is also called earthworm farming. Solid organic wastes, such as kitchen wastes, sludge, and dung, can be converted into nutrient-rich vermicompost used as high-quality organic fertilizer in organic farming.

21.5.5 Sanitary Landfill

There are various types of landfills, like municipal solid landfill and industrial landfill. A sanitary landfill is considered to be the best technique for solid waste management. Sanitary landfill, unlike open dumping, is managed by a local government or a state. A sanitary landfill is a land area or a large pit with its bottom covered with a sheet impervious to leakage of the leachate to protect against contamination of groundwater resources. The solid wastes are filled in the site earmarked for a sanitary landfill in layers and compressed to further solidify and reduce volume.

The alternate layering of wastes and soil helps accelerate the decomposition and production of methane, a toxic fuel gas. Methane gas production is controlled and the gas is not allowed to escape into and pollute the environment. This gas is rather used as fuel for the generation of electricity. A clay liner helps to isolate the landfill from the environment. There are, thus, two purposes of the sanitary landfill: (1) to protect the environment from solid and liquid wastes and (2) to use the decomposition product, methane, as a fuel, especially for electricity generation.

The sanitary landfill is not a simple technique. It involves specific engineering designs to protect the environment and use the methane gas produced to generate electricity. When an intended landfill comes to its closure, which usually takes decades-long time, it is sealed with clay and a new landscape designed with landscape engineering comes into being for other uses.

21.5.6 Recovery and Recycling

The recovery or recycling process involves taking discarded but useful items for the next use. When a product is brought into its initial use and then discarded, it is again converted into a useful product and is again brought into use. This process involves separation, collection, processing, marketing, and using the material which was once discarded as waste. Wastes containing papers, bottles, tin cans, aluminum vessels, etc. can be managed in a way that these materials are recovered for further use after their cleaning and sterilization and/or are recycled or even converted into other usable products.

Recovery and recycling is an eco-friendly and economic method of solid waste management. The main advantages derivable from this technique include the following:

1. Environmental protection through proper handling of the solid wastes
2. Reduction of the waste volume and burden of the waste material
3. Conservation of natural resources by reducing the need for raw material
4. Making waste management an economic process

21.6 The 3 Rs Rule: Reduce, Reuse, and Recycle

The 3 Rs, that is, reduce, reuse, and recycle, are regarded as a rule of waste management. Waste generation has become a natural habit of human societies. The 3 Rs rule suggests how can we maintain a balance between waste generation and its management and also govern our lifestyles regarding the use of natural resources:

- Reduce unnecessary consumption to reduce generating more and more waste.
- Reuse everything you can or give usable things to others to use.
- Recycle the recyclable wastes.

When we follow the rule of reduction of raw materials, it manifests in a corresponding reduction in the production of waste. For example, a reduction in the use of metals will help reduce the mining of ores, and this would greatly help reduction in waste generation enacted during mining activities. Reduce the consumption of paper, and there will be a substantial decrease in the cutting of trees and environmental pollution. If your plate has only as much food as would be necessary for you to eat, this would prevent the wastage of the leftover food.

There are many wastes that, instead of being discarded as waste, can be used again and again.

For example, glass bottles and containers can be reused many times. Waste paper can be used for making seed containers.

Recycling every recyclable material is ecologically the most crucial aspect of waste management. Biodegradation helps us convert organic matter into some highly valuable forms. For example, biodegradation of animal excreta, crop residues, leaf litter, etc. converts them into compost or vermicompost through which nutrients are recycled into the soil, resulting in the augmentation of soil fertility and crop productivity. Some nonbiodegradable wastes can be converted into useful products by physical means. For example, iron scraps and aluminum cans can be transformed into usable products.

Following the 3 Rs rule, we can save energy, raw material, land space, money, labor, and time on waste management and can substantially decrease environmental pollution. The 3 Rs, thus, are of critical ecological, environmental, economic, and ethical importance.

21.7 Summary

Control of the generation, storage, collection, transfer, processing, and disposal of solid wastes is what is called solid waste management. Solid waste management involves a set of technologies applicable for harnessing energy out of the garbage. There are six functional components in the whole process of solid waste management: identification of waste, on-site handling and storage, waste collection (collection from different sources), waste transfer, waste processing (sorting of reusable and recyclable wastes), and disposal (e.g., at landfill site). There are many reasons for the generation of solid wastes, for example, popollution (population and pollution), economic affluence and modern use-and-throw lifestyles, waste-generating technologies, etc.

Accumulation of waste material coupled with poor solid waste management can cause various problems to the environment, humans, and other organisms. Various sorts of effects can range from spoilage of aesthetic values to physicochemical and biological ones. The consequences might even assume the proportions of an epidemic or a pandemic. Spoilage of aesthetic looks of the environment has bad impacts on the human mind as well as on public health. Improper dumping of waste causes widespread environmental pollution. Solid wastes can change the physicochemical and biological properties of the soils and decline their productivity. Pollutants of the wastes can affect community structure in terrestrial, aquatic, and soil ecosystems. The health of the ecosystems and all the organisms is severely affected by a variety of toxic pollutants in the wastes. Improper dumping of solid wastes in or near residential areas and in vacant lands elsewhere is bound to create unhygienic conditions leading to the outbreak of various diseases, notably gastroenteritis, cholera, and jaundice. Wastes also provide breeding refuge for rodents that can carry the bacteria *Yersinia pestis*, the causal agents of plague.

Many wastes can also be converted into precious products. Several methods of solid waste management are applied for different purposes. Incineration, an industrial process of controlled burning of the wastes at very high temperatures (about 850 °C) in the presence of oxygen, is used to significantly reduce the volume of the wastes so that the final product, the ash, needs very limited land area for disposal. In pyrolysis, solid wastes are chemically decomposed by heat in the absence of oxygen. Usually carried out under pressure and temperatures up to 430 °C, the pyrolysis reduces the solid wastes into solid residues enriched in carbon, chemicals, and gases. Composting involves the degradation of organic wastes, especially animal excreta, by microorganisms in the presence of oxygen, and in the process, the organic waste is converted into compost and CO_2. Compost is of vital significance in improving and maintaining soil fertility and in increasing and sustaining crop productivity. Vermicomposting involves the use of earthworms to convert organic wastes, such as kitchen wastes, sludge, and dung that can be converted into nutrient-rich vermicompost used as high-quality organic fertilizer in organic farming. A sanitary landfill managed by a local government or a state is a land area or a large pit with its bottom covered with a sheet impervious to leakage of the leachate to protect against contamination of groundwater resources. The alternate layering of wastes and soil helps accelerate the decomposition and production of methane, a toxic fuel gas often used as fuel for electricity generation. When a product is brought into its initial use and then discarded, it is again converted into a useful product and is again brought into use. This process involves separation, collection, processing, marketing, and using the material which was once discarded as waste.

Recovery and recycling is an eco-friendly and economic method of solid waste management. Following the 3 Rs rule, that is, reduce, reuse, and recycle, we can save energy, raw material, land space, money, labor, and time on waste management and can substantially decrease environmental pollution.

21.8 Exercises

21.8.1 Multiple-Choice Questions

1. Solid waste management involves
   (a) Collection
   (b) Treatment
   (c) Disposal
   (d) All of the above
2. How many functional components are there in the process of solid waste management?
   (a) 2
   (b) 4
   (c) 5
   (d) 6
3. Which of the following does not produce waste?
   (a) Agriculture
   (b) An educational institute
   (c) Film industry
   (d) None of the above
4. What is used in the production of plastic?
   (a) Mercuric chloride
   (b) Vinyl chloride
   (c) Carbon tetrachloride
   (d) Lead oxide
5. Municipal solid waste (MSW) emanates from
   (a) Agriculture sector
   (b) Mining operations
   (c) Private houses, commercial establishments, and institutions
   (d) Construction sites
6. How many groups can the biomedical wastes be divided into, according to WHO?
   (a) 9
   (b) 7
   (c) 5
   (d) 4
7. Which of the following industries contributes to producing metals as waste?
   (a) Mining
   (b) Electroplating
   (c) Skiing
   (d) None of the above
8. The most serious environmental effect the hazardous wastes pose is
   (a) Contamination of groundwater
   (b) Increased land area for disposal
   (c) Air pollution
   (d) Insect killing
9. What is the characteristic of aerobic bacteria?
   (a) They are nourished by organic matter.
   (b) They oxidize organic matter.
   (c) They perpetuate in free oxygen.
   (d) They do as above.
10. The liquid material that passes through compacted solid waste in the process of waste management is known as
   (a) Kerosene oil
   (b) Petroleum
   (c) Leachate
   (d) Stone oil
11. The 3 Rs relating to solid waste management are
   (a) Reduce, reuse, and recycle
   (b) Reduce, replace, and recycle
   (c) Reuse, replace, and recover
   (d) Reject, remove, and rewind
12. What is used as fuel at a landfill?
   (a) Methane
   (b) CO_2
   (c) Leachate
   (d) H_2
13. Which of the following helps reduce the volume of solid waste to a significant extent?
   (a) Recycling
   (b) Incineration
   (c) Landfill
   (d) Composting

14. What is the process by which solid waste is burnt at very high temperatures?
    (a) Recycling
    (b) Recovery
    (c) Incineration
    (d) Sanitary landfill
15. At the end of its service life, that is, after it ends up waste, a product is converted into another useful product. This process is called
    (a) Recycling
    (b) Recovery
    (c) Reuse
    (d) Replacement
16. Why is burning waste in the open not advisable?
    (a) It needs applications of modern technology.
    (b) It raises several environmental and health-related issues.
    (c) It needs a lot of space.
    (d) It is not a cost-effective practice.
17. Which of the following is a fundamental way of reducing waste on an individual level?
    (a) Burning
    (b) Recycling
    (c) Source reduction
    (d) Reuse
18. Which of the following practices of waste management contributes to increasing agricultural productivity?
    (a) Composting
    (b) Vermiculture
    (c) Composting and vermiculture
    (d) None of the above
19. The decomposition of organic waste at high temperature in the absence of oxygen is known as
    (a) Incineration
    (b) Open burning
    (c) Carboxylation
    (d) Pyrolysis
20. When we throw away television sets, computers, and cell phones, the waste accumulated is called as
    (a) e-Waste
    (b) Trash
    (c) Recycled waste
    (d) Gray waste

Answers: 1-d; 2-d; 3-d; 4-b; 5-c; 6-d; 7-b; 8-a; 9-d; 10-c; 11-a; 12-a; 13-b; 14-c; 15-a; 16-b; 17-c; 18-c; 19-d; 20-a

21.8.2 Short-Answer Questions

1. How would you define solid waste management?
2. Differentiate between urban wastes and industrial wastes.
3. What is the difference between biodegradable and nonbiodegradable waste?
4. What is biomedical waste?
5. How is soil affected by waste?
6. What is vermiculture?
7. What is the main advantage of incineration?
8. What do you mean by 3 Rs?
9. What is meant by recycling wastes?
10. What are the advantages of compost?

21.8.3 Long-Answer Questions

1. How would you classify wastes? Explain with examples.
2. Discuss the impacts of waste accumulation.
3. Discuss various methods of solid waste management.
4. Why should we follow and pursue the 3 Rs philosophy in life?
5. Write short notes on the following:
    (a) Vermicomposting
    (b) Pyrolysis
    (c) Sanitary landfill

References

Guerrero LA, Maas G, Hogland W (2013) Solid waste management challenges for cities in developing countries. Waste Manag 30(1):220–232. https://doi.org/10.1016/j.wasman.2012.09.008

Mishra AR, Mishra SA, Tiwari AV (2014) Solid waste management – case study. Int J Res Adv Technol 2(1):396–399

22 Climate Change Mitigation

Climate change mitigation, without any doubt, is the most critical issue facing our contemporary world. So far, as we have discussed in a previous chapter, human activities are held to be the major cause of climate change. The magnitude of climate change has already captured the momentum, and despite adopting the most effective measures, it seems pretty difficult to attain a climate pattern like that of the pre-climate change era in a reasonable time length. However, it is not an impossible task to accomplish. The world is replete with knowledge, skill, and strategies to halt and reverse the global warming trend and restore global climate order. What is urgently needed is to implement workable strategies.

22.1 Three-Dimensional Strategies to Deal with Changing Climate

Climate change is going on; it is a reality that cannot be denied. That way it will go or what gravity it will assume, we cannot predict precisely, although many speculations based on scientific studies are being made all over the world. We, therefore, need to be cautious about and prepared for living with it. Climate change is gradually intensifying as indicated by unabatedly increasing atmospheric CO_2 levels. There can be no shortcut to climate change mitigation. It might take decades, even centuries, to resume the stable climate pattern of the pre-industrial age. Therefore, we need to adapt to this reality. Our ultimate goal has to be to bring down the current atmospheric CO_2 concentration to a level that could help the planet's climate system to a state of normalcy.

22.1.1 Preparedness

What shall we do if climate change assumes critical proportions? We cannot immediately afford to be dependent on long-term climate strategies, that is, adaptation and mitigation. In the event of climate change up to a level that matters, we shall have to be dependent on instant remedial measures. Preparedness is a short-term strategy, like disaster management. Some questions that need to be responded to cope with climate change are the following:

- **Food and nutrition security:** Do we have enough food storage to meet our people's requirements if agriculture fails for a few years?
- **Water security:** What is the status of our freshwater resources and public water supply systems during recurring drought conditions?
- **Health security:** Do we have enough of everything necessitated to provide adequate health cover for the people, including neutraceuticals, vaccines, and lifesaving medicines

V. Singh, *Textbook of Environment and Ecology*, https://doi.org/10.1007/978-981-99-8846-4_22

to manage the usual and newly emerging diseases in the changed environments?

- **Food production security:** Are our agricultural and food supply systems designed to function according to the changed environmental circumstances?
- **Dwelling security:** Are the facilities in our dwelling systems (inhabited houses in cities, towns, villages, workplaces, etc.) for saving people from adverse climatic conditions, etc. during climate extremes up to mark?
- **Life's dynamism:** Are our transport systems suitably developed enough for people's movement in extreme weather?
- **Ecological security:** What are the ecological measures to be adopted so that the long-term measures – namely, adaptation and mitigation – are not badly affected but are accelerated during the adoption of preparedness strategies?

A preparedness strategy can be evolved based on the answers to the questions posed above. There seems to be no consensus on a global basis regarding preparedness to combat climate crises. A few countries have declared a "state of emergency" which is certainly the right step toward formulation strategies to deal with the changing circumstances. The whole world, however, needs to be prepared with all preparedness to cope with impending climate crises while accelerating the operationalization of adaptation and mitigation processes.

22.1.2 Adaptation

Evolving adaptation strategies are imperative for coping with climate change on a long-term basis. These strategies ensure the survival and intactness of life support systems against an adverse environmental background. Adaptation mechanisms also ensure an appropriate living standard and an appropriate pace of socioeconomic development.

Life on Earth has evolved in tune with its adaptability to the ongoing environmental conditions and so have our socioeconomic systems. Resilience and sustainability of the current development state and the current socioeconomic system seem to be at stake if the environmental conditions are changed drastically. Therefore, we shall essentially need to evolve mechanisms that could help us maintain the present status of life and life systems even under altered environments.

Our modern agriculture built around Green Revolution technology is vulnerable to climate change. Evolved in line with adaptation mechanisms, agriculture can sustain itself against the environmental extremes emerging and emerge along the intensification of climate change processes. A paradigm shift in food production systems adaptable to the dismal scenario to emerge amid climate crises as suggested below (after Singh 2020) will nurture our best expectations in the era of climate change:

- **Agro-ecosystems/farming systems:** Development of agriculture based on agroecosystems or farming systems rather than on just cultivated lands. This would be an ecologically sound, fairly resilient, and sustainability-oriented food production system with forests, cultivated land, livestock, etc. integrated into the form of an agroecosystem or a farming system.
- **Development of agroforestry:** Agricultural practices involving only crop annuals (shallow-rooted plants) are extremely vulnerable to adverse environmental conditions. Agroforestry – with the integration of trees and other woody perennials with cultivated crop annuals – is more resistant to drought conditions and a more sustainable system of food production.
- **Increased level of biodiversity/agrobiodiversity in agroecosystems:** A system with a high degree of biodiversity, as a rule of nature, is more resilient to biotic and abiotic stresses, more sustainable, and more productive.
- **Altered cropping patterns and crop sequences:** Cultivation of shallow-rooted food crops (e.g., cereals, pseudocereals, vegetables, etc.) with deep-rooted ones (e.g., pulses and oilseeds) is more promising in the event of water scarcity and arid conditions.

- **Landraces and drought-resistant varieties:** The so-called high-yielding varieties of food crops developed by plant breeders are mostly water-guzzling and, therefore, would not survive drought conditions. There are many landraces and traditional crop varieties with high water use efficiency and the inherent capability of thriving under rain-fed farming conditions. Such crop varieties and landraces are still in practice in traditional systems in marginal areas, such as in mountain areas. Such traditional agricultural systems evolved through trials and errors over centuries need to be protected and promoted as they have a high degree of adaptation to the conditions borne out of climate change.
- **No-till systems:** When the land is plowed, it loses its moisture, and its carbon emission rates are also increased. No-till systems or agriculture based on minimum tillage would be highly promising from the viewpoint of soil and water conservation and minimum carbon emissions.
- **Principles of ecological agriculture:** Ecological agriculture is a healthy, vibrant, regenerative, and sustainable agriculture. Operationalization of the three principles of ecological agriculture, namely, living soil, biodiversity, and cyclic nutrient flows (Singh et al. 2014; Singh 2020), would lead to the development of ecological agriculture.

22.1.3 Mitigation

Mitigation of climate change is a long-term strategy that involves local, regional, and international policies, programs, projects, missions, strategies, tactics, and intended goals. The mitigation strategies are based on three premises: (i) carbon sequestration enhancement through technological means and biological phenomena, (ii) minimization of carbon emissions, and (iii) alternative energy sources or clean energy sources with no or minimum carbon emissions.

In addition, we shall inevitably need to inculcate a philosophy embracing ecological justice, pursue and follow environmental ethics, and have due respect for environmental laws. Vital for restoring the planet's climate order and saving our living planet, these have been discussed under the following subheads.

22.2 Carbon Sequestration

Carbon sequestration mechanisms involve the removal of atmospheric CO_2 to be captured, secured, stored elsewhere, and/ or used for some specific purpose. In the context of climate change mitigation, the purpose of carbon sequestration is to strike a global carbon balance so that the enhanced global warming effect of CO_2 could be nullified or minimized.

22.2.1 Geological Carbon Sequestration

The process of storing carbon in underground geologic formations or rocks is what geological carbon sequestration is. It is a long-term carbon storage in nature. Carbon dioxide is captured from the sites of emissions – such as natural gas processing, power plants, cement, steel, and chemical production industrial units – and artificially injected into porous rocks of the Earth. This technique of carbon sequestration is of value till thc fossil fuel energy sources responsible for excessive carbon emissions are utilized.

22.2.2 Carbon Capture, Utilization, and Storage

There could be many ways of carbon sequestration. The whole process enacted artificially in a bid to minimize carbon emissions is referred to as carbon capture, utilization, and storage (CCUS). This concept and the technologies involved are not new. The concept was first of all implemented in 1972. At present, 21 large-scale commercial CCUS projects are running worldwide with a capacity to capture up to 40 Mt. CO_2 per year. More than 50% of the CO_2 in the world is captured in the USA alone (Fig. 22.1). Out of these,

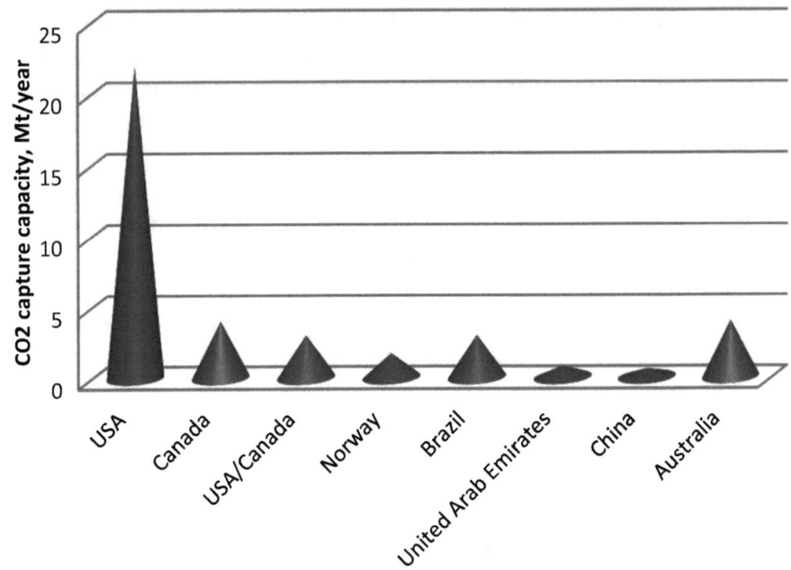

Fig. 22.1 Operational large-scale CCUS projects across the globe in 2020. (Source: Derived from IEA's (2020) data)

nine are in the USA, four in Canada, two in Norway, and one each in the USA/Canada (combined), Brazil, Saudi Arabia, United Arab Emirates, China, and Australia, according to International Energy Agency (IEA), an intergovernmental energy organization based in Paris, France. The Century Plant to capture carbon emission from natural gas processing based in the USA is the largest CCUS project with a CO_2 capture capacity of 8.4 Mt. per year, followed by the Shute Creek Gas Processing Facility, also based in the USA, with a capacity of 7.0 Mt. CO_2 annually from the natural gas processing plant. As many as 30 new ones have been recently announced. The story of the CCUS, however, has largely been one of the unmet expectations. Its impact on global warming has been almost negligible, according to IEA (2020).

The major object of the CCUS projects is to capture, store, and reuse CO_2 emissions from various polluting plants/industrial units. CO_2 is used in the production of chemicals and fuels. Thus, the carbon emissions do not return to the atmosphere where they play role in blocking the Earth's heat from escaping into space, thus causing global warming. The CCUS can be retrofitted to existing power and industrial plants, which, according to IEA (2020), are likely to still emit eight billion tons (Gt) of CO_2 in 2050. In the economic sectors where other technology options are limited, such as manufacturing of iron and steel, chemicals, chemical fertilizers, synthetic fuels, cement, etc., CCUS can be effectively implemented to tackle carbon emissions efficiently. Combined with bioenergy, CCUS can be used in sequestering CO_2 from the atmosphere which could be instrumental in striking a carbon balance. However, such efforts have not so far been brought into practice to an appreciable extent, and therefore, CCUS contributions to strike carbon balance have largely been negligible.

22.2.3 Photosynthesis or Biological Carbon Sequestration

Carbon dioxide is fixed into living organic matter through photosynthesis and chemosynthesis. Photosynthesis is the major phenomenon of the biosphere striking a carbon balance by removing atmospheric CO_2 and storing the same in plant biomass and soils. Since photosynthesis is carried out by the chlorophyll-containing plants, algae, and cyanobacteria in the presence of sunlight using CO_2 from the atmosphere and water from the soil or the hydrosphere, the rate of organic matter accumulation will be proportional to the land area covered by green vegetation and quantitative presence of algae and cyanobacteria in hydrosphere and soils.

Forests, grasslands, rangelands, and other terrestrial parts sequester about 25% of the atmospheric CO_2. Organic matter trapping carbon keeps on transferring from plants to the soils through fallen leaves and dead plant parts. In the soils, carbon is stored as soil organic carbon (SOC). The soils of the world contain 2–3 times more carbon than plant biomass. Soil organic carbon is a product of photosynthesis. Thus, should the adverse environmental and anthropogenic factors be kept off, the amount of SOC in the soils would be dependent on the rates of photosynthesis performed by the green vegetation above the soil surface and partly by the cyanobacteria present in the soil environment.

Soils also store carbon in its inorganic form – as carbonates. Dissolved in water, CO_2 percolates through soils and, mixed with calcium and magnesium minerals, forms carbonates. This process however takes thousands of years. In arid soils and deserts, carbonates form caliche, the sedimentary rocks containing calcium carbonate binding clay, sand, gravel, silt, etc. Carbon in the soil organic matter (SOM) is stored in the soil for decades, but carbon in the carbonates can stay for as many as 70,000 years.

Oceans absorb about 25% of the carbon emissions released as a result of human activities. Carbon is physically dissolved in water and is also captured by photosynthesizers in marine and fresh waters. Oceans serve as the largest carbon sinks on Earth. Cold waters have more capacity to absorb CO_2 than warmer ones. It is why the polar regions are more efficient carbon sinks than the warmer regions of the Earth. The physical dissolution of CO_2 is a two-way process. When CO_2 is released from the oceans into the atmosphere, it creates a positive atmospheric flux. A negative flux refers to the oceans absorbing CO_2.

The mantra of enhancing carbon sequestration in the planet's natural terrestrial ecosystems lies in the fortification of photosynthesis. For this to practically happen, most of the Earth's terrestrial space must flourish with photosynthesizers, especially green plants. Protection, conservation, and augmentation of natural forests, grasslands, and rangelands should be the priority of the strategy of climate change mitigation. Plantation of trees and other woody perennials in open spaces, cultivated lands, roadsides, riverbanks, and anywhere they can survive should be the second priority. The performance of our socioeconomic systems is rooted in Earth's ecosystems. The higher the photosynthetic efficiency of the ecosystems, the greater the degree of socioeconomic systems' sustainability. An appropriate land use system comprising the largest proportion of the land for performing the highest possible rates of photosynthesis should be an essential feature of our socioeconomic strategy.

Agricultural practices are responsible for soil erosion, land degradation, carbon emissions, biodiversity loss, and environmental pollution to a considerable extent. Our agriculture needs to be ecologically sound and environmentally safe and must function as a vibrant carbon sink, like natural ecosystems. Development of such agricultural systems is possible as well as a necessity to be incorporated into the strategies to avert climate change. This has been discussed earlier in this chapter.

Pollution-free marine ecosystems and freshwater lotic and lentic ecosystems have a greater capacity to absorb atmospheric CO_2 and provide an appropriate environment for boosting photosynthesis – thus enhancing their capacities to serve as efficient carbon sinks. Sound management of the Earth's water ecosystems must be an essential component of the strategies to confront and reverse climate change processes.

22.3 Minimizing Carbon Emissions

We are living life by imparting our carbon footprints, that is, by causing the emission of greenhouse gases. The larger the carbon footprint, the greater the damage done to the environment. Since greenhouse gases are concerned with global warming, the carbon footprint has a direct bearing on the ongoing climate change. If we are destined to halt the global temperature rise at

1.5 °C or less, the world would need to reduce its per capita footprint. The current carbon footprint varies from country to country and is too high: 18.3 tons per year of the USA and 8.2 tons per year in China, for example. The Deep Decarbonization Pathways Project calculated that the average per capita footprint needs to be reduced to 1.87 tons per year by 2050 to restrict temperature rise to 2 °C.

We have to be dependent on conventional sources of energy until they are completely replaced by inexhaustible and clean energy sources. Till then we have to develop technologies and engineering designs to minimize carbon emissions. In addition, we need to use high-quality/low-carbon fuels that emit less carbon rather than impure ones. Through implementation of new technology and energy-efficient machinery/engines along with low carbon fuels, we can substantially reduce carbon emissions. Such measures have been discussed in the Chap. 16 on Air Pollution. We also need to make such efforts to reduce environmental pollution that contributes to enhancing global warming.

Of course, technical measures are not enough to significantly reduce carbon emissions and global warming. We need to bring emission-reducing changes in our lifestyles, which would be phenomenal in curbing global warming. Some of the suggestions to reduce carbon footprint are as follows:

1. Eat low on the food chain/be vegetarian.
2. Reduce your food wastage.
3. Compost domestic organic wastes.
4. Use clean energy.
5. Avoid fast fashion.
6. Plant a garden.
7. Eat local and organic foods.
8. Save as much energy as you can.
9. Be dependent on public transport.
10. Use the least possible paper.
11. Contribute to protecting, conserving, and augmenting natural resources.
12. Plant trees wherever possible.
13. Maintain biodiversity in your surroundings.
14. Avoid pollution.
15. Participate in the programs creating awareness about climate change.
16. Pursue and practice eco-philosophy and environmental ethics.

The question is not only of reducing footprint but also of living carbon-negative lives. And that is possible. Bhutan is the only carbon-negative country in the world. The others can also afford to be like Bhutan. And why not? Let every doorstep function as a carbon sink. Let every human person adopt carbon negative lifestyle. That is possible if there are enough trees and enough space stocked with photosynthesizers, when per capita carbon fixation through photosynthesis is far greater than per capita carbon emissions.

22.4 Alternative Energy Sources

Conventional energy sources hold the key to all problems relating to global warming and climate change. Their replacement with alternative energy sources is one of the key solutions to climate change without compromise with an appropriate pace of socioeconomic development.

Alternative energy generally refers to energy sources other than fossil fuels – coal, petroleum, and natural gas. However, here our emphasis is on the energy resources that yield no pollutants that can be replenished and are inexhaustible. The alternative energy sources that the world needs to develop and be dependent on are solar energy, wind energy, geothermal energy, hydroelectric energy, tidal energy, biomass energy, hydrogen energy, and nuclear energy. All these forms have been discussed in the Chap 12 on Energy Resources.

Among the alternative energy sources, the category of renewable ones that are exhaustible – for example, fuel wood and biogas – also produce greenhouse gases and other pollutants. Such energy sources emanating from living sources – plants and animal wastes – will produce GHGs during their biodegradation even if they are not used for energy production. Therefore, such wastes from living sources can be used for energy production and taken advantage of.

Nuclear energy is also often categorized as an alternative energy source. However, it is neither an inexhaustible nor a renewable source as it is obtained from mined elements, such as uranium and thorium. Further, the processing of nuclear energy involves too many risks to life. However, since a very small amount of nuclear material is used in generating huge amounts of energy, it is also often categorized as sustainable energy.

Some of the energy sources that emanate from renewable sources of inexhaustible types can also generate pollution in the process of their implementation. For example, the way geothermal power is processed can generate pollution. Hydroelectric energy is a renewable energy, but its generation process by damming a river with a high dam produces environmental pollution apart from creating many other social, ecological, and environmental problems. If the hydroelectric energy is generated by using methods not leading to environmental disruptions, for example, employing run-of-the-river technology, it would be environmentally safe.

From the above account of alternative energy sources emerges another category of energy referred to as clean energy. Clean energy is the energy that is generated by using methods that do not emit GHGs and other pollutants. Therefore, while developing alternative energy sources for the replacement of conventional fossil fuel energy, our priority must be to go in for clean energy. Clean energy is the best bet for climate change mitigation.

22.5 Eco-philosophy

A civilization inculcates a philosophy and is broadly guided by it. Philosophy denotes more than love for wisdom. The philosophy that has been evolving in various corners of the world since the times of Thales of Miletus and Socrates is the analytical philosophy. It is cool, neutral, and unconcerned with evolutionary passages of life. This analytical – or conventional – philosophy does not have evolutionary content. It is anthropocentric, not eco-centric. It has not helped us maintain our ecological integrity. It has failed in evolving pathways toward a sustainable and happy future.

The ancient Indian philosophy emanating from Vedas, Upanishads, Ramayana, and Mahabharata is a kind of philosophy in which nature speaks ecstatically in its entirety. This philosophy generates a fertile ground of wisdom and is evolutionary. It drives humanity toward karma of nature's respect and conservation and builds a future that must be better than the present. In 1974, Professor Henryk Skolimowski (1930–2018) (Fig. 22.2) created a new philosophy which is a beautiful blend of Indian and Western philosophies. He gave his philosophy a new and vital dimension that nurtures our roots of survival and sustainability: the ecological dimension. He named his philosophy eco-philosophy. Eco-philosophy is a meaningful fusion of ecology and philosophy.

Skolimowski describes the world as a unity of multifarious, mutually interdependent manifestations (Małecka and Stark 2018). Actions, as Grandpierre (2018) says, play a crucial role in the way we conceive ourselves, life, and the universe. Philosophy is not, and it should not be, just theoretical content reflecting concepts. It must come up with the power of generating actions. It must

Fig. 22.2 Professor Henryk Skolimowski: The father of eco-philosophy

have the potential to transform actions into creativity. It must have an agenda of positive movement toward the sources that build up, take care of, and sustain life. It must be the source of wisdom that nurtures new hopes and that makes people responsible for their world and the whole cosmos. In our contemporary world, ecology cannot be meaningful without philosophy. And philosophy will stay cool, inert, incomplete, and somewhat less meaningful without ecology. When both unite into oneness, a synergy is created: ecology infuses life, more life, and more life, into living Earth, and philosophy affixes sanctity and new values to life and takes care of the whole ecosphere and the ecology structures. Ecology deals with all species, and philosophy deals with human persons valuing all living forms of life. Ecology is for greening the Earth. Philosophy is for greening the human mind. Ecology writes the story of a sustainable living planet. Philosophy writes the glory of Mother Earth (Singh 2019).

The conventional philosophy or the analytical philosophy, which has been ruling over the human psyche for centuries, did not serve society well. The eco-philosophy, on the other hand, is evolutionary and is committed to society and the living planet. Attributes of the Skolimowskian philosophy as compared to those of the analytical philosophy are presented in Table 22.1.

Eco-philosophy embraces five basic tenets, namely:

1. The world is a sanctuary.
2. Reverence for life is our guiding value.
3. Frugality is a precondition for inner happiness.
4. Spirituality and rationality do not exclude each other but complement each other.
5. To heal the planet we must heal ourselves.

With multiple attributes woven together (Fig. 22.3), eco-philosophy conceives the world as a sanctuary and inculcates a reverential attitude toward life. Modern human life is founded on the culture of wastage. Wastage saps out what is most precious in life – our inner happiness. The higher the amount of wastage, the higher the status of life: this kind of lifestyle is categorically deplored by eco-philosophy by its tenet "frugality is a precondition for inner happiness." Frugality means waste without grace. Spirituality is a must for our integrity that translates into ecological integrity. Our being spiritual and our being rational are complementary aspects of life. Ecological healing of the planet depends on our healing. A neutral, inactive, ill-fated, wretched, greedy, and violent person or a society cannot undertake the karma of healing the planet. Therefore, our healing is a prerequisite for the planet's healing (Skolimowski 1981, 1992, 2000).

Eco-philosophy nurtures vital values that guide us to take up the pathways to an ecological renaissance and strike an ecological balance with a sense of ecological justice translated into ecological actions (Fig. 22.4).

Table 22.1 Characteristics of Skolimowskian eco-philosophy compared to analytical philosophy

	Analytical philosophy	Eco-philosophy
1.	Piecemeal (analytical)	Comprehensive
2.	Pursuing information	Pursuing wisdom
3.	Environmentally and ecologically oblivious	Environmentally and ecologically conscious
4.	Related to the economics of material progress	Related to the economics of the quality of life
5.	Politically indifferent	Politically alive
6.	Socially unconcerned	Socially concerned
7.	Mute about individual responsibility	Vocal about individual responsibility
8.	Intolerant to transphysical phenomenon	Tolerant to transphysical phenomenon
9.	Health mindless	Health mindful
10.	Language-oriented	Life-oriented
11.	"Objective" (detached)	Committed
12.	Spiritually dead	Spiritually alive

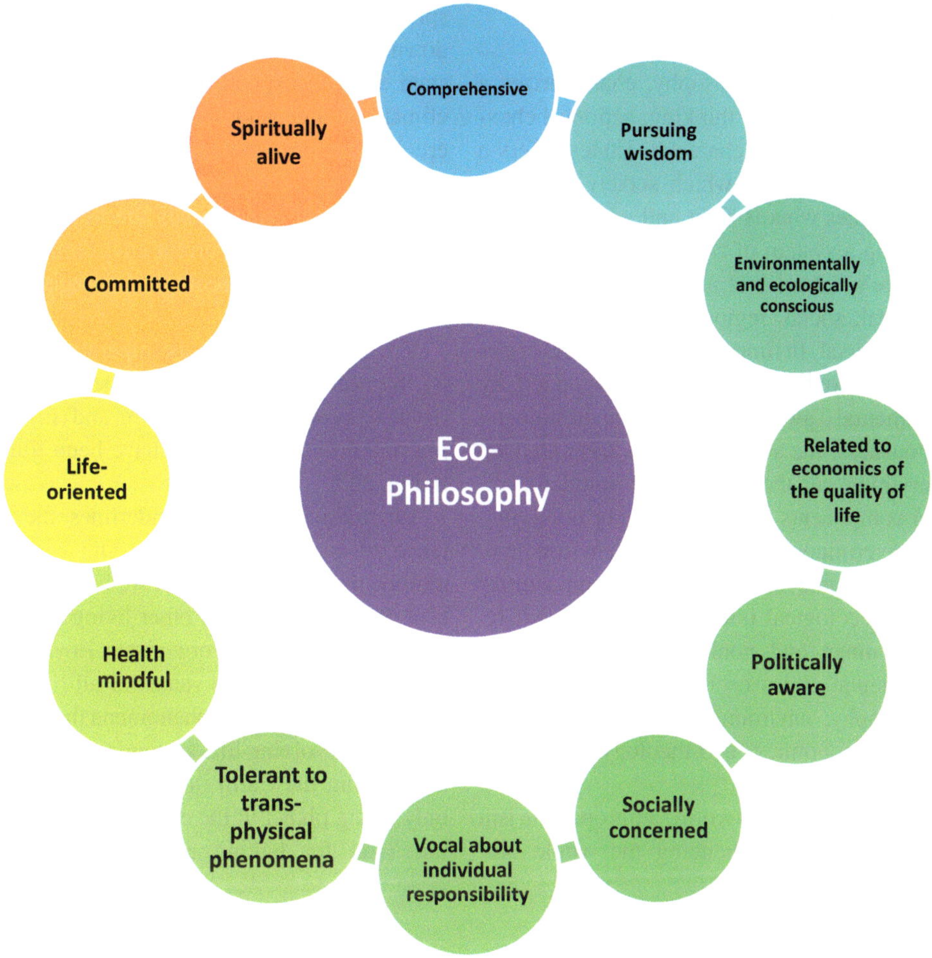

Fig. 22.3 Multiple attributes of eco-philosophy united in harmony

Fig. 22.4 Eco-philosophy contributing to the development of an ecologically balanced planet

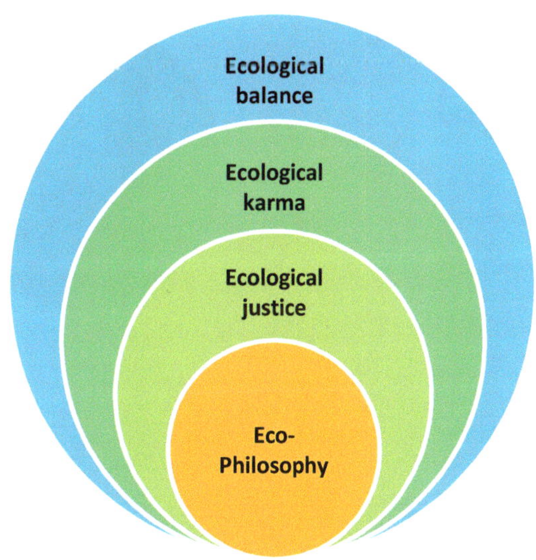

22.6 Environmental Ethics

Ethics, a branch of philosophy, characterizes a person based on his/her thinking, actions, behavior, moral values, and character. Ethics offers a value system based on which society perceives and determines what is right and what is wrong. It builds up a system of cultural values motivating people's behavior at all levels – individual, institutional, social, regional, international, and global/universal. In times of environmental crises necessitated to be appropriately responded by environmental awareness and environmental consciousness, the scope of ethics has extended to everything out there in the environment: all-natural components whether living or nonliving, all that determine the quality of the environment, all that have come through evolution and contribute to the ecological integrity of the biosphere, and all human and nonhuman forms of life. Environmental ethics exert a vital influence on the geography, environmental sociology, environmental economics, eco-theology, and environmental laws.

In our contemporary world, anthropocentrism is overwhelmingly influencing the whole biosphere, jeopardizing all ecosystems and breaking down the ecological integrity of the planet. This state of the biosphere is gradually precipitating climate change and species extinction. At this crucial juncture, ecological healing of the Earth emerges as an imperative for humanity. Environmental ethics holds the key to bringing the planet out of the current environmental mess and reestablish its ecological integrity (Singh 2017).

Environmental ethics is imperative to set out by the needs of (i) positive influence on nature, (ii) new knowledge about nature, and (iii) expanding moral concerns. These have been illustrated in Fig. 22.5.

Environmental ethics underlines the importance of eco-centric thinking and eco-friendly attitude that, in turn, suggest we live on the Earth as a part of life, like all other living species, not as a master of all other organisms. Environmental ethics instills in us the values of all life, all species, and all ecological phenomena that have contributed to evolving life, sustaining the biosphere, and turning the Earth into a living planet – the only living planet in the entire cosmos known so far. Vital for halting ecological disaster and

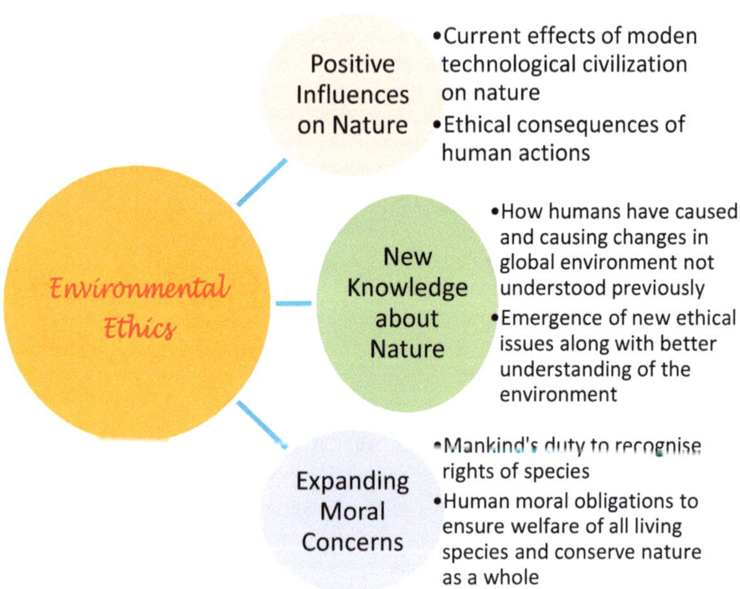

Fig. 22.5 Elements that give rise to the necessity for environmental ethics

restoring ecological balance, environmental ethics helps cultivate, adopt, and practice the values like the following:

1. The Earth belongs to all living beings, not just to the human species and the limited natural resources belong to all, not just meant for exploitation by human beings.
2. Socioeconomic development should keep pace with the ecological regeneration of natural resources; it should not be at the cost of the welfare of all living species.
3. A healthy, vibrant, and sustainable socioeconomic system rests on healthy, vibrant, and ecologically sustainable ecosystems; ecosystem degradation results in a collapsed economy.
4. The existence and well-being of human beings are dependent on the existence and well-being of other species.
5. We must respect overall biodiversity – the variety of ecosystems, species, and genotypes – and should not rely on monocultures in our food production systems.
6. We must hold a reverential attitude toward nature and nature's components – mountains, forests, rivers, lakes, trees, seeds, soils – so that we refrain from polluting our environment and conserve all natural resources the Earth has been flourishing with.
7. We must practice and pursue ecological philosophy, or eco-philosophy, reverberating with environmental ethics and pregnant with life-enhancing values.

22.7 Environmental Laws

An environmental law encompasses concerns about the protection of the environment. This involves regulations, statutes, and local, national, and international legislation and treaties. While designing a framework for the protection of the environment as a whole, or a part, or an aspect of it, an environmental law explains the legal consequences of environmental degradation/disaster toward individuals, and private, or government organizations.

The environmental laws cover pollution control, environmental protection principles, and natural resource conservation/sustainability. Each aspect of the environment covers various legal measures, as illustrated in Table 22.2. These are not strictly fixed norms. Other environmental aspects can be added as per the circumstances and needs of the hour. Further, all environmental laws do not prevail in all countries. Except for bilateral, trilateral, or international agreements/protocols/treaties, the environmental laws are largely country-specific.

International environmental law is a body of international law related to environmental protection often through bilateral or multilateral agreements. It rests on three sources, namely, (i) international treaties, (ii) customary international law, and (iii) judicial decisions of international courts.

The Centre for International Environmental Law (CIEL) has been using the power of law to protect the environment since 1989. It has its

Table 22.2 Comprehensive environmental regulations encompassing a range of environmental aspects

Pollution control	Environmental protection principles	Natural resource conservation/resource sustainability
• Air quality • Water quality • Safety from chemicals • Contamination cleanup • Waste management	• Precautionary principles • Prevention • Equity • Transparency • Sustainable development • Public participation • The polluter pays principle	• Water resources • Forest resources • Mineral resources • National parks, sanctuaries, and biosphere reserves • Wildlife • Species protection • Fish and game • Land use • Environmental impact assessment

headquarters in Washington DC and Geneva. CIEL has a team of attorneys, policy experts meant to extend legal counsel, policy research, advocacy, and capacity building in the fields such as environmental health, climate, energy, etc.

22.7.1 International Environmental Agreements

International environmental agreements, also popular as Environmental Protocols and Environmental Treaties, have proliferated in global scenarios in recent decades. Largely inspired by the 1972 Stockholm Conference on Human Environment, the 1987 Brundtland Commission Report *Our Common Future*, and the 1992 Earth Summit, several international agreements serve as a form of treaties binding in international environmental law with a specific purpose of attaining an environmental goal, often within a timeframe. International environmental treaties, protocols, or agreements are primarily developed by United Nations bodies/agencies often on the occasion of an international conference/summit. The agreements – the intergovernmental documents – can be between two countries (bilateral environmental agreement), three countries (trilateral environmental agreement), or more than three countries (multilateral environmental agreement), or it can be intended for all the countries of the world.

The first-ever international environment law came into being in 1857 as a result of a German agreement meant for regulating the water from Lake Constance between Switzerland and Austria. Kim (2013) counts as many as 747 multilateral environmental agreements witnessed by the world between 1857 and 2012. What international environmental agreements have flooded the human world, however, have evolved in the wake of global environmental awareness following the publication of Rachel Carson's eye-opening *Silent Spring* in 1962 and then a series of environmental conferences.

The international environmental agreements have spurred revolutionary changes in the laws affecting the global environment. These have implications for the national environmental laws in those countries which have ratified them. However, their importance is considered for the entire globe. Not all agreements pertain to all countries. Some of them involve a few countries or a region to attain particular goals.

Some of the notable international environmental conventions and treaties having bearing on the global environment are the following:

- United Nations Conference on Human Environment, Stockholm, Sweden, 5–16 June 1972
- United Nations Conference on New and Renewable Sources of Energy, Nairobi, Kenya, 10–21 August 1981
- Vienna Convention for the Protection of the Ozone Layer, Vienna, Austria, 22 March 1985
- The Montreal Protocol on Substances that Deplete Ozone, Montreal, Canada, 26 August 1987
- UN Convention on the Law of the Sea, Montego Bay, Jamaica, 10 December 1982
- UN Framework Convention on Climate Change (UNFCCC), New York, USA, 30 April to 9 May 1992
- Convention of Biological Diversity (CBD), Rio de Janeiro, Brazil, June 1992
- Convention on the Law of the Non-Navigational Uses of International Water Courses, New York, USA, 21 May 1997
- Kyoto Protocol, Kyoto, Japan, 11 December 1997
- Cartagena Protocol on Biosafety under CBD, 29 January 2000
- Stockholm Convention on Persistent Organic Pollutants, Stockholm, Sweden, 22 May 2001
- International Tropical Timber Agreement, 2006, Geneva, Switzerland, 27 January 2006
- Amendment to Annex-B of the Kyoto Protocol to the UNFCCC, Nairobi, Kenya, 17 November 2006
- Eleventh Conference of Parties (COP) to the UN Convention on Biological Diversity, Hyderabad, India, 1–19 October 2012
- Paris Agreement under the United Nations Framework Convention on Climate Change (Paris Climate Agreement or Paris Agreement), Paris, France, 12 December 2015

The Convention on Biological Diversity, the Montreal Protocol, the Kyoto Protocol, and the Paris Agreement are among the most talked about international treaties which have had a phenomenal impact on the global environment. A brief account of these treaties, which can be regarded as milestones in the pathways to climate change mitigation, is presented under the following subheads.

22.7.1.1 Convention on Biological Diversity

The Convention on Biological Diversity (CBD), often referred to as Biodiversity Convention, is an international legal instrument. Work on the treaty was concluded in Nairobi in May 1992 with the adoption of the Nairobi Final Act by the Nairobi Conference for the Adoption of the Agreed Text of the Convention of Biological Diversity. Then it was tabled for signatures by the parties at the Earth Summit in Rio de Janeiro on 5 June 1992. The treaty entered into force on 29 December 1993.

CBD encompasses all three forms of biological diversity: ecosystem, species, and genetic diversity. It has been ratified by 196 nations. This multilateral treaty has three main goals:

1. Conservation of biological diversity
2. Sustainable use of its components
3. Fair and equitable sharing of benefits arising from genetic resources

Biodiversity is the basis of sustainable development. The overall objective of CBD, thus, is to develop national strategies for the conservation and sustainable utilization of biodiversity. Human actions leading to sustainable development are an inherent spirit of the CBD. The CBD has two supplements:

1. **Cartagena Protocol on Biosafety to the Convention on Biological Diversity** was adopted on 29 June 2000 in Cartagena, Columbia, and entered into force on 11 September 2003. This is an international treaty concerning the governance of the movement of the living modified organisms (LMOs) developed using the tools of biodiversity across the nations.
2. **The Nagoya Protocol on Access to Genetic Resources and the Fair and Equitable Sharing of Benefits Arising from their Utilization to the Convention on Biological Diversity** was adopted on 29 October 2010 in Nagoya, Japan, and entered into force on 12 October 2014. This facilitates CBD's third objective, namely, fair and equitable sharing of the benefits to emanate from the sustainable utilization of genetic resources.

The Secretariat of CBD is located in Montreal, Canada. Its main function is to extend assistance to all the nations (parties) toward the implementation of CBD.

22.7.1.2 Montreal Protocol

The Montreal Protocol on Substances that Deplete Ozone – or simply, the Montreal Protocol – came into force to reduce ozone layer depletion by reducing the production of ozone-depleting substances. This international treaty was signed on 26 August 1987. The Vienna Convention for the Protection of the Ozone Layer in 1985 was the inspiring event to develop an international treaty aimed at the protection of the stratospheric ozone layer. Came into force on 26 August 1989, the Montreal Protocol has been revised nine times: London (1990), Nairobi (1991), Copenhagen (1992), Bangkok (1993), Vienna (1995), Montreal (1997), Australia (1998), Beijing (1999), and Kigali (2016).

The treaty takes into account ozone-depleting substances, especially halogenated compounds containing chlorine and bromine. N2O is not included in the list of compounds.

The Montreal Protocol has been ratified by 196 countries. This Treaty has the credit of being the first universally ratified Treaty. And so is the Vienna Convention. This has also earned the credit of being the most successful international agreement.

22.7.1.3 Kyoto Protocol

The Kyoto Protocol was adopted in Kyoto city of Japan on 11 December 1997 and brought into effect on 16 February 2005. In the Kyoto Protocol, the United Nations Framework Convention on Climate Change (UNFCCC) aimed at reducing atmospheric concentrations of six GHGs, namely, CO_2, CH_4, N_2O, HFCs, PFCs, and SF_6. The Protocol is based on the common-but-differentiated responsibilities arguing that since the developed countries have historically been more responsible for GHG emissions, they are more obliged to reduce current rates of GHG emissions.

There were originally 192 countries as the parties to the Protocol. The first phase of the commitment to reducing GHG emissions was from 2008 to 2012 in which all the 36 developed countries participating in this phase reduced carbon emissions, but the world, on the whole, registered an emission increase of 32% from 1990 to 2010. The second phase of the commitment, the Doha Amendment to Kyoto Protocol, began in 2012 with 37 countries (34 of them ratifying it) with binding targets of GHG emission reductions. Despite the withdrawal of many countries from the second phase of commitment, 147 countries have accepted the Doha Agreement that came into force on 31 December 2020.

22.7.1.4 Paris Agreement

The Paris Climate Agreement, a legally binding international treaty, was evolved at UNFCCC's COP 21 on 12 December 2015 and entered into force on 4 November 2016. It has been adopted by 197 countries. The main aims of the Paris Agreement are the following:

1. Hold the rise of global temperature to well below 2 °C, preferably 1.5 °C, compared to the temperature of the pre-industrial age.
2. Increase the capabilities to adapt to the adverse impact of climate change and foster climate resilience and low carbon-based development without affecting food production processes.
3. Ensure flows of funds in a pathway toward low GHG emissions and climate-resilient development.

The countries of the world are expected to reach the peak of carbon emissions as soon as possible to attain a carbon-neutral world by the mid-century. Implementation of the treaty requires socioeconomic transformation based on the "best available science."

The function of the Paris Agreement is designed based on a five-year cycle for which the countries are required to submit their plans for tackling climate change by reducing GHG emissions. This plan of individual countries is known as Nationally Determined Contributions (NDCs). The NDCs from all parties have been submitted to work toward the realization of the Paris Agreement goals. The countries of the world are also working on their own long-term low greenhouse gas emission development strategies – LT-LEDS. The LT-LEDS reflects a country's long-term plan, development priorities, and vision. However, unlike NDCs, LT-LEDS is not mandatory for a country to submit.

22.8 Summary

Climate change is gradually intensifying as indicated by unabatedly increasing atmospheric CO_2 levels. Our ultimate goal has to be to bring down the current atmospheric CO_2 concentration to a level that could help the planet's climate system to a state of normalcy. To deal with climate change, our strategy focus should be on preparedness, adaptation, and mitigation. Carbon sequestration mechanisms involve the removal of atmospheric CO_2 to be captured, secured, stored elsewhere, and/or used for some specific purpose. This can be carried out through geological carbon sequestration, carbon capture, utilization and storage, and biological carbon sequestration (photosynthesis). We need to bring emission-reducing changes in our lifestyles, which would be phenomenal in curbing global warming.

Replacement of conventional energy sources (fossil fuels) with alternative clean energy sources (solar energy, wind energy, geothermal energy, hydroelectric energy, tidal energy, biomass energy, and hydrogen energy) offers one of the key solutions to climate change without compromise with an appropriate pace of socioeconomic development. Clean energy is the energy that is generated by using methods that do not emit GHGs and other pollutants. In 1974, Professor Henryk Skolimowski created a new philosophy known as eco-philosophy. Eco-philosophy is a meaningful fusion of ecology and philosophy that instills in the human mind the values like ecological ethics, ecological justice, and ecological karma and develops human love for nature and its conservation and paves the way to the ecological renaissance of the planet. Environmental ethics exert vital influence on the geography, environmental sociology, environmental economics, eco-theology, and environmental laws. The environmental laws cover pollution control, environmental protection principles, and natural resource conservation/sustainability. International environmental law is a body of international law related to environmental protection often through bilateral or multilateral agreements. It rests on three sources, namely, (i) international treaties, (ii) customary international law, and (iii) judicial decisions of international courts.

International environmental agreements, also popular as Environmental Protocols and Environmental Treaties, have proliferated in global scenarios in recent decades. Some of the notable international environmental conventions and treaties phenomenal for global environment are United Nations Conference on Human Environment, Stockholm (1972); United Nations Conference on New and Renewable Sources of Energy, Nairobi (1981); Vienna Convention for the Protection of the Ozone Layer, Vienna (1985); the Montreal Protocol on Substances that Deplete Ozone, Montreal (1987); UN Convention on the Law of the Sea, Montego Bay (1982); UN Framework Convention on Climate Change (UNFCCC), New York (1992); Convention of Biological Diversity (CBD), Rio de Janeiro (1992); Convention on the Law of the Non-Navigational Uses of International Water Courses, New York (1997); Kyoto Protocol, Kyoto (1997); Cartagena Protocol on Biosafety under CBD (2000); Stockholm Convention on Persistent Organic Pollutants, Stockholm (2001); International Tropical Timber Agreement, 2006, Geneva (2006); Amendment to Annex-B of the Kyoto Protocol to the UNFCCC, Nairobi (2006); Eleventh Conference of Parties (COP) to the UN Convention on Biological Diversity, Hyderabad, India (2012); Paris Agreement under UNFCCC (Paris Climate Agreement or Paris Agreement), Paris (2015); etc.

22.9 Exercises

22.9.1 Multiple-Choice Questions

1. Climate change mitigation involves
   (a) Carbon sequestration
   (b) Minimization of carbon emissions
   (c) Alternative energy sources
   (d) All of the above
2. Excessive plowing of the soil results in
   (a) Increased carbon emissions
   (b) Increased soil moisture loss
   (c) Increased moisture loss as well as increased carbon emissions
   (d) None of the above
3. What is the full form of CCUS?
   (a) Climate Change in the USA
   (b) Carbon Capture, Utilization, and Storage
   (c) Carbon Capture in the USA
   (d) Climate Change through Undisturbed Soil
4. Which country captures most of the carbon through CCUS technology?
   (a) USA
   (b) Russia
   (c) Canada
   (d) France
5. How much percentage of the CO_2 released by human activities is absorbed by the seas?
   (a) 10%
   (b) 15%
   (c) 25%
   (d) 80%.
6. Carbon in the soil organic matter (SOM) is stored in the soils for decades, but carbon in the inorganic carbonates can stay for as many as
   (a) 1000 years
   (b) 2000 years
   (c) 7000 years
   (d) 70,000 years

7. Which of the following statements is wrong?
   (a) Cold waters have more capacity to absorb CO_2 than warmer ones.
   (b) Hot waters have more capacity to absorb CO_2 than cold waters.
   (c) The polar regions are more efficient carbon sinks than the warmer regions of the Earth.
   (d) When CO_2 is released from the oceans into the atmosphere, it creates a positive atmospheric flux.
8. When the CO_2 is released from the oceans into the atmosphere, it creates
   (a) A positive atmospheric flux.
   (b) A negative atmospheric flux.
   (c) Either a negative or a positive atmospheric flux.
   (d) A flux depending upon the number of living animals in the oceans.
9. A negative atmospheric CO_2 flux means that
   (a) Oceans are releasing CO_2.
   (b) Oceans are absorbing CO_2.
   (c) Atmospheric temperature is increasing.
   (d) Ocean temperature is increasing.
10. Which of the following countries imparts the largest carbon footprint?
   (a) Russia
   (b) The Netherlands
   (c) USA
   (d) Brazil
11. Which food habit of mankind is most appropriate to assist the climate change mitigation processes?
   (a) Vegetarian
   (b) Meat eating
   (c) Fish eating
   (d) Omnivory
12. Which of the following countries is said to be a carbon-negative one?
   (a) India
   (b) Bhutan
   (c) Nepal
   (d) Myanmar
13. The energy generated by using methods that do not emit GHGs and other pollutants is
   (a) Clean energy
   (b) Clean energy
   (c) Exhaustible energy
   (d) Biomass energy
14. Who did create eco-philosophy in 1974?
   (a) Henryk Skolimowski
   (b) James Lovelock
   (c) E.P. Odum
   (d) Arne Naess
15. "The World is a Sanctuary" is the basic tenet in the
   (a) Deep ecology
   (b) Eco-philosophy
   (c) Gaia theory
   (d) Theory of fertilizing the universe
16. Environmental ethics exerts a vital influence on
   (a) Geography
   (b) Environmental sociology
   (c) Environmental economics
   (d) All of the above
17. Where are the offices of the Centre for International Environmental Law (CIEL) located?
   (a) New Delhi, India
   (b) Amsterdam, Netherlands
   (c) Washington DC, USA, and Geneva, Switzerland
   (d) Nairobi, Kenya
18. The 1972 Stockholm Conference focused on
   (a) New and renewable energy
   (b) Acid rain
   (c) Human environment
   (d) Ozone layer depletion
19. Cartagena Protocol and Nagoya Protocol are the supplements to the
   (a) Kyoto Protocol
   (b) Montreal Protocol
   (c) Paris Agreement
   (d) Convention on Biological Diversity (CBD)
20. Which international treaty is the Doha Amendment associated with?
   (a) Montreal Protocol
   (b) Paris Climate Agreement
   (c) Convention on Biological Diversity (CBD)
   (d) Kyoto Protocol

Answers: 1-d, 2-c, 3-b, 4-a, 5-c, 6-d, 7-b, 8-a, 9-b, 10-c, 11-a, 12-b, 13-a, 14-a, 15-b, 16-d, 17-c, 18-c, 19-d, 20-d

22.9.2 Short-Answer Questions

1. Name the three strategies needed to deal with climate change.
2. What do you mean by carbon sequestration?
3. What food habits would be helpful to climate change mitigation?
4. What is clean energy?
5. Mention the four tenets of eco-philosophy.
6. "The world is a sanctuary." What do you mean by this tenet of eco-philosophy?
7. What do you mean by international environmental law?
8. Mention the names of any four international treaties.
9. Where is the Secretariat of the Convention on Biological Diversity (CBD) located?
10. Name any four Environmental Protection Acts of India.

22.9.3 Long-Answer Questions

1. Write an account of the carbon capture, utilization, and storage (CCUS) techniques.
2. What are your views on reducing greenhouse gas emissions?
3. How can alternate energy sources be vital in dealing with climate change? Suggest appropriate examples.
4. Describe the main features of the Paris Climate Agreement.
5. Write short notes on the following:
   (a) Biological carbon sequestration
   (b) Eco-philosophy
   (c) Environmental ethics

References

Grandpierre A (2018) The fundamental biological activity of the universe. In: Smith W, Smith J, Verducci D (eds) Eco-phenomenology: life, human life, post-human life in the harmony of the cosmos. Analecta Husserliana (the yearbook of phenomenological research), vol CXXI. Springer, Cham. https://doi.org/10.1007/978-3-319-77516-6_10

IEA (2020) CCUS in clean energy transitions. IEA, Paris. https://www.iea.org/reports/ccus-in-clean-energy-transitions

Kim RE (2013) The emergent network structure of the multilateral environmental agreement system. Glob Environ Chang 23(5):980–991. https://doi.org/10.1016/j.gloenvcha.2013.07.006

Małecka A, Stark K (2018) Henryk Skolimowski's eco-philosophy as a project of living philosophy. In: Smith W, Smith J, Verducci D (eds) Eco-phenomenology: life, human life, post-human life in the harmony of the cosmos. Analecta Husserliana (the yearbook of phenomenological research), vol CXXI. Springer, Cham. https://doi.org/10.1007/978-3-319-77516-6_27

Singh V (2017) Environmental ethics and ecological integrity. The Times of India, July 25, 2017

Singh V (2019) Fertilizing the universe: a new chapter of unfolding evolution. Cambridge Scholars Publishing, London. 285 pp

Singh V (2020) Environmental plant physiology: botanical strategies for a climate smart planet. Taylor and Francis (CRC Press), Boca Raton. 216 pp

Singh V, Shiva V, Bhatt VK (2014) Agroecology: principles and operationalisation of sustainable mountain agriculture. Navdanya, New Delhi. 64p+viii pp

Skolimowski H (1981) Eco-philosophy: designing new tactics for living. Marion Boyars, London

Skolimowski H (1992) Living philosophy: eco-philosophy as a tree of life. Penguin/Arkana, London/New York/Ringwood/Toronto/Auckland

Skolimowski H (2000) Philosophy for a new civilization. Gyan Publications, New Delhi. 345 pp

Environment, Development, and Sustainability

23

Environment, development, and sustainability are the three phenomenal terms that pervaded the latter half of the twentieth century and, because of their vital interconnectedness for the welfare of the world, are likely to prevail in all the times to come. The environment is the crux of the living planet. The term "development" basically denotes socioeconomic welfare, which, to sustain, depends on the state of the environment. The term "sustainability" is rooted in the environment and implies the state of the things flowing through "development" and how long in the future the development attributes would sustain without adverse impact on the environment.

23.1 Global Environmental Challenges

The rise in greenhouse gas emissions is projected to occur all over the world, especially in countries registering steady growth rates. As a consequence, climate change indicators are emerging with greater magnitude and intensity. The fear of the breach of holding global temperature rise well below 2 °C is stalked in the minds of climate scientists and activists. The countries most vulnerable to climate change are those which overwhelmingly depend on climate-sensitive sectors, such as agriculture and fisheries. Their adaptive capacity is often low, especially because of the vulnerable livelihoods derived from biotic resources prone to climate change. Persistent poverty prevailing in some of the countries adds to climate miseries. There are certain links between climate change, poverty, and food security risks.

Nature's biodiversity is continuously depleting. The rate of species extinction is estimated to have increased 1000-fold over the natural rate, which is one of the most serious threats to the life of the planet. Biodiversity loss is resulting in the suppression of ecosystem functioning and is being increasingly realized in terms of curtailment in ecosystem services.

Substantial anthropogenic impact on the natural ecosystem can be realized from the fact that a considerable proportion of the net primary productivity has been converted through altered land uses, such as for agricultural production and animal production. Again, a pretty large land area modified as cities, towns, airports, railway stations, roads, industries, etc. has been reduced to the level of zero primary productivity. Modern agriculture practices have devoured most of the traditional crop genetic resources. The narrow genetic base in the world's agriculture creates an extremely vulnerable situation. The overall biodiversity loss translating into suppressed ecosystem functioning especially affects the communities directly relying on ecosystem services.

Global resource extraction rates have been increasing over the years, and they have been

V. Singh, *Textbook of Environment and Ecology*, https://doi.org/10.1007/978-981-99-8846-4_23

outweighing resource regeneration rates as well as gains in resource efficiency. For example, in 1980, the extraction rates from ecosystems and mines were 40 billion tons which increased to 58 billion tons in 2005. Over this period, world economic output (GDP) registered an increase of 110% against the resource extraction rates of 45% (EEA 2019). This difference reveals a clear-cut decoupling of global resource extraction and economic growth.

The ever-intensifying pressure of economic development coupled with economic growth competition among nations has driven the world to a state where it would need 1.7 Earths to support it to meet its current demands of renewable natural resources, according to Global Footprint Network (GFN). The term "economic development" in the era of liberalization, privatization, and globalization (LPG) haplessly implies using more and more resources and increasing more and more carbon emissions.

Progress in the area of environment-related health is generally expected with the advances in medical sciences. However, in 2004, as many as 8.9% of deaths were recorded in the world due to five types of risks, namely, (i) unsafe water, sanitation, and hygiene (3.2%), (ii) indoor smoke from solid fuels (3.3%), (iii) urban outdoor pollution (2.0%), (iv) lead exposure (0.2%), and climate change (0.2%) (Fig. 23.1). Apart from these deaths, 8.5% people fell victim to DALYs (disability-adjusted life years) in the same year (Fig. 23.2).

Resource overexploitation, biodiversity loss, soil degradation, environmental disruptions, and

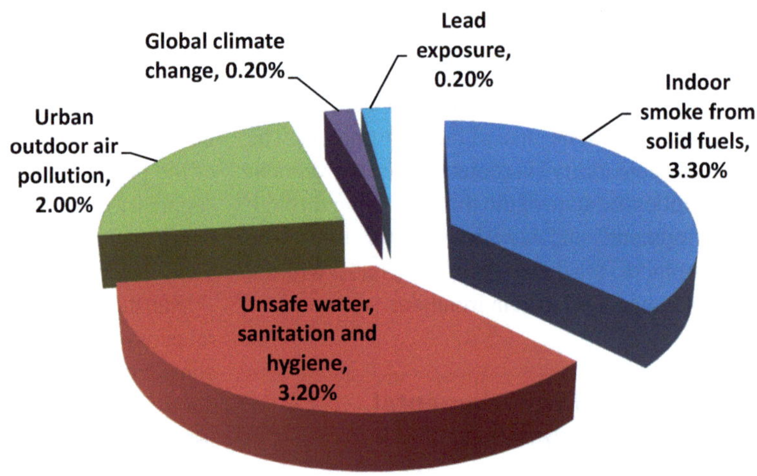

Fig. 23.1 The percentage of deaths attributable to five environmental risks, 2004. (Source: EEA 2019)

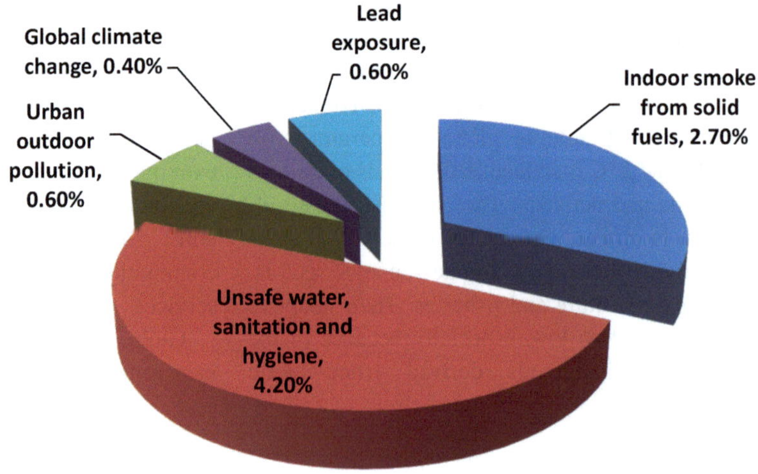

Fig. 23.2 The percentage of disability-adjusted life years (DALYs) attributable to five environmental risks, 2004 (EEA 2019)

climate change are all intertwined factors, each factor triggering all other factors. Environmental challenges are all set to multiply risks to food, water, energy, and health security at a global scale. The gravest ever and continuously intensifying environmental crises also seem to hide the Sixth Mass Extinction in their folds.

23.2 What Is Sustainable Development?

Economic growth costs natural resources and deteriorates the environment. Global warming gradually precipitating into climate change is also an outcome of the conventional economic development model. Such a development is, in fact, unsustainable. Unsustainable development leads to natural resource degradation, environmental disruption, sociocultural inequity, and a gloomy future.

Sustainable development, on the other hand, is the one that ensures natural resource conservation, environmental safety, sociocultural equity, and a happy future. The term "sustainable development" was first of all used in 1980 in the World Conservation Strategy (IUCN-UNEP-WWF 1980). However, the concept and paramount need for sustainable development, first of all, appeared in the wake of the 1987 release of *Our Common Future*, a highly celebrated report of the World Commission on Environment and Development (WCED), popularly called the Brundtland Commission Report in respect of the WCED Chair, Gro Harlem Brundtland.

Today's well-recognized definition of sustainable development has been derived from the Brundtland Report: *Development that meets the needs of the present without compromising the ability of future generations to meet their own needs.*

Sustainable development integrates the environment, society, culture, and economy. In other words, the well-being of the environment, society, culture, and economy on a sustained basis are the major concerns in the strategies aimed at sustainable development.

23.2.1 Sustainability

Sustainability, although often used as synonymous with sustainable development, is somewhat different from the latter. While sustainable development refers to many processes and pathways to attain intended goals (e.g., sustainable agriculture, sustainable forestry, sustainable economy, sustainable society, sustainable lifestyles, sustainable production, sustainable consumption, etc.), sustainability generally refers to the human ability to live constantly in tune with the biosphere's capacities.

The whole planetary life is bound by a common evolutionary phenomenon. Human existence is not separable from the existence of all other living species. Therefore, human well-being on Earth emanates from that of the entire biodiversity prospering in the biosphere. Sustainability, in essence, is an ecological phenomenon. The sustainability of the human race, of human society, of a national economy, or the world is rooted in ecological sustainability. As Singh (2019) puts it: "There can be no sustainability without ecological sustainability of its own."

Sustainability has now assumed the status of science: science of sustainability. It is relatively a new discipline of science that helps us understand nature's complexity, biosphere's capacities, species' interconnectedness within an ecosystem, human development interventions, and ecological, environmental, social, cultural, and economic issues and provides us opportunities for evolving strategies vital for fulfilling all needs of the present and future generations while ensuring ecological balance and maintenance of biosphere's capacities. Sustainability is also an aspect of the living philosophy, especially of eco-philosophy, that guides us to design, develop, and evolve lifestyles fostering ecological affluence for the well-being of mankind together with the well-being of all living forms.

23.2.2 Sustainable Future

A sustainable future is difficult to define, as no one knows what kind of future is there in the stock of time. The future is about time, not about the human race. The future is to be out there irrespective of human will. The question is: future with or without humanity? When we attempt to define a sustainable future, it must essentially involve the human element as an integral part of it. A sustainable future is – and it has to be – from a human perspective. As Singh (2019) defines, *a sustainable future would be witnessed by a healthy, happy, and vibrant humanity*.

What would, then be a likely scenario of a sustainable future? It would be a scenario filled with all hopes for a sustainable life: ecological affluence (ecological balance, blossoming biodiversity, healthy environment, and healthy resource base) and healthy, happy, and vibrant humanity amidst it. When the human species is at the helm of all affairs on the living planet, we can infer that a sustainable future is not to be on its own; it has to be created. And humanity has to be an architect of a sustainable future.

23.2.3 The World Commission of Environment and Development (WCED)

The United Nations General Assembly, in its resolution 38/161 of 19 December 1983, "Process of Preparation of the Environmental Perspective to the Year 2000 and Beyond" among other things, welcomed the establishment of a special commission which later on was known as World Council on Environment and Development (WCED). The purpose of WCED was to prepare a report on environment and the global problems.

In December 1983, the then UN Secretary Javier Perez de Cuellar invited Gro Harlem Brundtland to chair the WCED. Brundtland was the first female and the youngest ever three-time Prime Minister of Norway (1981, 1986–1989, and 1990–1996). She has also held the prestigious position of Director General of the World Health Organization (1998–2003). After the appointment of Gro Harlem Brundtland as the Chairperson, WCED became popular as the Brundtland Commission.

23.2.4 Our Common Future

After working for four years, WCED was officially dissolved in December 1987 after the release of its report, popular as the Brundtland Commission Report, titled *Our Common Future* in October 1987. The *Our Common Future* soon became a highly acclaimed report and one of the landmark documents spelling out the new and mind-igniting concept of Sustainable Development.

The *Our Common Future* underlined the dire fact that many crises facing the planet are interlocking crises that are elements of a single crisis of the whole and placed environmental issues firmly on the global political agenda. The targets of the Brundtland Commission Report were finding out pathways to sustainable development through multilateralism and interdependence of nations. Some of the immediate outcomes on account of the groundwork contained in the *Our Common Future* were the organization of the 1992 Earth Summit, the adoption of Agenda 21, the Rio Declaration, and the constitution of the Commission for Sustainable Development.

23.2.5 Education on Sustainability

Education on sustainability or sustainable development is increasingly becoming popular at higher education levels all over the world. Many universities have adopted sustainability in their curricula. Sustainability education is also an integral part of many global frameworks and conventions. Article 6 of UNFCCC for climate change mitigation, Article 13 of the Convention of Biological Diversity (CBD), Sendai Framework for Disaster Risk Reduction, etc. are some of the crucial examples.

23.3 Environment-Development-Sustainability Linkages

The environment of the planet is the basis of survival, material progress, and sustainability of the human race. The environment includes both abiotic and biotic components. When we think of the abiotic or physical dimension of the environment, our emphasis is on the quality of this component. There is a definite composition of all the components of the physical environment – for example, of the atmosphere, hydrosphere, and lithosphere – in which all living organisms survive, nurture, and sustain themselves. Substantial alteration in the original composition of the environment leads to its quality deterioration. The biotic component includes all the varieties of plants, animals, and microorganisms that can thrive and which all are of vital value for human development.

Development is a process enacted through human actions, often interaction with environmental resources, aimed at creating positive change or adding up physical, environmental, social, cultural, and economic elements. Development is an inherent need of people to survive, sustain, and use their inherent and evolved capabilities to reach and enjoy an ever higher level of their living status.

Changes for betterment in lifestyles through development are not bound to a periphery. Changes are and have to be continuous. Development is dynamic and it has to be successional, always destined to protect, regenerate, conserve, and enrich natural resources. Such kind of ecological-conscious and conservation-oriented development is a sustainable development ensuring a happy and sustainable future. If the development processes are haphazard, unwise, and driven by greed rather than by need, the destruction of resources and environmental pollution and many other sociocultural tensions would be a natural outcome, and such a resource-eclipsing development would be an unsustainable one. An unsustainable development would not help us carry on for a long time.

23.4 Ecological Sustainability

There can be many levels of sustainability, ranging from a population of a species to the whole biosphere (e.g., sustainability of the elephants in tropical wildlife, of a forest ecosystem, or the living planet). There can be many forms and dimensions of sustainability (e.g., agricultural sustainability, economic sustainability, sustainability of future generations, etc.). Ecological sustainability is the basis of all levels, all kinds, and all dimensions of sustainability. Sustainability, in essence, is rooted in ecological sustainability. In other words, sustainability emanates from ecological sustainability, or there can be no sustainability without ecological sustainability of its own.

Sustainability is not something static; it is a dynamic process. It denotes the transformation of a simple or less complex, vulnerable, and unsustainable system into a complex, resilient, and sustainable system. Biodiversity is the basis of sustainability. The higher the measure of biodiversity, the higher the level of sustainability. Green vegetation constituting photosynthetic autotrophs, or the producers, supporting the populations of all organisms in an ecosystem – herbivores and carnivores – is the basis of biodiversity in a terrestrial ecosystem. The phenomenon that nurtures the basis of biodiversity – that is, the core of sustainability – is photosynthesis. Thus, photosynthesis is the basis of sustainability. Complementary to the green plants performing photosynthesis are green algae and cyanobacteria which also add to the efficiency of photosynthesis in the soil and the aquatic ecosystems. Chemosynthesis is another mechanism operating by bacteria in the soil that assists the phenomena of biodiversity and subsequent sustainability. Nitrogen fixation in soil and aquatic ecosystems performed by nitrogen-fixing bacteria is a natural way of nourishing plants with nitrate ions necessary for plants to synthesize their proteins, the building blocks in the structure of all organisms. Nitrogen fixation, thus, also helps enhance biodiversity and ecological sustainability. Net photosynthesis, chemosynthesis, and nitrogen fixation lead to enhanced biodiversity and strike an ecological balance, ecological integrity, resilience, and ecological sustainability (Fig. 23.3).

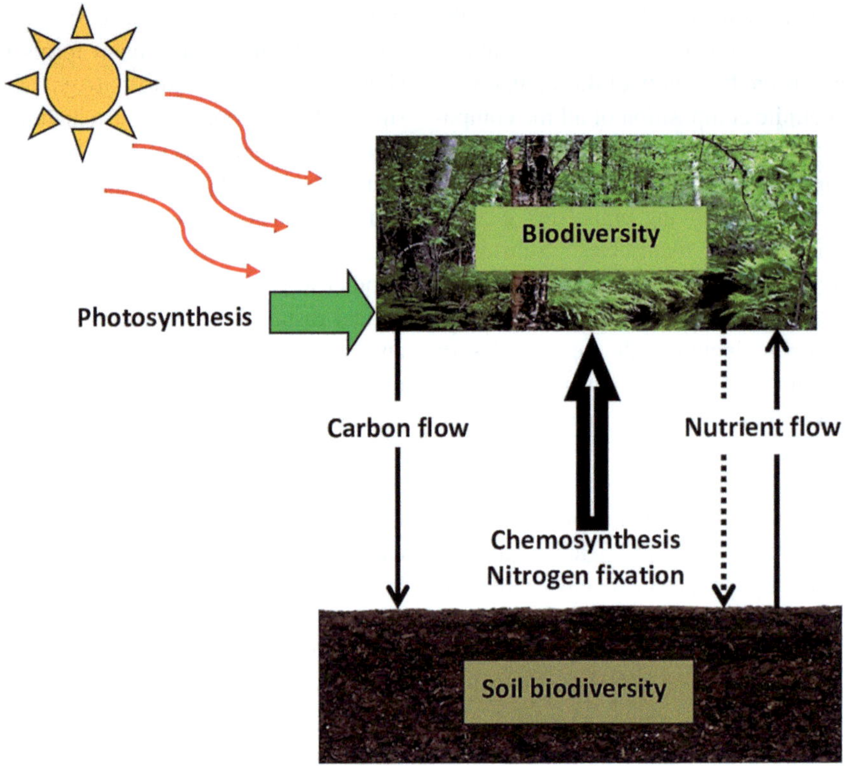

Fig. 23.3 The ecological resilience and sustainability of an ecosystem: biodiversity boosted through photosynthesis, chemosynthesis, and nitrogen fixation

23.5 Strategies for Sustainable Development

As discussed above, a sustainable development strategy builds upon a healthy and vibrant ecological foundation. This is the primacy of the strategy aimed at attaining sustainability. Sustainability is to be attained by employing human ecological activities. Therefore, the first and foremost need for a sustainable world is a sustainable society.

23.5.1 Sustainable Society

A sustainable society is a building block of a sustainable world, which has to be an essential goal to attain. A sustainable society is an architect of sustainable development. Essential features of a sustainable society include its eco-centric approach to every walk of life. This society adopts, pursues, and implements vital principles emanating from a living philosophy founded on ecological philosophy nurturing ecological consciousness, ecological ethics, ecological justice, ecological karma (actions), and ecological vision.

A sustainable society is invariably a socially just society. Social justice helps peace and happiness prevails within a society. Such a society depends on a viable economy, rather than on an exploitative economy.

A sustainable society trusts in intergenerational and intragenerational equity. The former emphasizes minimizing any adverse impact on natural resources and the environment for future generations, while the latter suggests minimizing the wealth gaps within and between the nations during a development process.

A sustainable society depends upon the carrying capacity of a system, regenerative production systems, minimum waste production, waste recycling, waste utilization, clean energy resources, environmental health, public hygiene, value education, gender equity, and on creative lifestyles. Such a society is driven is an ecologically conscious society committed to the conservation and enhancement of natural resources.

23.5.2 Sustainable Agriculture

Foods produced through agricultural practices are not only necessary for survival but also for physical, intellectual, psychological, and emotional development and sustenance. A sustainable future for humanity evolves through sustainable agriculture.

Modern agriculture being largely practiced all over the world is not sustainable as it does not fulfill ecological rules. Sustainable agriculture evolves on an ecological axis. There is an agriculture-ecology axis on which a sustainable food production system is developed. The ecological roots of sustainable agriculture lie in an ecosystem, a living system empowered to transform light into life – through photosynthesis.

There are four traits of sustainable agriculture. It has to be (i) ecologically sound, (ii) regenerative, (iii) economically viable, and (iv) socially just. Among these, the first two, ecologically sound and regenerative ones, are ecological attributes, while the latter two, economically viable and socially just ones, are the social attributes.

The traits of sustainable agriculture mentioned above embrace certain indicators, some of them invisible (Fig. 23.4). Operationalization of sustainability can be possible based on these sustainability indicators.

A sustainable agricultural system has many components in linkages with each other and integrated into a self-regulating unit of nature providing essential products of vital human needs: foods, fodder, fiber, fertilizer, etc. (Singh 2020). The uncultivated land covered with forests, rangelands, grazing lands, etc. is the key component of a farming system. This vegetated area is a natural storehouse of nutrients; helps to conserve soil, water, and biodiversity; traps atmospheric CO_2 and functions as a carbon sink; contributes to maintaining an appropriate microclimate; and serves to enhance the adaptation capabilities of an agroecosystem.

An agroecosystem is an arbitrarily defined unit of nature with constant energy flows and chemical cycling, involving uncultivated land areas (forests, grasslands/grazing lands), cultivated land, and livestock in organic linkages with each other, woven into a complex unitary whole, functionally oriented to produce foods and other life-supporting products – such as feed, fiber, fuel, fertilizers, etc. – and provide vital ecological functions. Agroecosystem crosses over or overlaps diverse ecosystems designed and managed by a farming community for the production of fundamental life-sustaining products (Singh 2020). The other natural attributes of an agro-ecosystem approach are the following:

- A unification of different solar-powered ecosystems into a single functional unit.
- Protecting and conserving some areas for thriving natural biodiversity for. ecological balance necessary for the ecological integrity of agricultural systems
- Self-containment of a farming system: essential inputs are produced within the system.
- Shifting partial pressure for food, feed, and other life-support products in uncultivated areas.
- Increasing the level of biodiversity to have bearing on the whole ecosystem.
- Diversification of economic activities.
- Resource conservation way of development and succession.
- Increasing resilience and sustainability.
- Increasing sociocultural cohesion through common property resource management in a community-based farming system, especially in some areas, such as in India, practicing traditional agriculture.
- Giving a base to food and livelihood security and sovereignty.

In the wake of climate change, we need to impart an additional dimension of sustainability to agriculture. A sustainable agricultural system should have a high degree of adaptation capabil-

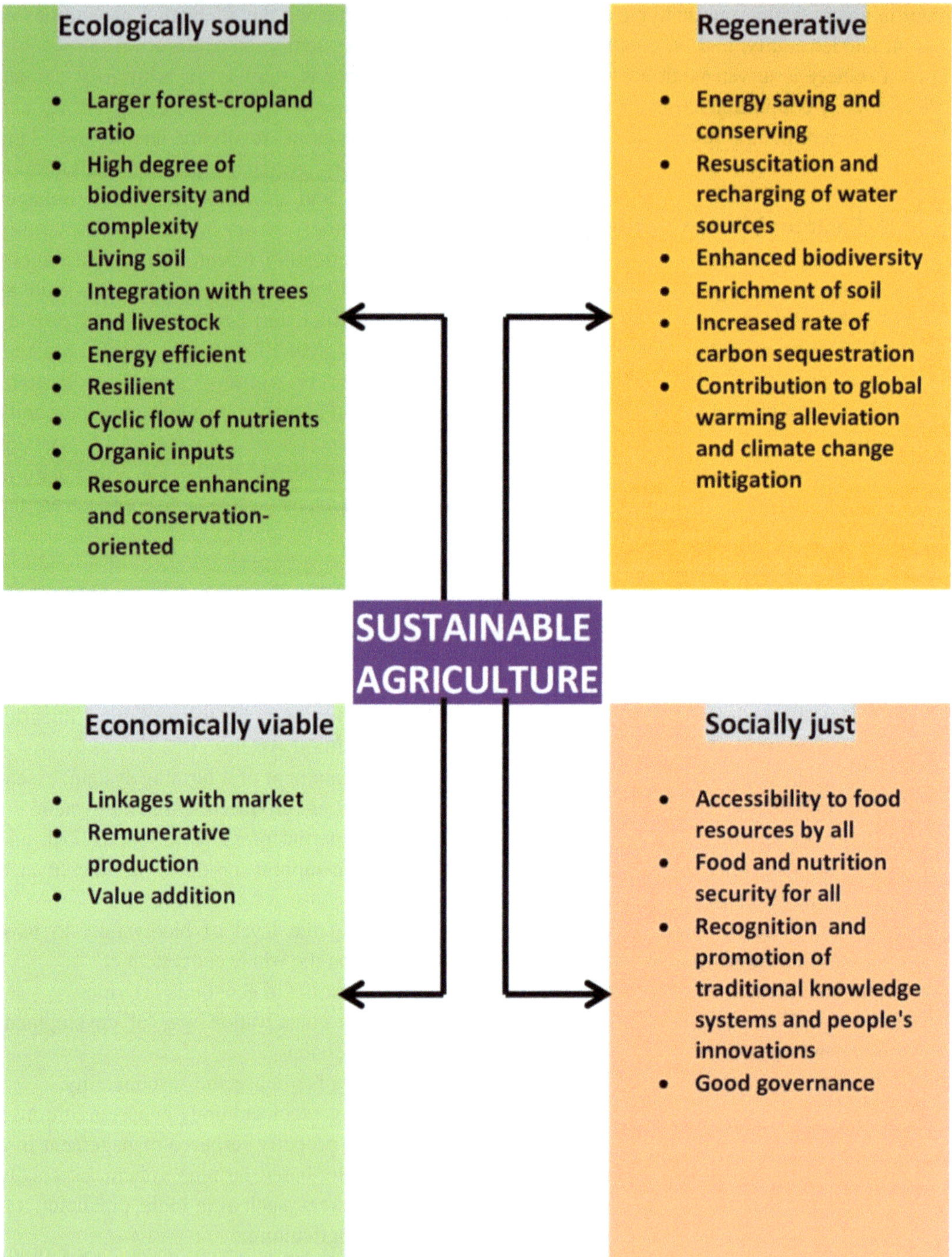

Fig. 23.4 Indicators of sustainability for characteristics of sustainable agriculture

ity against climate change. The adaptation mechanisms can be operationalized through the management of available natural resources in an agro-ecosystem. These resources are typically found in traditional agricultural systems in some of the areas, such as in the mountain areas.

23.6 Global Policies on Sustainable Development

The concept of Sustainable Development introduced in the WCED Report *Our Common Future* has been articulated into various global, regional, and national policies and programs. The 2030 Agenda for Sustainable Development, which was adopted by all the member states of the United Nations, has been an important blueprint for initiating programs vital for global development and peace. The call for action by all participating member nations came to the fore as the 17 Sustainable Development Goals (SDGs) intended to reach for. The major agendas of the SDGs include improvement in health and education and poverty and inequality reduction, spurring economic growth and at the same time preserving forests and oceans and tackling climate change.

Also known as Global Goals, the 17 SDGs are integrated, which means that actions taken to achieve one goal would have bearing on other goals. Developmental processes must strike a balance between social, economic, and environmental sustainability. Reaching the SDGs goals requires a partnership of the national governments, the private sector, civil society, and citizens.

Inspired by the outcome of the WCED, a series of international policy decisions/agreements emerged that became key documents of international policies. Let us have a look into historical developments in this context:

- 1992: Adoption of Agenda 21 at Earth Summit in Rio de Janeiro regarding the comprehensive plan of action for sustainable development to improve human lives and protect the environment
- 2000: Millennium Declaration at the Millennium Summit in New York spelling out eight Millennium Development Goals (MDGs)
- 2002: Johannesburg Declaration on Sustainable Development and the Plan of Implementation adopted at the World Summit on Sustainable Development in South Africa
- 2012: Adoption of "The Future We Want" to launch a process to develop a set of SDGs to build upon MDGs and to establish the UN High-Level Political Forum on Sustainable Development at the United Nations Conference on Sustainable Development (Rio + 20)
- 2013: Setting up of a 30-member Open Working Group to develop proposals on SDGs by UN General Assembly
- 2015: Adoption of the 2030 Agenda for Sustainable Development with 17 SDGs at the UN Sustainable Development Summit
- 2015: Sendai Framework for Disaster Risk Reduction
- 2015: Addis Ababa Action on Financing for Development
- 2015: Paris Agreement on Climate Change

The High-level Political Forum for Sustainable Development, a central UN Platform, serves to follow up and review the SDG processes. The Division for Sustainable Development Goals (DSDGs) at the UN Department of Economic and Social Affairs (UNDESA) extends support, including capacity building, towards the processes leading to the achievement of SDGs.

23.7 Summary

The environment is the crux of the living planet. The term "development" basically denotes socio-economic welfare, which, to sustain, depends on the state of the environment. The term "sustainability" is rooted in the environment and implies the state of the things flowing through "development" and how long the development attributes would sustain without adverse impact on the environment.

Environmental crises are driving climate change. Climate change indicators are emerging with greater magnitude and intensity. Nature's biodiversity is continuously depleting. The rate of species extinction is estimated to have increased 1000-fold over the natural rate, which is one of the most serious threats to the life of the planet. The narrow genetic base in the world's agriculture creates an extremely vulnerable situation. The overall biodiversity loss translating into suppressed ecosystem functioning especially affects the communities directly relying on ecosystem services. Environmental challenges are all set to multiply risks to food, water, energy, and health security at a global scale. Economic growth costs natural resources and deteriorates the environment.

Sustainable development is the one that ensures natural resource conservation, environmental safety, sociocultural equity, and a happy future. Today's well-recognized definition of sustainable development has been derived from the Brundtland Report: *Development that meets the needs of the present without compromising the ability of future generations to meet their own needs*. The whole planetary life is bound by a common evolutionary phenomenon. Human well-being on Earth emanates from that of the entire biodiversity prospering in the biosphere. Sustainability, in essence, is an ecological phenomenon. The sustainability of the human race, of human society, of a national economy, or of the world is rooted in ecological sustainability. Sustainability is also an aspect of the living philosophy, especially of eco-philosophy, that guides us to design, develop, and evolve lifestyles fostering ecological affluence for the well-being of mankind together with the well-being of all living forms of the Earth.

Development is an inherent need of people to survive, sustain, and use their inherent and evolved capabilities to reach and enjoy an ever higher level of their living status. There can be many forms and dimensions of sustainability (e.g., agricultural sustainability, economic sustainability, sustainability of future generations, etc.). Ecological sustainability is the basis of all levels, all kinds, and all dimensions of sustainability. Sustainability, in essence, is rooted in ecological sustainability. Sustainability is not something static; it is a dynamic process. It denotes the transformation of a simple or less complex, vulnerable, and unsustainable system into a complex, resilient, and sustainable system. Biodiversity is the basis of sustainability. The phenomenon that nurtures the basis of biodiversity – that is, the core of sustainability – is photosynthesis. Thus, photosynthesis is the basis of sustainability. Sustainability is to be attained by employing human ecological activities. Therefore, the first and foremost need for a sustainable world is a sustainable society. A sustainable society is an architect of sustainable development. A sustainable society trusts in intergenerational and intragenerational equity. Foods produced through agricultural practices are not only necessary for survival but also for physical, intellectual, psychological, and emotional development and sustenance. A sustainable future for humanity evolves through sustainable agriculture. There are four traits of sustainable agriculture. It has to be (i) ecologically sound, (ii) regenerative, (iii) economically viable, and (iv) socially just. In the wake of climate change, we need to impart an additional dimension of sustainability to agriculture. A sustainable agricultural system should have a high degree of adaptation capability against climate change.

The concept of Sustainable Development introduced in the WCED Report *Our Common Future* has been articulated into various global, regional, and national policies and programs.

23.8 Exercises

23.8.1 Multiple-Choice Questions

1. The rate of species extinction, in comparison to the natural rate, is estimated to have increased
   (a) 1000 times
   (b) 50 times
   (c) 20 times
   (d) 10 times
2. According to Global Footprint Network (GFN), how many Earths are needed to meet the requirements of the world at the current rate of resource extraction?
   (a) 1.4
   (b) 1.7
   (c) 2
   (d) 4

3. The term "sustainable development" was first of all used in 1980 in the
   (a) UN General Assembly
   (b) World Commission on Environment and Development
   (c) World Conservation Strategy
   (d) Stockholm Conference
4. World Commission on Environment and Development (WCED) was established in
   (a) 1972
   (b) 1980
   (c) 1983
   (d) 1989
5. Who was appointed as the Chairperson of the World Commission on Environment and Development?
   (a) Gro Harlem Brundtland
   (b) Javier Perez de Cuellar
   (c) Kurt Waldheim
   (d) Tatiana Valovaya
6. The World Commission on Environment and Development (WCED) was popularly called as the
   (a) Sustainable Development Commission
   (b) UN Development Commission
   (c) World Conservation Strategy
   (d) Brundtland Commission
7. The title of the WCED Report was
   (a) Our Common Future
   (b) Sustainable Development
   (c) Development and Sustainability
   (d) World Environment and Development
8. In which year was the WCED Report released?
   1. 1893
   2. 1987
   3. 1990
   4. 1992
9. The World Commission on Environment and Development Report is also known as the
   (a) Sustainable Development Report
   (b) Sustainability Report
   (c) Brundtland Commission Report
   (d) UN Development Report
10. The immediate outcome on account of the groundwork contained in the *Our Common Future* was the organization of the
    (a) 5 June 1987 as the World Environment Day
    (b) 1992 Earth Summit at Rio de Janeiro
    (c) Earth Day on 22 April every year
    (d) 16 October as the World Sustainable Development Day
11. The adoption of Agenda 21, the Rio Declaration, is about the
    (a) Constitution of the Commission for Sustainable Development.
    (b) Constitution of the Millennium Development Goals
    (c) Organization of Rio + 20 in 2015
    (d) Organization of the Climate Summit every year
12. The Brundtland Commission Report finds out pathways to sustainable development through
    (a) New technological developments
    (b) UN institutional managements
    (c) Multilateralism and interdependence of nations
    (d) Specific needs of individual nations
13. Sendai Framework is for
    (a) Biodiversity conservation
    (b) Sustainable forestry
    (c) Climate change mitigation
    (d) Disaster risk reduction
14. What form of sustainability forms the core of all forms of sustainability?
    (a) Social sustainability
    (b) Economic sustainability
    (c) Ecological sustainability
    (d) Industrial sustainability
15. What is the basis of ecological sustainability?
    (a) Biodiversity
    (b) Croplands
    (c) Agroforestry
    (d) Livestock

16. Which of the following is ecologically the most stable one?
    (a) Cultivated land
    (b) Natural forest
    (c) Grazing land
    (d) Agroforestry system
17. A sustainable agricultural system cannot be
    (a) Ecologically sound
    (b) Economically viable
    (c) Economically exploitative
    (d) Regenerative
18. How many Sustainable Development Goals (SDGs) does the 2030 Agenda for Sustainable Development have?
    (a) 5
    (b) 10
    (c) 14
    (d) 17
19. Which of the following events took place in the year 2000?
    (a) Johannesburg Declaration on Sustainable Development and the Plan of Implementation at the World Summit on Sustainable Development in South Africa
    (b) Setting up of a 30-member Open Working Group to develop proposals on SDGs by UN General Assembly
    (c) Sendai Framework for Disaster Risk Reduction
    (d) Millennium Declaration at the Millennium Summit in New York
20. The Division for Sustainable Development Goals (DSDGs) works with the
    (a) United Nations Framework Convention on Climate Change (UNFCCC)
    (b) UN Department of Economic and Social Affairs (UNDESA)
    (c) Food and Agriculture Organization (FAO)
    (d) Intergovernmental Panel on Climate Change (IPCC)

Answers: 1-a, 2-b, 3-c, 4-c, 5-a, 6-d, 7-a, 8-b, 9-c, 10-b, 11-a, 12-c, 13-d, 14-c, 15-a, 16-b, 17-c, 18-d, 19-d, 20-b

23.8.2 Short-Answer Questions

1. What is the definition of sustainable development as per the document of the World Council on Environment and Development?
2. What is the difference between sustainable development and sustainability?
3. What do you mean by a sustainable future?
4. Name the Report of the Brundtland Commission.
5. Who was the chairperson of the World Commission on Environment and Development and the three-time Prime Minister of Norway?
6. What was the immediate outcome that emanated from the Brundtland Report?
7. What do you mean by a viable economy?
8. Name the four traits of sustainable agriculture.
9. What was the outcome of the 2030 Agenda for Sustainable Development?
10. What do you know about the Millennium Development Goals (MDGs)?

23.8.3 Long-Answer Questions

1. Explain the environment-development-sustainability linkages.
2. Why is sustainable development imperative in our times? Elaborate with your reasoning.
3. Write an essay on the statement: “There can be no sustainability without ecological sustainability of its own.”
4. What is sustainable agriculture? How can you design and develop a sustainable agricultural system?
5. Write short notes on the following:
    (a) Global environmental challenges
    (b) Sustainable society
    (c) Global policies on sustainable development

References

EEA (European Environment Agency) (2019) The European Environment – state and outlook 2020: knowledge for transition to a sustainable development. European Environment Agency, Copenhagen. 499 pp

IUCN-UNEP-WWF (1980) World conservation strategy: living resource conservation for sustainable development. IUCN, Gland. 77 pp

Singh V (2019) Fertilizing the universe: a new chapter of unfolding evolution. Cambridge Scholars Publishing, London. 285 pp

Singh V (2020) Environmental plant physiology: botanical strategies for a climate smart planet. Taylor and Francis (CRC Press), Boca Raton. 216 pp

Glossary

3Rs (Reduce, Reuse, Recycle) A waste management principle aimed at minimizing waste generation and maximizing resource utilization.

Abiotic Components Nonliving factors in an ecosystem, including inorganic and organic substances, and climate regime.

Abyssal Zone The deep and dark zone of the open ocean, characterized by high pressure and low temperatures and inhabited by unique species.

Acid Rain Rainfall or precipitation with high levels of acidic components, formed when industrial emissions like SO_2 and NO_x react with atmospheric moisture, leading to environmental damage.

Acoustic Environment The surrounding soundscape or auditory environment in a particular location.

Activated Sludge Process A method of secondary treatment in which wastewater is mixed with a culture of microorganisms in an aeration tank. The microorganisms break down organic matter, converting it into carbon dioxide, water, and microbial biomass.

Adaptation Strategies and measures taken to adjust to changing climate conditions and minimize the negative impacts of climate change on ecosystems and human societies.

Aeolian Relating to or caused by wind.

Aesthetic Values The significance of beauty and visual appeal in forests and natural landscapes.

Afforestation The establishment of forests by planting trees in an area.

Age Structure The proportion of individuals of different age groups in a population.

Aggregation The multiplication of organisms of an established species to thrive better in a new environment.

Agriculture The practice of cultivating plants, raising animals, and producing food, fiber, and other goods for human consumption.

Agrobiodiversity The diversity of crops, livestock, and other organisms involved in agricultural ecosystems.

Agroecology The study of ecological processes applied to agricultural systems, focusing on sustainable and ecological farming practices.

Agroforestry A land use system that combines agriculture and forestry, integrating woody perennials with annual crops and livestock.

A-Horizon The horizon that contains clay, silt, sand, and organic matter percolating from the O-horizon.

Aichi Biodiversity Targets A set of 20 ambitious targets adopted during the 10th COP to the CBD in Aichi, Japan, guiding global biodiversity conservation efforts.

Air Pollutants Harmful substances released into the atmosphere, including gases, liquid droplets, solid particles, or biological agents, that contaminate the air we breathe.

Air Pollution The presence of harmful substances in the air, either naturally occurring or human-generated, in quantities that exceed the natural capacity of the atmosphere to disperse and dilute them.

Albedo Reflectivity of a surface; lower albedo means more absorption of sunlight.

Allogenic Succession Succession caused by external factors, where a community is replaced by a new one due to changes in the external environment.

V. Singh, *Textbook of Environment and Ecology*, https://doi.org/10.1007/978-981-99-8846-4

Alpha Diversity The diversity of species within a specific location or habitat.

Altitudinal Gradients Changes in biodiversity with increasing elevation.

Amazonia Refers to the Amazon rainforest and its surrounding areas, covering parts of several South American countries, including Brazil, Peru, Colombia, and Venezuela.

Ammonification The process by which organic nitrogen is converted into ammonia by decomposers.

Anaerobic Digestion Biological process where microorganisms break down organic matter in the absence of oxygen, producing biogas.

Anoxia A condition where an area or body of water lacks oxygen, which can lead to the death of aquatic organisms.

Antagonism A form of symbiosis where one interacting species is harmed.

Anthropocene Extinction The current ongoing mass extinction primarily driven by human activities.

Anthropocentric A perspective that places human beings at the center of the universe or considers human needs and interests as the primary focus.

Anthropocentrism A worldview that places humans at the center of the universe and considers human interests as the primary focus, often disregarding the impacts on other living beings and ecosystems.

Anthropogenic Ecosystem Ecosystems transformed or created by human intervention.

Apex Predators Top carnivores that are not preyed upon by other carnivores and occupy the highest position in an ecosystem's food chain.

AQI Air quality index – a numerical scale used to communicate air quality to the public based on pollutant levels.

Aquaculture The farming of aquatic organisms such as fish, shellfish, and aquatic plants, usually in controlled environments like ponds or tanks.

Aquatic Ecosystem Ecosystems located in water.

Aquifers Water bodies of porous rocks or sediment saturated with groundwater.

Arable Land Land suitable for growing crops.

Aranya Culture Refers to the forest culture, highlighting the significance of forests in shaping knowledge, wisdom, and cultural ethos.

Atmosphere Mixture of gases surrounding Earth, including nitrogen, oxygen, carbon dioxide, and trace gases.

Autecology The study of individual species in relation to the environment.

Autogenic Succession Succession caused by internal factors, where a community modifies its environment, leading to its replacement with new communities.

Autotrophic Succession Succession that begins in an inorganic environment and involves the dominance of autotrophic organisms like plants.

Autotrophs (Producers) Photosynthetic chlorophyllous plants and chemosynthetic bacteria that produce food for other organisms.

Bathyal Zone The twilight zone of the open ocean, located between the euphotic zone and the abyssal zone.

Bedrock The solid rock that lies beneath the soil, providing a stable foundation for the Earth's surface.

Benthic Zone The bottom-most zone of a lake or pond, consisting of soil and sediments rich in organic matter, supporting decomposers.

Beta Diversity The turnover or change in species composition between different habitats or locations.

B-Horizon The horizon containing humus, clays, and other mineral nutrients transported from the A-horizon.

Bioaccumulation The process in which certain pollutants, such as heavy metals and POPs, accumulate in organisms and increase in concentration as they move up the food chain.

Biodiesel Diesel consisting of esters of long-chain fatty acids, derived from vegetable oils and animal fats, used as a cleaner alternative to conventional diesel.

Biodiversity Hotspots Biologically rich areas facing critical levels of habitat loss and species endangerment.

Biodiversity The variety of life on Earth at various levels of organization, including genetic diversity, species diversity, and ecosystem diversity.

Biofilms Microbial films that form on surfaces submerged in water, consisting of diverse communities of microorganisms that can contribute to the breakdown of organic matter.

Biogas Energy Production of fuel gases using anaerobic bacterial decomposition of organic waste.

Biogas A mixture of gases (mainly methane) produced from the decomposition of organic matter in the absence of oxygen, often used as a renewable energy source.

Biogeochemical Cycles Natural processes through which elements and compounds are transferred and recycled between living organisms and their environment.

Biological Magnification The process by which certain pollutants become increasingly concentrated and magnified as they move up the food chain in ecosystems affected by water pollution.

Biomass Energy Energy obtained from organic materials, such as agricultural and forestry residues, animal waste, and dedicated energy crops, through processes like combustion or anaerobic digestion.

Biomass The total mass of living organisms, including plants, animals, and microorganisms, in a given area or ecosystem.

Biome The set of flora and fauna living in a habitat and occupying certain geography.

Biomolecule A large molecule essential for life, including proteins, nucleic acids (DNA and RNA), lipids (fats), and carbohydrates.

Bio-oil A dark brown liquid produced from pyrolysis, used in diesel engines and gas turbines to produce electricity.

Biosphere Reserves Special areas reconciling biodiversity conservation with sustainable development through designated zones.

Biosphere Sphere of life, encompassing parts of the lithosphere, hydrosphere, and atmosphere, supporting all living organisms.

Biotic Community Another term for a community, referring to the living organisms of diverse species within an ecological unit.

Biotic Components Living beings within an ecosystem, including plants, animals, and microorganisms.

Biotic Potential The inherent power of population growth under ideal environmental conditions.

Biotope A habitat shared by many species.

Black Foot Disease A disease prevalent in regions with high levels of arsenic in groundwater, causing skin lesions, gangrene, and cancer due to chronic exposure to arsenic.

BOD (Biochemical Oxygen Demand) A measure of the amount of oxygen consumed by microorganisms in the biological degradation of organic matter present in water, used to assess organic pollution in water bodies.

Botanical Gardens Facilities conserving plant species through cultivation and management of living collections.

Bt Crops Crops that have been genetically modified to express a toxin from the bacterium *Bacillus thuringiensis* (Bt), which provides resistance against certain pests.

BTEX Benzene, toluene, ethylbenzene, and xylene – toxic compounds often found in contaminated environments.

Buffer Zones Areas designed to separate different land uses or activities to reduce conflicts and impacts.

C_3 Plants A type of plants that use the C_3 photosynthetic pathway, which is less efficient under high temperatures and high CO_2 levels compared to C_4 plants.

Captive Breeding Programs Breeding endangered species in controlled environments for potential reintroduction into the wild.

Carbon Capture, Utilization, and Storage (CCUS) A technology that involves capturing CO_2 emissions from industrial processes, converting them into useful products or storing them underground to reduce GHG emissions.

Carbon Cycle The biogeochemical cycle involving the movement of carbon among different reservoirs on Earth.

Carbon Footprint The total amount of greenhouse gas emissions produced directly or indirectly by an individual, organization, product, or activity, usually measured in CO_2 equivalent (CO_2e).

Carbon Negative A state in which an individual, organization, or activity removes more carbon dioxide from the atmosphere than it emits, effectively reducing the overall carbon footprint.

Carbon Reservoir A natural or artificial storage location for carbon-containing substances.

Carbon Sequestration The process of capturing and storing carbon dioxide to mitigate its impact on the atmosphere.

Carbon Sink A natural or artificial reservoir that absorbs and stores carbon dioxide from the atmosphere.

Carnivores Animals that feed on other animals.

Carnivores Organisms that consume other animals as their primary source of energy.

Carrying Capacity The maximum population size an environment can sustainably support without causing damage or depletion of resources.

Cartagena Protocol on Biosafety An international protocol focusing on the safe handling and use of genetically modified organisms.

CBD Convention on biological diversity, an international agreement aimed at conserving biodiversity and sharing benefits from genetic resources.

Centers of Origin of Crop Plants Geographic regions where specific cultivated plants originated and evolved over time.

Char The solid residue obtained from pyrolysis, containing noncombustible matter and carbon, used in the manufacture of activated carbon and soil amendment.

Chemosynthesis The process by which some bacteria and other organisms produce energy from the oxidation of inorganic substances, rather than through photosynthesis.

C-Horizon The lowermost portion of the soil profile, comprising weathered plant material.

CITES Convention on International Trade in Endangered Species – an international agreement to protect endangered species from overexploitation.

Clean Energy Energy derived from sources that produce little or no greenhouse gas emissions and minimal environmental pollution, such as solar, wind, and hydroelectric power.

Clementsian Theory A theory of succession proposed by Frederic Clements, emphasizing the progression toward a stable climax community.

Climate Change Long-term changes in temperature and weather patterns due to human activities, primarily the emission of greenhouse gases.

Climate Crisis A situation where significant and harmful changes in the Earth's climate system occur due to human activities, primarily related to greenhouse gas emissions.

Climate Stability The consistency and predictability of climate patterns over time.

Climatic Factors Atmospheric factors creating conditions for life, including light, temperature, water, humidity, and winds.

Climax Community The relatively stable and self-sustaining community that develops after ecological succession reaches its final stage.

CNG Compressed natural gas, a clean-burning alternative fuel for vehicles derived from natural gas.

Coactions The interaction between two species influencing each other's growth and survival.

Coal A combustible black or brownish-black sedimentary rock that is a fossil fuel used primarily for electricity generation and industrial processes.

COD (Chemical Oxygen Demand) A measure of the amount of oxygen required to chemically oxidize the organic and inorganic compounds present in water, used to assess pollution in water bodies, particularly in industrial and wastewater treatment processes.

Cognitive Functions Mental processes and abilities related to thinking, reasoning, memory, attention, and problem-solving.

Cold Deserts Deserts with extremely low temperatures, where precipitation occurs mainly in the form of snow and vegetation is adapted to harsh climatic conditions.

Colloids Tiny particles suspended in a medium, such as soil, which do not settle but remain dispersed.

Combustion The process of burning a substance, usually with the release of heat and light, and often accompanied by the production of carbon dioxide and other gases.

Community Ecology The study of interactions among various organisms (plants, animals, and microorganisms) and their populations occupying a common space and living in interaction and adjustment with each other, constituting a community.

Competition A natural phenomenon where individuals or populations compete for limited resources.

Composting The degradation of organic wastes by microorganisms in the presence of oxygen, resulting in compost and CO_2, which improves soil fertility.

Conservation Biology A field of biology focused on the preservation and protection of biodiversity and ecosystems.

Conservation The protection and preservation of natural resources to ensure their sustainable use and prevent depletion.

Consumerism A culture that encourages excessive consumption and acquisition of goods and services.

Contaminants Harmful or undesirable substances that pollute or taint the environment.

Contour Ploughing Plowing across the slope following its elevation contour lines to reduce the formation of gullies and rills during heavy rainfall.

Cropping Intensity The number of crop cycles or seasons in a year on a specific piece of land.

Crude Density The number or biomass of individuals per unit area.

Cryopreservation Preservation of genetic material at extremely low temperatures.

Cryosphere The frozen water on Earth, including snow, ice, glaciers, ice sheets, and permafrost.

Cyclic Succession A characteristic of a well-established community, where cyclic changes occur in its structure, often influenced by seasonal variations.

Dam Removal The process of dismantling dams to restore the natural hydrology and ecological balance of rivers.

Decomposers Microorganisms that break down complex organic matter into simpler inorganic components, playing a crucial role in nutrient recycling.

Decomposition The process in which complex organic matter is broken down into inorganic raw materials, releasing nutrients for use by plants.

Demography The study of population dynamics, including birth rates, death rates, immigration, and emigration.

Dendrothermal Energy Energy derived from plants, such as fuel wood and charcoal, obtained through energy plantations.

Denitrification The process by which nitrates and nitrites are converted back into molecular nitrogen.

Depletion The reduction or exhaustion of natural resources due to excessive consumption or exploitation.

Desalination The process of removing minerals and salts from seawater to make it fit for consumption and other purposes.

Desert Reclamation The process of converting deserts into fertile lands through various techniques and interventions.

Desertification The process by which fertile land becomes desert due to natural or human-induced factors, such as deforestation, drought, or improper agriculture.

Desiccation The process of drying out or becoming dehydrated.

Detritivores/Scavengers Small animals that feed on the dead bodies of other organisms.

Detritus Food Chain A food chain that starts with dead organic matter, called detritus, and involves energy flow through detritus-feeding organisms.

Detritus Organic matter resulting from decomposition, providing energy for decomposers and contributing to nutrient cycling.

Development The process of positive change and progress, involving physical, social, cultural, economic, and environmental elements, aimed at improving living conditions and well-being.

Devil's Bargain An agreement or deal that may seem advantageous but is likely to have negative consequences in the long run.

Disclimax A degraded state of an ecosystem, resulting from mismanagement and leading to reduced productivity and biodiversity.

Dispersion Patterns The spatial arrangement of individuals within a population, including random, clumped, and regular patterns.

Dissolved Oxygen (DO) The amount of oxygen gas dissolved in water, necessary for supporting the respiration and survival of aquatic organisms.

Dominance Refers to the influential species or group of species in a community that control its dynamics and interactions, often based on trophic roles.

Draught Animal Power The use of domesticated animals, such as cattle or buffaloes, to provide power for agricultural activities like plowing and tilling.

Dredging Extracting minerals from water sources using buckets and draglines.

Dynamic Natural Resource A resource that is constantly changing and replenishing, like water in the hydrological cycle.

Ecesis The continuing causes of succession, relating to physicochemical or edaphic factors and their responses to species.

Eco-centric A perspective that emphasizes the intrinsic value of all living beings and ecosystems and promotes the well-being of all the entire ecological community, rather than focusing solely on human interests.

Ecological Balance The state of equilibrium achieved through interactions among structural components and functional attributes of an ecosystem.

Ecological Balance The state of equilibrium and harmony within an ecosystem, where all components are in proportion and function cohesively.

Ecological Climax The final stage of ecological succession where a stable, self-sustaining community is reached.

Ecological Density The number and biomass of individuals per unit area of the habitat they occupy.

Ecological Healing The process of restoring and maintaining ecological balance and integrity on the planet.

Ecological Health The condition of an ecosystem with balanced biodiversity and functional processes, promoting its ability to sustain life.

Ecological Integrity The state of ecological balance and health within an ecosystem, ensuring all components function together harmoniously.

Ecological Interactions The relationships and connections between different organisms within an ecosystem.

Ecological Justice The fair and equitable distribution of environmental resources and benefits among all members of society and consideration of environmental impacts on vulnerable communities.

Ecological Processes Natural processes and interactions that occur within ecosystems, affecting biodiversity and ecological balance.

Ecological Pyramids Graphic representations of ecological parameters such as energy content, biomass, and the number of organisms at various trophic levels in an ecosystem.

Ecological Regeneration The natural renewal and restoration of ecosystems over time.

Ecological Security Measures and strategies to safeguard ecosystems, biodiversity, and natural resources to ensure their resilience and ability to cope with climate change and other environmental challenges.

Ecological Stability The ability of a community to withstand environmental disturbances and maintain its structure and function over time.

Ecological Succession The process of change in the structure and composition of an ecological community over time, leading to the establishment of a stable ecosystem.

Ecological Turnover Ratio The ratio of respiration (R) to biomass (B), which influences species diversity.

Ecology Study of relationships between organisms and their environment.

Eco-philosophy A philosophical approach that integrates ecological awareness and values, emphasizing the interconnectedness of all living beings and the importance of respecting and conserving nature.

Ecosystem Balance The state of harmony and stability achieved through interactions between structural components and functional attributes.

Ecosystem Complexity The level of diversity and interactions within an ecosystem.

Ecosystem Diversity The variety of habitats, communities, and ecological processes occurring in a given region.

Ecosystem Ecology A branch of ecology that studies interactions among populations, communities, and their physical environment.

Ecosystem Efficiency The degree of compatibility between ecosystem structures and functions, with improvements in one influencing the other.

Ecosystem Functioning The processes and interactions that maintain the balance and health of an ecosystem.

Ecosystem Functions The functional attributes of an ecosystem that determine its state and are influenced by its structure and operating factors.

Ecosystem Mechanisms Processes that drive changes in the genetic makeup of populations over generations.

Ecosystem Productivity The rate at which an ecosystem produces organic matter, typically measured as the amount of biomass (living matter) produced by photosynthesis.

Ecosystem Stability The capacity of an ecosystem to maintain its structure and function in the face of disturbances.

Ecosystem Structure The physical organization of an ecosystem, characterized by the arrangement of biotic and abiotic components.

Ecosystem Turnover Changes in species composition and diversity between different ecosystems or habitats.

Ecosystem A community viewed in relation to its environment, comprising the interactions between living organisms and their physical surroundings.

Edaphic Factors Soil-related factors, including soil type, texture, structure, composition, pH, water, minerals, and organic matter.

Effluent The partially treated wastewater that flows out of the settling tank during secondary treatment.

Electrokinetic Remediation A technique that uses electric currents to remove or neutralize contaminants in the soil.

Electrolysis The process of using an electric current to induce a chemical reaction.

Elements Fundamental substances that cannot be broken down into simpler substances.

Elixir A magical or medicinal potion believed to have the power of prolonging life or turning metals into gold; in this context, it refers to water as a life-giving substance.

Endemic Species Species that are found only in specific geographic regions and nowhere else.

Energy Flow The movement of energy through an ecosystem via various trophic levels, following the laws of thermodynamics.

Energy Loss (Respiration) The release of energy during metabolic processes.

Energy Resources Natural sources of energy that can be harnessed and utilized to perform work or generate power, such as fossil fuels (coal, petroleum, natural gas), renewable sources (solar, wind, hydro, geothermal, tidal), and nuclear energy.

Energy Trapping The process of capturing solar energy through photosynthesis.

Energy The capacity to do work or produce an effect, typically in the form of heat, light, or motion.

Engine A machine that converts energy into mechanical power or motion, often used to power vehicles or industrial processes.

Environment Everything that surrounds everything, the medium containing all living beings.

Environmental Ethics A branch of philosophy that defines moral principles and values guiding human behavior toward the environment, encompassing both living and nonliving components.

Environmental Remediation The process of cleaning up polluted water bodies and restoring ecosystems to minimize environmental impacts.

Equine Referring to horses, mules, and donkeys.

Erosion Control Strategies and measures to prevent or reduce soil erosion, typically involving land management practices and conservation efforts.

Erosion The process of wearing away or removal of soil, rock, or other materials from the Earth's surface by natural agents like wind, water, or ice.

Erosivity The ability of erosive agents, like water and wind, to cause soil erosion.

Essential Nutrients Nutrients that are required for the proper functioning and growth of living organisms.

Euphotic Zone The well-lit upper layer of the open ocean, supporting abundant plant and animal life due to sufficient light penetration.

Eutrophication A process in aquatic ecosystems where excessive nutrients, primarily nitrogen and phosphorus, lead to the overgrowth of algae and aquatic plants, disrupting the ecosystem's structure and functioning.

Exhaustible Limited in quantity and can be depleted over time.

Exosphere Outermost atmospheric layer, where gases transition into space.

Exotic Species Non-native species introduced into new ecosystems, often causing negative impacts on native species and ecosystems.

Ex Situ Conservation Preservation of species outside their natural habitats, often in controlled environments.

External Digestion Decomposers breaking down organic matter using enzymes secreted into the surrounding medium.

Extinction Rate The rate at which species are disappearing from the Earth over a specific time period.

Facilitation The process by which pioneer species modify the environment to make it suitable for later successional species.

Food Chain The transfer of food energy from one organism to another in a linear sequence, showing the flow of energy in an ecosystem.

Food Pyramid A graphical representation of different food groups and their recommended proportions in a balanced diet.

Food Security The state of having reliable access to sufficient, safe, and nutritious food to meet the dietary needs and food preferences for an active and healthy life.

Food Web Patterns Patterns of feeding relationships among organisms within an ecosystem, forming complex interconnections.

Food Web A network of interconnected food chains that depicts the complex feeding relationships among various organisms in an ecosystem.

Forest Resources Refers to the various assets, materials, and ecological benefits derived from forests, including timber, fuel wood, biodiversity, and ecosystem services.

Fossil Fuels Hydrocarbons formed from the remains of plants and animals buried millions of years ago, including coal, petroleum, and natural gas.

Freshwater Water with low salt content, suitable for drinking and various human needs.

Frugality Practicing simplicity and restraint in consumption to reduce waste and environmental impact.

Fuel Cells Electrochemical cells that convert the chemical energy of a fuel into electricity, with hydrogen as a common fuel source.

Gamma Diversity The total species diversity within a large geographic region or entire landscape.

Gaseous Cycles Biogeochemical cycles with nutrient reservoirs in the atmosphere.

GDP Gross domestic product, a measure of the economic output of a country.

Gene and Seed Banks Specialized facilities storing genetic material from diverse species for conservation and research.

Genetic Diversity The variation within species, including diversity of genes, alleles, and genetic traits.

Genetic Drift Random changes in gene frequencies within a population due to chance events.

Genetic Pollution The unintended spread of genetically modified organisms and their genes into the environment, potentially causing ecological disruptions.

Genetically Modified Organisms (GMOs) Organisms whose genetic material has been altered through genetic engineering techniques, usually to introduce specific desirable traits.

Geosphere The solid part of the Earth, including the rocks, minerals, and landforms.

Geothermal Energy Energy derived from the heat stored in the Earth's interior, used for electricity generation or heating and cooling purposes.

Glacial Ice Ice formed from the compression of snow over many years, found in glaciers.

Glacier Melt The process of ice melting from glaciers, often accelerated due to rising temperatures and climate change.

Global Warming The gradual increase in the Earth's average temperature due to the buildup of greenhouse gases in the atmosphere, primarily caused by human activities.

Globalization The process of increased interconnectedness and interdependence among countries, societies, and economies on a global scale.

Grazing Food Chain A type of food chain that starts with photosynthetic producers and involves energy flow through herbivores to carnivores.

Green Revolution A period of significant agricultural advancements in the twentieth century that led to increased crop productivity through the use of high-yielding varieties, irrigation, and modern agricultural technologies.

Green Transportation Environmentally friendly transportation systems, like electric vehicles and public transport, that help combat air pollution.

Greenhouse Gases (GHGs) Gases in the atmosphere that trap heat and contribute to the greenhouse effect.

Gross Primary Productivity (GPP) The total rate of energy capture and biomass production by producers.

Ground Subsidence The sinking of land due to mining or other activities.

Groundwater Water that is stored beneath the Earth's surface in aquifers.

Gully Erosion The formation of deep channels or gullies in the landscape due to the concentrated flow of water.

Habitat Destruction The alteration or destruction of natural habitats due to human activities such as deforestation, urbanization, and agriculture.

Habitat Fragmentation The division of a large intact habitat into smaller segments, often caused by roads, railways, and other human developments.

Habitat Gradients Biodiversity variations across different types of habitats or ecosystems.

Habitat The area or place where a species lives, grows, and reproduces, providing a suitable environment for populations.

Hazardous Air Pollutants (HAPs) Substances known or suspected to cause serious health effects, including heavy metals, benzene, formaldehyde, and dioxins.

Heavy Metals Metallic elements with high atomic weights that can be toxic to living organisms in excessive amounts.

Herbivores Organisms that primarily feed on plants and algae as their source of energy.

Heterogeneous Community A diverse and varied community with different species, characteristics, or elements.

Heterotrophic Succession Succession that begins in a predominantly organic environment and involves the dominance of heterotrophic organisms like fungi and animals.

Heterotrophs (Consumers) Organisms, mainly animals, that obtain their food from producers or other consumers.

Holistic Development Comprehensive and balanced development that considers physical, intellectual, ethical, aesthetic, emotional, and psychological well-being.

Human Impact The influence of human activities on ecosystems, affecting trophic levels and ecological processes.

Hydroelectric Projects Infrastructure projects that generate electricity through the use of flowing or falling water.

Hydroelectricity Electricity produced from the kinetic energy of flowing water, typically in hydropower plants.

Hydrogen Energy Energy derived from hydrogen, which can be separated from water through various processes

Hydrological Cycle Also known as the water cycle, it is the continuous process of water circulation on Earth through evaporation, condensation, and precipitation.

Hydropower Electricity generated from the kinetic energy of flowing or falling water, typically using dams and turbines.

Hydrosphere All the water on Earth's surface, including oceans, lakes, rivers, and groundwater.

Hydrothermal Community Organisms living near deep-sea hydrothermal vents, relying on chemosynthesis.

Hydrothermal Vents Openings on the ocean floor through which hot, mineral-rich fluids and gases escape into the surrounding seawater.

Importance Values A measure of the significance of each species in relation to the entire community.

Incineration An industrial process of controlled burning of waste at high temperatures, reducing waste volume and generating ash.

Indoor Air Pollutants Pollutants originating from various sources within buildings, such as tobacco smoke, VOCs, biological contaminants, and radon gas.

Indoor Air Pollution Pollution caused by harmful substances indoors, such as smoke from traditional biomass fuels, which can be mitigated through clean cooking solutions.

Induced Succession Succession that is human-induced, typically through human intervention and management of ecosystems.

Industrial Age The period characterized by the rise of industrialization and the use of machinery for production and manufacturing, usually associated with the eighteenth and nineteenth centuries.

Infiltration The process of water seeping into the soil and replenishing groundwater.

Inhibition The process by which early successional species modify the environment to hinder the establishment of later successional species.

Initiating Causes Factors that initiate succession, including climate and biotic interactions.

Inorganic Compound A chemical compound that does not contain carbon-hydrogen (C-H) bonds.

Inorganic Pollutants Nonliving substances originating from industrial discharges, mining operations, and agricultural activities include heavy metals, acids, salts, and other chemical compounds.

In Situ Conservation Preservation of species and ecosystems within their natural habitats.

Integrated Pest Management An approach to pest control that uses a combination of techniques to minimize environmental impact.

Interspecific Competition Competition between populations of different species.

Intraspecific Competition Competition among members of the same species.

Intrinsic Rate of Natural Increase (r max) The specific growth rate in a population with a stationary age distribution.

Intrinsic Value The worth and rights of biodiversity independent of its usefulness to humans.

Invasive Species Non-native species that invade and establish themselves in an ecosystem, often causing disruptive effects on native species and their environment.

Inversion Layers Meteorological phenomenon where a layer of warm air traps cooler air near the Earth's surface, exacerbating air pollution in regions where they occur.

IPBES Intergovernmental Science-Policy Platform on Biodiversity and Ecosystem Services, assessing the state of biodiversity and ecosystems.

IPCC Intergovernmental Panel on Climate Change – a scientific body assessing climate change and its impacts.

IUCN Red List The International Union for Conservation of Nature's categorization of species based on their conservation status, ranging from extinct to least concern.

IUCN International Union for Conservation of Nature – an organization working on conservation and sustainable use of natural resources.

Kerogens Organic matter in sedimentary rocks that can be converted into hydrocarbons through heating.

Keystone Species A species that plays a crucial role in affecting many other species within an ecosystem, often exerting a disproportionate impact on community structure.

Kinetic Energy Energy possessed by a moving object, in this context, water or wind.

Land The solid surface of the Earth, including all natural features and resources, utilized for various human needs and activities.

Latitudinal Gradients Changes in biodiversity from the poles to the equator.

Lignite A type of coal with low carbon content.

Limnetic Zone The open water area of a lake or pond where sunlight can penetrate, supporting photosynthesizers like phytoplankton.

Lithosphere Outer crust of Earth, comprising continents and basins, interacting with the hydrosphere and atmosphere.

Littoral Zone The near-shore area of a lake or pond where solar radiation penetrates, providing an appropriate environment for macrophytes to grow.

Livelihood Security The assurance of sustainable income and resources to support a person or community's basic needs and well-being.

Livelihood Systems The diverse strategies and activities undertaken by individuals and communities to earn a living and support their well-being.

Lotic Ecosystems Ecosystems of flowing water, such as rivers and streams.

LPG Agriculture Refers to agriculture in the era of liberalization, privatization, and globalization, where agricultural practices are influenced by free markets and biotechnology-driven approaches.

LPG Liquefied petroleum gas, a flammable mixture of hydrocarbon gases used as a fuel for heating, cooking, and vehicles.

Machine A mechanical device designed to perform specific tasks by converting energy into useful work.

Macronutrients Nutrients that are required in large quantities by organisms for their proper growth and development. In the context of foods, they include carbohydrates, fats, and proteins.

Malnourishment A condition where a specific component of the diet, such as proteins, minerals, or vitamins, is deficient or absent, leading to health-related problems.

Marine Protected Areas (MPAs) Sections of oceans, seas, and coastlines designated to conserve marine biodiversity and ecosystems.

Mass Extinction Events Cataclysmic events in Earth's history that resulted in significant loss of biodiversity.

Mass Extinction A widespread and rapid decrease in the biodiversity on Earth, leading to the loss of a large number of species over a relatively short geological time period.

Matter Physical substance that occupies space and has mass, comprising atoms and molecules.

Megadiversity Countries Nations recognized for exceptionally high levels of biodiversity.

Membrane Filters Filters that use a thin, semi-permeable membrane to separate particles and microorganisms from water.

Mesosphere Atmospheric layer above the stratosphere, where meteorites burn up.

Metabolism The set of chemical reactions that occur within living organisms to maintain life, including energy production and utilization of nutrients.

Metabolism The set of chemical reactions that occur within living organisms to maintain life, growth, and energy production.

Metallic Minerals Minerals containing metals like gold, copper, iron, etc.

Metallurgical Processes Techniques used to extract metals from ores.

Methane Clathrate Also known as methane hydrate, it is a solid ice-like compound in which methane molecules are trapped within a lattice of water molecules. It is stable at high pressures and low temperatures, typically found in deep ocean sediments and permafrost regions.

Methanogens A group of bacteria that produce methane during anaerobic decomposition.

Methemoglobinemia Also known as blue baby syndrome, a disease occurring when infants ingest water contaminated with high levels of nitrate or other pollutants, interfering with the oxygen-carrying capacity of red blood cells.

Microclimate The localized climate conditions within a small area or habitat, often influenced by factors like vegetation and topography.

Microhabitat The immediate surroundings and physical factors of an individual plant or animal.

Micronutrients Nutrients that are required in small quantities by organisms for various physiological functions. In the context of foods, they include vitamins and minerals.

Millennium Development Goals (MDGs) Eight global development goals established by the United Nations to reduce poverty and improve living conditions by 2015.

Minamata Disease A neurological disorder caused by long-term exposure to methylmercury, a highly toxic form of mercury, named after the city of Minamata, Japan, where a severe outbreak occurred due to industrial wastewater discharge.

Mineralization The release of nutrients into the environment during decomposition.

Minerals Naturally occurring substances found in Earth's crust, used in various applications.

Monoculture Cultivation or plantation of a single type of crop or tree species over large areas.

Monospecific Populations Populations consisting of individuals of only one species.

Montreal Protocol An international treaty aimed at phasing out the production and use of ozone-depleting substances to protect the ozone layer.

Mortality The death rate or the number of individuals dying in a given period or per unit of time.

Motor An electrical or mechanical device that converts electrical or other forms of energy into mechanical motion.

Mulching The practice of covering the soil surface with a layer of organic or inorganic material to conserve moisture, control temperature, and suppress weed growth.

Multifunctionality The ability of an agroecosystem to provide multiple benefits, such as food production, environmental conservation, and cultural values.

Mutualism A form of symbiosis where both interacting species benefit.

Nagoya Protocol An international protocol promoting fair and equitable sharing of benefits from the utilization of genetic resources.

Natality The birth rate or the number of offspring produced per female per unit of time.

National Parks Large tracts of land dedicated to conserving natural landscapes and ecological processes.

Natural Gas A fossil fuel consisting mainly of methane, used for heating, cooking, and electricity generation and as a fuel for vehicles.

Natural Heritage The collection of natural resources, including biodiversity, that contribute to the identity and history of a region or planet.

Natural Phenomena Natural events or occurrences that take place in the environment, such as rain, photosynthesis, and pollination.

Natural Resources Elements found in the Earth's environment that have economic value and are essential for human survival, well-being, and economic growth.

Net Primary Productivity (NPP) The rate at which energy or organic matter is stored by producers after accounting for respiration and maintenance.

Neutraceuticals Products derived from food sources with potential health benefits beyond basic nutrition.

Nitrification The process of converting ammonia into nitrites and nitrates by nitrifying bacteria.

Nitrogen Cycle The biogeochemical cycle involving the conversion of nitrogen between various compounds.

Nitrogen Fixation The conversion of atmospheric nitrogen into a usable form by certain bacteria, essential for plant growth and ecosystem functioning.

Noise Pollution The presence of excessive and unwanted sounds in the environment that disrupt the natural balance and cause annoyance, discomfort, or health issues.

Nomads People or communities who move from place to place, usually in search of food and resources for their livelihoods.

Nonmetallic Minerals Minerals without metallic properties, like asbestos, granite, etc.

Nonrenewable Energy Sources Energy sources that are finite and cannot be replenished naturally within a reasonable time frame, such as fossil fuels and nuclear energy.

Nonrenewable Energy Energy sources derived from finite resources that cannot be replenished once they are depleted, such as fossil fuels.

Nonrenewable Resources Natural resources that cannot be replenished or take a very long time to replenish, such as fossil fuels (coal, oil, natural gas) and minerals.

Nuclear Accidents Catastrophic events involving nuclear reactors that can lead to radioactive releases and widespread environmental contamination.

Nuclear Energy Energy produced from the controlled release of nuclear reactions, usually involving uranium or thorium, to generate electricity.

Nuclear Fission A nuclear reaction in which the nucleus of an atom splits into smaller nuclei, releasing a large amount of energy.

Nuclear Fusion A nuclear reaction in which two light atomic nuclei combine to form a heavier nucleus, releasing a tremendous amount of energy.

Nuclear Power Plant A facility that uses nuclear reactions to produce electricity.

Nudation The initial stage of succession where a bare area or rock becomes available for colonization.

Nutrient Cycles The processes through which essential elements and compounds like carbon, nitrogen, and phosphorus are circulated and recycled within ecosystems.

Nutrient Cycling The process of recycling inorganic nutrients within an ecosystem, facilitated by decomposers.

Nutrient Flows The movement of nutrients within an ecosystem, involving the uptake, recycling, and transfer of nutrients through various components.

Nutrients Substances that provide nourishment and are essential for the growth, maintenance, and functioning of living organisms.

Ocean Ecosystem The largest ecosystem on Earth, covering most of the Earth's surface and characterized by stable saline water, photic and aphotic zones, and varying temperatures.

Odum's Theory Concepts proposed by Eugene Odum, including the emphasis on ecosystem development and the role of energy flow in succession.

O-Horizon The organic horizon, consisting of fallen leaves, twigs, bark, and other plant parts.

Oikos Greek word for "home," from which "ecology" is derived.

Oil Spills Accidental or deliberate oil spills from offshore drilling, transportation vessels, or oil refineries can cause significant water pollution, coating the water surface and impacting marine ecosystems and wildlife.

Open-Pit Mining A type of surface mining where minerals are extracted from open pits or quarries.

Ores Mineral deposits of economic significance from which elements are extracted.

Organic Farming Agricultural practices that avoid synthetic chemicals and promote natural processes for soil fertility and pest control.

Organic Pollutants Substances derived from living organisms or their byproducts, such as sewage, agricultural runoff, and organic chemicals, that can lead to oxygen depletion in water bodies and harm aquatic organisms.

Outbreak A sudden increase in the occurrence of a disease in a specific geographic area or population.

Overnutrition The consumption of an excessive amount of food, leading to health issues such as obesity and related diseases.

Ovine Referring to sheep and goat.

Oxygen Cycle The biogeochemical cycle that involves the production, utilization, and movement of oxygen in the Earth's atmosphere, hydrosphere, lithosphere, and biosphere.

Ozone Layer A layer in the stratosphere containing ozone (O_3) that protects against UV radiation.

Ozone Treatment A method of advanced oxidation that uses ozone gas to degrade persistent organic compounds and remove contaminants from water.

Ozone-Depleting Substances (ODS) Industrial chemicals, such as CFCs and halons, which contribute to the thinning of the ozone layer.

PAH Polycyclic aromatic hydrocarbon – a group of organic compounds with high environmental persistence and toxicity.

Pandemic A worldwide outbreak of a contagious disease affecting a large number of people.

Parasites Organisms that derive their food directly from other living organisms.

Particulate Matter (PM) Tiny solid or liquid particles suspended in the air, classified based on their size, with PM_{10} (particles with a diameter of 10 micrometers or less) and PM2.5 (particles with a diameter of 2.5 micrometers or less) being of particular concern.

Pathogens Bacteria, viruses, and other microorganisms from human and animal waste that can contaminate water sources and cause waterborne diseases.

Pedodiversity The diversity of soil types and characteristics found in different regions.

Perennial Rivers Rivers that flow throughout the year.

Permafrost Soil or rock that remains frozen for at least two consecutive years.

Permissible Ambient Noise Levels The acceptable or safe levels of noise pollution in different areas, as defined by regulatory authorities.

Persistence The natural tendency of an ecosystem to endure and maintain stability.

Petroleum A naturally occurring liquid found beneath the Earth's surface, consisting of hydrocarbons, used primarily as fuel and to produce various products.

Petroplants Alternative and renewable sources of liquid fuel that can replace diesel and petrol, derived from specific plant species.

Philosophy The study of fundamental questions about existence, knowledge, values, reason, mind, and reality, often guiding human beliefs, behavior, and understanding of the world.

Phosphorus Cycle The biogeochemical cycle that involves the production, utilization, and movement of phosphorus in the Earth's atmosphere, hydrosphere, lithosphere, and biosphere.

Photochemical Smog Air pollution formed when sunlight reacts with pollutants like NOx and VOCs, resulting in the formation of harmful chemicals, contributing to haze and poor air quality.

Photosynthesis The process by which autotrophs convert solar energy into food using carbon dioxide and water in the presence of solar radiation.

Photosynthetic Efficiency The ability of plants to convert sunlight into energy through photosynthesis.

Photovoltaic (PV) Cells Devices that convert sunlight directly into electricity through the photovoltaic effect.

Phytoplankton Microscopic algae or protists floating in water bodies, serving as primary producers in marine ecosystems.

Phytoremediation The use of plants to remove, degrade, or stabilize pollutants in the soil.

Pioneers The first group of organisms or species to establish in a new environment during primary succession.

Pollinators Animals, usually insects or birds, that assist in the transfer of pollen from one flower to another, enabling plant reproduction.

Pollution The presence of harmful substances or pollutants in the environment, often caused by human activities.

Polyspecific Populations Populations consisting of individuals of more than one species.

Population Density The number of individuals of a species per unit area or volume of the environment.

Population Ecology A branch of ecology that studies species populations, their dynamics, and their interactions with the environment.

Population Growth Rate The rate at which a population's size and density change over time.

Population Growth The increase in the number of individuals in a population over time.

Population Pollution (Popollution) The concept linking population growth to increased consumption and waste generation.

Population Size The number of individuals of a species in a particular area at a specific time.

Power The ability or capacity to do work or influence the behavior of others.

Precipitation Any form of water that falls from the atmosphere to the Earth's surface, including rain, snow, sleet, and hail.

Preparedness Short-term strategies and plans to cope with the immediate effects of climate change up to a critical level before long-term adaptation and mitigation measures can take full effect.

Primary Pollutants Pollutants directly emitted into the atmosphere in their harmful form, including PM, SO_2, NOx, CO, and VOCs.

Primary Producers Organisms capable of capturing solar energy and converting it into chemical energy through photosynthesis.

Primary Productivity The rate at which solar energy is captured and converted into organic compounds by producers (photosynthesizers).

Primary Succession Ecological succession that starts on a primitive substratum (land or water body) with no previous living organisms.

Producers Organisms, mainly plants and algae, that convert solar energy into biochemical energy through photosynthesis.

Productive Functions Forest functions related to providing economically valuable products like timber, fuel wood, food, medicines, etc.

Productivity The rate of biomass production in an ecosystem, expressed as mass per unit area per unit of time.

Profundal Zone The deeper zone of a lake or pond below the thermocline, where sunlight cannot penetrate and oxygen is limited.

Propagules Reproductive bodies of plants, such as seeds or spores.

Protective Functions Forest functions that contribute to ecosystem stability, soil conservation, water regulation, and biodiversity preservation.

Pseudocereals Non-grass plants that produce seeds used in a similar way to true cereals but do not belong to the Poaceae family.

Pyramid of Biomass A graphical representation of the total biomass (living organic matter) at each trophic level in an ecosystem.

Pyramid of Energy A graphical representation of the amount of energy trapped or utilized per unit area and time at different trophic levels in a food chain.

Pyramid of Numbers A graphical representation of the number of organisms at each trophic level in an ecosystem.

Pyrolysis A thermal decomposition process of organic waste in the absence of oxygen, producing energy-rich compounds like syngas and bio-oil.

Rainfed Areas Regions where agriculture relies primarily on rainfall for irrigation, rather than artificial water sources.

Ramayana and Mahabharata Two major epics of ancient Indian literature, Ramayana being the story of Lord Rama and Mahabharata depicting a great war and moral dilemmas.

Rangelands Large areas of open land used for grazing animals, usually in arid or semiarid regions.

Recyclable Wastes Materials that can be processed and reused to create new products, reducing the demand for raw materials.

Redox Reactions Chemical reactions involving the transfer of electrons between reactants.

Regulative Functions Forest functions that involve the absorption and release of gases, water, mineral elements, and radiation energy, regulating various environmental factors.

Renewable Energy Sources Energy sources that are naturally replenished and not depleted with use, such as solar, wind, hydro, and geothermal energy.

Renewable Energy Energy sources derived from natural processes that are continuously replenished, such as solar, wind, hydro, and geothermal energy.

Renewable Resources Natural resources that can be replenished or naturally replaced over time, such as solar energy, wind energy, and water.

Resilience The ability of an ecosystem to recover and return to its original structure after a disturbance.

Resistance The ability of a community or ecosystem to withstand and remain unchanged in the face of disturbances.

Resource Any material or asset that can be used to fulfill human needs or wants.

Retrogressive Succession A reversible succession that moves in a backward direction, taking the community back toward a less complex state after reaching the climax stage.

Revetment Planting vegetation on mine spoils to restore the land.

Rill Erosion The formation of small channels or rills in the soil due to the concentrated flow of water.

Rocket A vehicle or device that obtains thrust by expelling gas at high velocity, used for space exploration or propulsion.

Runoff Water that flows over the land surface and eventually into rivers or lakes.

Run-of-the-River A hydropower generation technique that utilizes the natural flow of the river without significant storage or damming.

Savanna A type of tropical or subtropical grassland ecosystem with scattered trees and shrubs.

Schizophrenia A mental disorder characterized by hallucinations, delusions, and disordered thinking and behavior.

Secondary Pollutants Pollutants formed through chemical reactions in the atmosphere, such as ground-level ozone, sulfuric acid, nitric acid, and secondary organic aerosols.

Secondary Productivity The rate of increase in the biomass of consumers per unit area and time.

Secondary Succession Ecological succession that occurs in an area previously occupied by a community that has been partially or completely destroyed.

Sedimentary Cycles Biogeochemical cycles with nutrient reservoirs in rocks.

Sedimentation The process of depositing sediment, including soil particles, in a new location, often caused by erosion.

Sediments Soil erosion, construction activities, and deforestation can result in sediment runoff, leading to water pollution, clouding water, reducing light penetration, and negatively impacting aquatic habitats.

Self-Regulating Entity An ecosystem's ability to maintain balance and stability.

Sequestration The process of capturing and storing a substance, such as carbon dioxide, in a stable form to prevent its release into the atmosphere.

Seral Communities The communities that replace one another during succession at a site.

Sewage Treatment The process of removing contaminants and pollutants from wastewater to make it safe for disposal or reuse, involving stages like primary, secondary, and tertiary treatment.

Sheet Erosion The removal of a thin layer of soil from a large area due to the flow of water.

Silo The vertical section of soil from the surface to the rock material, depicting all its horizons.

Skeletal Fluorosis A condition resulting from the consumption of fluoride-contaminated water, leading to the accumulation of fluoride in bones and causing joint pain, stiffness, and skeletal deformities.

Social Forestry A concept involving local people's participation in forestry activities and management for socioeconomic benefits.

Socioeconomic Relating to the interaction between social and economic factors, often used to describe the development and progress of societies.

Soil Biodiversity The variety of living organisms found within the soil, including bacteria, fungi, nematodes, earthworms, and more.

Soil Conservation Practices and methods aimed at preserving and protecting soil from degradation and erosion.

Soil Degradation The deterioration of soil quality and fertility due to human activities or natural processes.

Soil Erosion The process of detachment and transportation of soil particles from one location to another by agents such as wind, water, or human activities.

Soil Fertility The ability of soil to provide essential nutrients for plant growth and support healthy vegetation.

Soil Permeability The ability of soil to allow water and air to pass through its pores.

Soil The uppermost layer of the land, consisting of a mixture of organic matter, minerals, water, and air, essential for plant growth.

Solar Energy Energy derived from the sun's radiation, which can be converted into electricity or used for heating.

Solar Energy Radiant energy from the sun, captured by autotrophs during photosynthesis to convert into chemical energy.

Solid Waste Useless, unwanted, and undesirable materials generated from human lifestyles, posing environmental and health threats.

Sound Level Meter (SLM) A device used to measure sound pressure levels in decibels (dB).

Space-Based Solar Power A system of harnessing solar energy in outer space and transmitting it to Earth for use.

Species Composition The variety of species present in an ecosystem, which may vary from one ecosystem to another.

Species Displacement The replacement of native species by invasive or introduced species.

Species Diversity Index A measure used to determine ratios between species number and importance values within a community.

Species Diversity The variety of different species within a community, measured in terms of the number of species and their relative proportions.

Species Evenness The relative abundance of different species within a community.

Species Richness The number of different species present in a given area.

Spiritual Values The inherent desire for spiritual experiences and tranquillity that forests and nature provide.

Splash Erosion The detachment and transportation of soil particles by raindrop impact.

Stabilizing Causes Factors that support the prosperity of diverse species in appropriate environmental conditions, contributing to community stability.

Standing Crop The amount of inorganic nutrients present in an ecosystem at a particular time, representing part of the nonliving matter.

Standing State The amount of inorganic nutrients present in an ecosystem at a particular time, representing part of the nonliving matter.

Stratification Patterns Vertical layering of a community, representing different strata or levels within the ecosystem.

Stratification The presence of multiple layers or strata within an ecosystem, each inhabited by different species.

Stratosphere Layer above the troposphere, containing the ozone layer that filters harmful UV rays.

Stratospheric Ozone Layer A high concentration of ozone (O_3) molecules in the stratosphere that protect life on Earth by absorbing harmful UV radiation from the sun.

Strip Mining A technique using shovels and bulldozers to extract minerals from shallow deposits.

Subsurface Mining Mining that involves digging horizontal channels to reach deeper mineral deposits.

Succession The process of ecological change in a community over time, involving the replacement of one group of species by another

Sulfur Cycle The biogeochemical cycle that involves the production, utilization, and movement of sulfur in the Earth's atmosphere, hydrosphere, lithosphere, and biosphere.

Surface Mining Mining that takes place at or near Earth's surface.

Sustainability The capacity to endure and maintain balance and health over time, often used in the context of environmental conservation and resource management.

Sustainable Agriculture Farming practices that are environmentally friendly, socially responsible, and economically viable in the long term.

Sustainable Development Goals (SDGs) Seventeen global goals adopted by the United Nations in 2015, aimed at addressing various social, economic, and environmental issues by 2030.

Sustainable Development Development that meets the needs of the present without compromising the ability of future generations to meet their needs.

Sustainable Future A future characterized by the responsible use of resources, protection of the environment, and equitable development, ensuring the well-being of current and future generations.

Sustainable Livelihoods Livelihood strategies that meet the needs of the present generation without compromising the ability of future generations to meet their own needs.

Sustainable Society A society that prioritizes environmental protection, social justice, and economic viability for a balanced and equitable future.

Sustainable Use The utilization of natural resources in a manner that maintains ecological balance and supports long-term well-being.

Symbiosis A relationship in which two or more species live together in close association.

Synecology The study of organisms' groups in relation to the environment.

Syngas Synthesis gas, a mixture of combustible gases like CO, H_2, CH_4, and others, used as a fuel source.

Terracing The technique of carving steps or terraces into the slopes of hills or mountains to create flat areas for farming and prevent soil erosion.

Terrestrial Ecosystem Ecosystems located on land.

Tertiary Treatment Also known as advanced or final treatment, this stage focuses on achieving a high level of water quality by removing any remaining impurities and preparing the water for reuse or release into sensitive environments.

Thermal Energy Energy in the form of heat.

Thermochemical Technologies Technologies that extract energy from waste through processes like combustion, gasification, and pyrolysis.

Thermodynamics The study of energy transformations in physical and chemical systems, guiding energy flow in ecosystems.

Thermosphere Thickest atmospheric layer, with high temperatures and ionized gases.

Tidal Energy Energy harnessed from the gravitational pull of the moon and the sun on the Earth's oceans, resulting in the rise and fall of tides, which can be converted into electricity.

Tolerance The ability of species to survive and thrive in a variety of environmental conditions.

Top Consumers Apex predators, such as lions and tigers, that occupy the highest position in the food chain.

Toxin A poisonous substance produced by living organisms, capable of causing harm to other organisms.

Transesterification A chemical process used to produce biodiesel from vegetable oils or fats.

Transphysical Phenomenon Phenomenon or events that transcend the physical realm and cannot be explained by physical laws or scientific theories.

Transpiration The process by which water is taken up by plant roots and released into the atmosphere as water vapor through small openings (stomata) in the leaves.

Trickling Filter Process A method of secondary treatment where wastewater trickles over a bed of rocks or plastic media coated with microbial films (biofilms) that degrade organic matter as the water passes over them.

Trophic Efficiency The percentage of energy transferred from one trophic level to the next in a food chain.

Trophic Levels The hierarchical levels within a community, representing different groups of species based on their feeding relationships and energy transfer.

Trophic Organization The arrangement of organisms in an ecosystem based on their feeding habits, leading to different trophic levels.

Troposphere Lowest atmospheric layer, essential for life, weather systems, and water cycle.

UNCCD (United Nations Convention to Combat Desertification) An international treaty aimed at combating desertification, adopted in 1994 and effective since 1996.

Undernourishment A condition where a person receives less than 90% of the minimum dietary intake on a long-term basis, resulting in inadequate energy for an active and productive life.

Unidirectional Energy flow in ecosystems is always one-way, from producers to consumers, with no reverse flow.

Upanishads Philosophical texts that explore the concepts of reality and self.

Urbanization The process of increasing urban areas and population concentration in cities and towns.

Vector-borne Diseases Infectious diseases that are transmitted to humans or animals through the bites of vectors, such as mosquitoes, ticks, or fleas, which carry and transmit the disease-causing pathogens.

Vedas Ancient Hindu scriptures containing sacred knowledge and hymns.

Vermicomposting The use of earthworms to convert organic waste into nutrient-rich vermicompost, a high-quality organic fertilizer.

VOCs Volatile organic compounds – organic chemicals that easily evaporate into the air and contribute to air pollution.

Volatile Organic Compounds (VOCs) Organic compounds emitted from various sources, contributing to the formation of ground-level ozone and secondary organic aerosols.

Vulnerability to Extinction The susceptibility of a species to extinction due to various factors like population size, habitat specificity, and trophic level.

Waste-to-Energy The process of recovering energy from waste materials, converting them into useful forms of energy like electricity or heat.

Water Crisis A situation characterized by a lack of sufficient water resources to meet the needs of a region or population.

Water Pollution The contamination, degradation, or alteration of water bodies due to introduction of harmful substances or excessive amounts of synthetic pollutants.

Water Resources Refers to the various sources of water, including oceans, seas, lakes, rivers, etc., that are available for use.

Water Scarcity A condition where the demand for water exceeds the available supply.

Water Table The depth at which the ground beneath the surface is saturated with water.

Weathering The process by which rocks and minerals are broken down into smaller particles by physical, chemical, or biological means.

Wildlife Sanctuaries and Reserves Designated areas aimed at safeguarding wildlife and their habitats.

Wind Energy Energy harnessed from the motion of the wind to generate electricity.

Wind Farms Groups of wind turbines used for electricity generation.

Wind Power Electricity generated from the kinetic energy of wind using wind turbines.

Zonation Patterns Horizontal layering of a community, representing subcommunities existing amidst homogeneous environmental conditions.

Zoogenic Caused by animals.

Zooplankton Microscopic animals, including single-celled organisms and tiny crustaceans, floating in water bodies like ponds, lakes, and oceans.

Zoos and Wildlife Reserves Facilities providing a safe environment for endangered species, engaging in breeding and research.

MIX
Papier aus verantwortungsvollen Quellen
Paper from responsible sources
FSC® C105338

If you have any concerns about our products,
you can contact us on
ProductSafety@springernature.com

In case Publisher is established outside the EU,
the EU authorized representative is:
Springer Nature Customer Service Center GmbH
Europaplatz 3, 69115 Heidelberg, Germany

Printed by Libri Plureos GmbH
in Hamburg, Germany